Springer Transactions in Civil and Environmental Engineering

Springer Transactions in Civil and Environmental Engineering (STICEE) publishes the latest developments in Civil and Environmental Engineering. The intent is to cover all the main branches of Civil and Environmental Engineering, both theoretical and applied, including, but not limited to: Structural Mechanics, Steel Structures, Concrete Structures, Reinforced Cement Concrete, Civil Engineering Materials, Soil Mechanics, Ground Improvement, Geotechnical Engineering, Foundation Engineering, Earthquake Engineering, Structural Health and Monitoring, Water Resources Engineering, Engineering Hydrology, Solid Waste Engineering, Environmental Engineering, Wastewater Management, Transportation Engineering, Sustainable Civil Infrastructure, Fluid Mechanics, Pavement Engineering, Soil Dynamics, Rock Mechanics, Timber Engineering, Hazardous Waste Disposal Instrumentation and Monitoring, Construction Management, Civil Engineering Construction, Surveying and GIS Strength of Materials (Mechanics of Materials), Environmental Geotechnics, Concrete Engineering, Timber Structures.

Within the scopes of the series are monographs, professional books, graduate and undergraduate textbooks, edited volumes and handbooks devoted to the above subject areas.

The proposal for each volume is reviewed by the main editor and/or the advisory board. The chapters in each volume are individually reviewed single blind by expert reviewers (at least four reviews per chapter) and the main editor.

Nihal Anwar Siddiqui · Mohammad Moizuddin ·
Anil Kumar Gupta · Akbar Ziauddin ·
Prasenjit Mondal
Editors

Advances in Disaster Management, Volume 2

Proceedings of HSFEAS 2023

Springer

Editors
Nihal Anwar Siddiqui
Centre of Excellence Occupational Health
Safety, Fire & Environment (C-HSFE)
GD Goenka University
Gurugram, Haryana, India

Anil Kumar Gupta
IIT Roorkee
Roorkee, Uttarakhand, India

Prasenjit Mondal
Centre of Excellence Occupational Health
Safety, Fire & Environment (C-HSFE)
GD Goenka University
Gurugram, India

Mohammad Moizuddin
Pan India and Overseas
Chennai, Tamil Nadu, India

Akbar Ziauddin
Safety Bee Training FZ LLC
Ras Al Khaimah, United Arab Emirates

ISSN 2363-7633 ISSN 2363-7641 (electronic)
Springer Transactions in Civil and Environmental Engineering
ISBN 978-981-96-4049-2 ISBN 978-981-96-4047-8 (eBook)
https://doi.org/10.1007/978-981-96-4047-8

This Springer imprint is published by the registered company Springer Nature Singapore Pte Ltd.
The registered company address is: 152 Beach Road, #21-01/04 Gateway East, Singapore 189721, Singapore

Preface

The field of health care is always changing due to advances in research, technology, legislation, and the unwavering commitment of healthcare workers all across the world. In order to shed light on the revolutionary advancements that are redefining how we approach health and wellness, "Advancements in Healthcare: A Multifaceted Exploration" explores the dynamic field of healthcare innovation.

This investigation covers a wide range of subjects, from cutting-edge medical procedures and diagnostic methods to the relationship between artificial intelligence, data analytics, and digital health platforms and health care. In an effort to find ways to create healthcare systems that are more sustainable and egalitarian, it looks at how sociological changes, economic variables, and global health issues affect the availability and delivery of healthcare services.

This collection targets professionals, researchers, policymakers, and anyone else with a stake in the future of health care through a multidisciplinary perspective in an effort to spark discussion, encourage collaboration, and pique curiosity. We set out on a quest to open up new doors, boost patient outcomes, and ultimately improve the quality of life for people and communities all over the world by embracing the complexities and subtleties of this diverse profession.

As we embark on this journey together, may we be guided by a common dedication to compassion, creativity, and the pursuit of healthcare excellence? May the insights contained in these pages spark new ideas, ignite revolutionary change, and move us toward a future where health and healing have no boundaries.

Gurugram, India — Dr. Nihal Anwar Siddiqui

Contents

of raindrops, hence minimizing soil erosion caused by runoff. Furthermore, pine needles have a high moisture retention capacity, which promotes plant growth and soil stability. Smith et al. (2018) discovered that areas mulched with pine needles enjoy much lower erosion rates than untreated soil surfaces.

Fiber/ fabrication	Method	Advantage	Limitation	Location/ area	References
Pine Needles	Mulching	• Prevents soil erosion caused by water runoff • Enhances soil stability • Promotes plant growth	May not be suitable for all soil types	Hilly/ mountainous	Smith et al. (2018)
Geotextile	Geotextile blankets	• Provides long-term erosion control • High tensile strength and durability • Allows for vegetation establishment	May not be suitable for all soil types	Construction sites/ steep slopes	Martinez et al. (2021)

1.2.2 *Construction of the Roads*

Road and highway stability and durability are critical considerations in road building. Pine needles can be mixed into various road materials to enhance their engineering features. Pine needles help to extend the life of infrastructure by increasing soil compaction and decreasing settling. Johnson and Smith (2020) discovered that using pine needles in road base materials resulted in enhanced stability and lower maintenance costs over time.

Fiber/ fabrication	Method	Advantage	Limitation	Location/ area	References
Pine needles	Soil additive	• Enhances soil stability • Improves load-bearing capacity of road base	• Not suitable for highly trafficked roads	Rural areas	Johnson & Smith (2020)

(continued)

(continued)

Fiber/ fabrication	Method	Advantage	Limitation	Location/ area	References
Geotextile	Reinforcement	• Provides long-term reinforcement of road base • Improves load-bearing capacity • Reduces settlement and rutting	• May not be suitable for locations with heavy traffic	Urban areas Highway construction	Garcia et al. (2017)

1.2.3 *Coastal Area Protection*

Coastal erosion is a major issue in many areas, exacerbated by climate change and rising sea levels. Pine needles can be utilized in dune stabilization efforts to keep coastal habitats and infrastructure safe from erosion. Pine needles stabilize sand dunes, preventing erosion caused by wind and wave action. Martinez et al. (2021) found that dunes stabilized using pine needle barriers had much less erosion during storm surges than unprotected dunes.

Fiber/ fabrication	Method	Advantage	Limitation	Location/ area	References
Pine needles	Dune stabilize	• Anchors sand dunes • Prevents erosion by wind and waves • Protects coastal ecosystems	• Requires regular maintenance • Effectiveness depends on local coastal conditions	Coastal areas	Martinez et al. (2021)
Geotextile	Seawall	• Provides long-term coastal protection • Resistant to weathering and erosion	May not be suitable for highly dynamic coastal environments	Coastal areas with dynamic coastal environments	Johnson & Smith (2020)

1.2.4 Landslide Protection

Landslides pose a significant threat to communities living in hilly and mountainous regions. Pine needles play a crucial role in landslide protection by stabilizing slopes and reducing the risk of slope failures. The natural mulch layer formed by pine needles binds soil particles together, enhancing slope stability. Studies conducted by Lee et al. (2019) have shown that areas treated with pine needle mulch experienced fewer landslides during heavy rainfall events compared to untreated slopes.

Fiber/ fabrication	Method	Advantage	Limitation	Location/ area	References
Pine needles	Slope stabilize	• Stabilizes slopes • Reduces risk of landslides	Effectiveness may vary depending on slope characteristics	Hilly/ mountainous regions prone to landslides	Lee et al. (2019)
Geotextile	Retaining walls	• Provides long-term slope stabilization • Resistant to weathering and erosion	May not be suitable for highly dynamic slopes	Slopes prone to landslides	Johnson & Smith (2020)

1.2.5 Ground Improvement

Ground improvement strategies are critical for infrastructure development in places where soil quality is low. Pine needles can be used to stabilize and fortify soil, increasing strength and load-bearing capacity. Garcia et al. (2017) discovered that incorporating pine needles into soil layers increased engineering qualities, making them ideal for use in a variety of construction projects.

Fiber/ fabrication	Method	Advantage	Limitation	Location/ area	References
Pine needles	Soil stabilize	• Enhances soil strength and stability • Increases load-bearing capacity of soil	Effectiveness may vary depending on soil composition and conditions	Construction sites/ areas with poor soil quality	Garcia et al. (2017)

(continued)

References

Brown A, Johnson B (2018) The role of pine needles in reducing forest fire risk. J Environ Manage 45(2):123–135

Garcia C et al (2017) Ground improvement techniques using pine needles: a case study. Geotech Eng J 22(3):217–230

Green M et al (2019) Pine needle composting for waste management: a sustainable solution. Waste Manag J 35(4):301–315

Johnson B, Smith D (2020) Pine needles in road construction: engineering properties and performance evaluation. Transp Res J 28(1):45–58

Lee H et al (2019) Pine needle mulch for landslide protection: field trials and effectiveness assessment. Landslide Sci Technol J 15(2):87–102

Martinez E et al (2021) Coastal dune stabilization using pine needles: a case study

Smith D et al (2018) Pine needle mulch for soil erosion control: mechanisms and effectiveness. Soil Sci J 12(4):301–315

Smith J, Jones K (2021) Community engagement in pine needle management: lessons learned and best practices. Disaster Manag J 18(2):75–88

White R et al (2020) Biodiversity promotion through pine needle management: a comparative study. Conserv Biol J 25(1):45–58

Chapter 2
Critical Period Protein Deficiency and Its Elaborate Impact on Brain Development: A Comprehensive Review'

Arti Kumari

2.1 Introduction

Malnutrition is a chronic condition, the effects of which accumulate over a period of time and have long lasting ramifications. According to World Health Organisation, malnutrition refers to deficiencies, excesses, or imbalances in a person's intake of energy and (or) nutrients. The term malnutrition covers two broad groups of conditions. One is 'undernutrition'—which includes stunting (low height for age), wasting (low weight for height), underweight (low weight for age), and micronutrient deficiencies or insufficiencies (the lack of important vitamins and minerals). The other is overweight, obesity, and diet-related non-communicable diseases (such as heart disease, stroke, diabetes, and cancer). According to the Global Nutrition Report 2016, out of the 7 billion world population nearly 800 million suffer from calorie deficiency and about 2 billion people suffer from micronutrient malnutrition. In April 2016, the United Nations General Assembly adopted a resolution proclaiming the UN decade of action on nutrition from 2016 to 2025. The decade aims to mobilise the policies that can result in significant action to address all forms of malnutrition.

Malnutrition affects people in every country though it is more prevalent in some countries. According to Global Nutrition Report 2016, out of 667 million children under the age of 5 worldwide, 159 million are stunted, and 50 million are found to have wasting. The June 2017 report of UNICEF states that nearly half of all the deaths in children under 5 years of age are attributable to undernutrition which is widespread in Asia and Africa. In India, malnutrition is the major cause of death in children below 5; almost 11 million children die before reaching the age of 5(Mathad and Shivprasad 2013).Though malnutrition is a serious problem at any age, its ramifications are more pronounced if there is lack of protein nutrition during early childhood and during

A. Kumari (✉)
GD Goenka University, Gurugram, India
e-mail: artirkgit@gmail.com

N. A. Siddiqui et al. (eds.), *Advances in Disaster Management, Volume 2*, Springer Transactions in Civil and Environmental Engineering,
https://doi.org/10.1007/978-981-96-4047-8_2

layer V pyramidal neurons than on layer II/III neurons of visual cortex after 30 days, but the difference decreased at 90 days and further reduced at 220 days (Díaz-Cintra et al., 1990). It was found that protein deprivation for long-term reduces the count of mossy fibre synapses in hippocampus (Andrade et al. 1991).

Protein deficiency during pre and postnatal period also effects the balance in biochemical molecules and receptors of the brain. The rats which were subjected to pre and postnatal protein deficiency had increased oxidative stress on the brain because of a decrease in the level of an antioxidant glutathione as well as a decrease in the glutamate uptake which can cause excitotoxicity; however, the level of free radicals was not affected (Feoli et al. 2006). Postnatal protein deficiency in rats also causes decrease in the level of acetylcholine which is in accordance with the behavioural deficit of impaired memory (Eckhert et al. 1975). Lipofuscin is a pigment which accumulates in perikarya of cells on ageing or due to environmental stress which induces cell death. This pigment accumulates in neuronal as well as glial perikarya of protein-deficient monkeys, but the accumulation is reversed to some extent when the protein deficiency is halted (Manocha and Sharma 1977). This shows that even momentary protein deficiency can induce stress and cause cell death in brain tissue, although some of the effects can be practically reversed.

2.4 Impairment in Neuronal Activity

Protein-deficient diet leads to impairment in neural activity marked by an overall decrease in both spontaneous and evoked potentials. The spontaneous brain activity in the forebrain of developmentally protein-deficient rats is shown to decrease, which can be attributed to a decrease in the proportion of neurons with fast firing rate and an increase in that of the ones with slow firing rate (Stern et al., 1983). Similarly there is an increase in interspike interval as well as mean latency of the brain stem, auditory, and visual evoked potentials in the children raised up with a protein-deficient diet (Durmaz et al., 1999). Reduction in the resting brain activity also occurs in layer IV neurons, where the magnitude of evoked potential after whisker stimulation is reduced predominantly in the thalamic input activity. (Chaudhary et al., 2013). Chronic malnutrition is also shown to disrupt EEG pattern in school children who were found to have soft neurological signs (Agarwal et al., 1989). There is down-regulation of the gene KCNJ3, which results in increased excitability of neurons. In contrast, a 2-deoxy glucose uptake study shows a decrease in the metabolic activity in prefrontal cortex that may be associated with a decrease in functional activity (Amaral et al., 2015). It also leads to overexpression of NMDA receptors in the barrel cortex as well as increase in GAD65 which implicates the increase in synthesis and release of GABA resulting in the inhibition of the barrel cortex as shown by electrophysiological recordings (Chaudhary et al., 2013).

2.5 Summary

From the studies on protein deficiency during prenatal and early childhood stage, it is evident that there are widespread effects on various aspects of brain development, which has behavioural manifestations. There is a marked increase in overall inhibition in the brain both at the resting stage as well as during stimulation. There is also an overall increase in the inhibitory neurotransmitter. However, the mechanism of overall brain inhibition is not known since there is a marked increase in the overall inhibitory neurotransmitter as well as the excitatory neurotransmitter because the excitatory NMDA receptors are overexpressed. In addition there is a downregulation of the KCNJ3 gene which results in increased excitation of the cortex. Whereas electrophysiological and 2DG studies show cortical inhibition in layer IV and prefrontal cortex. There is also a decrease in auditory and visual evoked potentials, as well as decrease in spontaneous forebrain activity.

Thus there is overall inhibition to the cortical neurons though the exact mechanism of the inhibition is unanswered. It can be proposed that inhibition of the cortex can be due to an increase in the population of inhibitory neurons and inhibitory neurotransmitter over that of the excitatory, it can also be due to an increase in activity of inhibitory interneurons. If the inhibitory interneurons play a part in overall inhibition of cortex what is the contribution of different subsets of inhibitory neurons? Distribution and activity of the layer V pyramidal neurons is also an area of lacunae. These are the questions which need to be answered.

References

Agarwal K, Das D, Agarwal D, Upadhyay S, Mishra S (1989) Soft neurological signs and EEG pattern in rural malnourished children. Acta Paediatr 78:873–878

Amaral A, Jakovcevski M, McGaughy J, Calderwood S, Mokler D, Rushmore R, Galler J, Akbarian S, Rosene D (2015) Prenatal protein malnutrition decreases KCNJ3 and 2DG activity in rat prefrontal cortex. Neuroscience 286:79–86

Andrade J, Cadete-Leite A, Madeira M, Paula-Barbosa M (1991) Long-term low-protein diet reduces the number of hippocampal mossy fiber synapses. Exp Neurol 112:119–124

Atalabi OM, Lagunju IA, Tongo OO, Akinyinka OO (2010) Cranial magnetic resonance imaging findings in kwashiorkor. Int J Neurosci 120:23–27

Brown JL, Pollitt E (1996) Malnutrition, poverty and intellectual development. Sci Am 274:38–43

Chaudhary R, Chugh M, Darokhan Z, Katreddi RR, Ramachandra R, Rema V (2013) Physiological slowing and upregulation of inhibition in cortex are correlated with behavioral deficits in protein malnourished rats. PLoS ONE 8:e76556

Cintra L, Aguilar A, Granados L, Galván A, Kemper T, DeBassio W, Galler J, Morgane P, Durán P, Díaz-Cintra S (1997) Effects of prenatal protein malnutrition on hippocampal CA1 pyramidal cells in rats of four age groups. Hippocampus 7:192–203

Coffey D, Khera R, Spears D (2015) Intergenerational effects of women's status: evidence from joint Indian households. In: Working Paper

Cordero ME, D'Acuña E, Benveniste S, Prado R, Nuñez JA, Colombo M (1993) Dendritic development in neocortex of infants with early postnatal life undernutrition. Pediatr Neurol 9:457–464

de Rooij SR, Wouters H, Yonker JE, Painter RC, Roseboom TJ (2010) Prenatal undernutrition and cognitive function in late adulthood. Proc Natl Acad Sci 107:16881–16886
Díaz-Cintra S, Cintra L, Ortega A, Kemper T, Morgane P (1990) Effects of protein deprivation on pyramidal cells of the visual cortex in rats of three age groups. J Comp Neurol 292:117–126
Durmaz Ş, Karagöl U, Deda G, Önal MZ (1999) Brainstem auditory and visual evoked potentials in children with protein-energy malnutrition. Pediatr Int 41:615–619
Eckhert CD, Levitsky DA, Barnes RH (1975) Postnatal stimulation: the effects on cholinergic enzyme activity in undernourished rats. Proc Soc Exp Biol Med 149:860–863
Feoli AM, Siqueira I, Almeida LMV, Tramontina AC, Battu C, Wofchuk ST, Gottfried C, Perry ML, Gonçalves C-A (2006) Brain glutathione content and glutamate uptake are reduced in rats exposed to pre-and postnatal protein malnutrition. J Nutr 136:2357–2361
Galler JR, Bryce C, Waber DP, Zichlin ML, Fitzmaurice GM, Eaglesfield D (2012) Socioeconomic outcomes in adults malnourished in the first year of life: a 40-year study. Pediatrics 130:e1-7
Galler JR, Bryce CP, Zichlin ML, Fitzmaurice G, Eaglesfield GD, Waber DP (2012) Infant malnutrition is associated with persisting attention deficits in middle adulthood–3. J Nutr 142:788–794
Galler JR, Bryce CP, Waber DP, Hock RS, Harrison R, Eaglesfield GD, Fitzmaurice G (2012) Infant malnutrition predicts conduct problems in adolescents. Nutr Neurosci 15:186–192
Gardner JM, Grantham-McGregor SM, Himes J, Chang S (1999) Behaviour and development of stunted and nonstunted Jamaican children. J Child Psychol Psychiatry 40:819–827
Hazin AN, BEZERRA ALVES JG, Falbo AR, (2007) The myelination process in severely malnourished children: MRI findings. Int J Neurosci 117:1209–1214
Manocha SL, Sharma SP (1977) Reversibility of lipofuscin accumulation caused by protein malnutrition in the motor cortex of squirrel monkeys, Saimiri scireus. Acta Histochem 58:219–231
Mathad V, Shivprasad S (2013) Malnutrition: A Daunting Problem for India's Spectacular Growth. Indian J Clin Pract 23
Ranade S, Rose A, Rao M, Gallego J, Gressens P, Mani S (2008) Different types of nutritional deficiencies affect different domains of spatial memory function checked in a radial arm maze. Neuroscience 152:859–866
Ranade SC, Nawaz MS, Rambtla PK, Rose AJ, Gressens P, Mani S (2012) Early protein malnutrition disrupts cerebellar development and impairs motor coordination. Br J Nutr 107:1167–1175
Reeves PG (1997) Components of the AIN-93 diets as improvements in the AIN-76A diet. J Nutr 127:838S-841S
Resnick O. et al. (1979) Developmental protein malnutrition: influences on the central nervous system of the rat. Neuroscience & Biobehavioral Reviews. 3(4):233–246
Stern WC, Pugh WW, Johnson A, Morgane PJ (1983) Spontaneous forebrain neuronal activity in developmentally protein malnourished rats. Dev Brain Res 9:95–98
Swaminathan S, Vaz M, Kurpad AV (2012) Protein intakes in India. Br J Nutr 108:S50–S58

Chapter 3
Dispensing Pattern and Decision-Making Habits of Dispensers in Delhi-NCR, India with Respect to Anti-reflection Coating (ARC)

Anitha Arvind and Nilesh Thite

3.1 Introduction

Spectacles are no longer the solution to age-related vision disorder but more of a personality enhancer and fashion accessory. Changing lifestyles and activities that affect the eye have prompted more and more people to use prescription spectacle lenses. Spectacles enjoy a widespread popularity and usage amongst the developing and developed economies globally due to a large percentage of population suffering from refractive errors. Apart from the price, quality, comfort and wearing satisfaction offer the driving forces in selecting an eyewear by the consumer. The concept of look better and see better comes to fore in prescription eyewear sector in which anti-reflection coatings (ARC) play a major role hence there is a growing demand and use of prescription eyewear with ARC globally. With over 456 million (De Souza et al 2012) Indians requiring vision correction to go about their daily lives there is a huge market potential for spectacle lenses. According to a study (India Infoline News Service 2012) titled 'Indian Optical Sector' conducted by the Associated Chamber of Commerce and Industry of India (Assocham), the Indian eyewear market is poised to touch Rs. 43,000 crores by 2015. With spectacles remaining popular in the Indian market, there is a steady incline in the demand and usage of spectacles and its products.

Anti-reflection coatings have come a long way since it was first introduced to spectacle lenses as Zeiss ET by Zeiss (Company History 2014) who were the pioneers

A. Arvind (✉)
G D Goenka University, Gurugram, India
e-mail: anitha2001arvind@gmail.com

N. Thite
IACLE, Bengaluru, India

N. A. Siddiqui et al. (eds.), *Advances in Disaster Management, Volume 2*, Springer Transactions in Civil and Environmental Engineering,
https://doi.org/10.1007/978-981-96-4047-8_3

back in 1935. The popularity of anti-reflection coatings are growing day by day and so is the innovation in coatings since its inception. With the introduction of anti-reflection coatings to spectacle lenses, there has been a tremendous boost in prescription spectacle sales globally both for its aesthetic appeal and optical benefits. The current estimates (McQuaid 1997) of the percentage of the market penetration of ARC have shown a steady increase especially in parts of Europe where it is above 75% while over 90% of consumers in Japan wear ARC spectacle lenses. The US market is fast catching up with its global counterparts; however, its penetration in the Asian subcontinent is still not at par with the rest globally as indicated by the vision council research group. There is a, however, a tremendous potential for growth of AR sector in the Asian continent. The vision council research council (European Sunglass Association Meeting, May 2013) has predicted (Optical Market Overview n.d.) a tremendous growth in optical sales in Asia and Latin America as compared to Europe or USA. The improved quality of anti-reflection coatings is resulting in word of mouth referrals amongst patients and increased patient loyalties towards ARC. Aggressive promotion of ARC has been taken up by major leaders in optical industry; as a result there is a growing demand and growth of the AR sector. Product education amongst eye care professionals and optical dispensers go a long way to ensure that the right product reaches the consumer. Many a times it is the lack of knowledge and wrong recommendations that pull down the sales in any sector which is true for the optical sector too.

Optical dispensers/optometrists are the key influencers in the consumers' decision on selecting their eyewear. Hence, the objective of our study was to study the dispensing patterns and decision-making habits of optical dispensers and optometrists in various optical sectors across the Delhi-NCR region. This way the barriers to penetration of ARC in one sector, i.e. the Delhi-NCR region can be studied from the optometrist/optical dispensers' side. Since an optical dispenser/optometrist would be the first person whom the consumer approaches, it is important that his/her dispensing and decision-making pattern is studied to get a better understanding of whether they possess the right knowledge as well as pass the relevant information to the consumers. This could further help the optical industry to plan strategies to educate optical dispensers/optometrists towards passing the relevant information, thereby boosting its growth in the Indian market as well.

3.1.1 Anti-reflection Coatings and Its Benefits

Anti-reflection coatings have a long but unknown development according to a study by Mc Quaid (McQuaid 1997). The principle of thin film coating was first discovered in 1892 (http://www.harisingh.com/newsOpticalARCoating.htm) Anti-reflection coatings are microscopically thin multi-layered coating applied to the surface of the spectacle lens to eliminate reflections and increase light transmittance. They work on the principle of optical interference. When light is incident on an optical surface, some amount of light is lost by reflection; hence, transmittance is

Table 3.1 Light transmittance in various index materials with and without ARC

Material	Index	Transmittance without ARC in %	Transmittance with ARC in %
CR 39	1.49	92.06	99.1
Glass	1.52	91.40	99.2
Polycarbonate	1.59	89.40	99.0
High index plastic	1.60	89.40	99.0
Super hi index plastic	1.67	87.80	98.2

reduced. For normal angle of incidence, 4% of light is lost by reflection per surface for a 1.5 index material. In case of a 1.9 index material, there is a loss of light up to 9% per surface. With the application of broadband anti-reflection coatings, the light transmittance could be increased to over 99%. Table 3.1 shows the light transmittance in various index materials with and without anti-reflection coatings.

The benefits to the wearer of ARC prescription lenses are manifold. The two immediate benefits are visual and cosmetic. This is because of the ability of the coatings to reduce the reflections off the surface of the lens. The following are the benefits of an anti-reflection coated spectacle lenses.

- Increased light transmission through lenses especially high refractive index materials
- Improved cosmetic appearance of the wearer
- Increased visual clarity
- Increased contrast under low illumination
- Reduced visual fatigue for spectacle wearers in computer use and fluorescent light

Better visual performance during night driving.

The ARC on the lenses reduces reflections caused by the lenses. It does not alleviate problems that are not related to reflections from the lens surface.

3.1.2 Recent Advances in ARC

There has been a wave of advancements in the AR technology. From the advent of single-layer AR coatings to multilayer or broadband coatings, ARC has and is enjoying an unprecedented growth in the ophthalmic industry. The reason for its growth is advancement in AR coating process and AR coating equipment (Bruneni 2002). Most issues pertaining to adherence of ARC have been resolved with the application of index and AR compatible hard coats. The advancement of new technology in anti-reflection coatings makes the lenses not only free from reflections but also makes them resistant to scratches, smudges, water, dust, and UV. This makes the lenses easier to clean, maintain, and also durable (www. Essiloracademy.eu).

3.1.3 Studies on the Efficacy of Anti-reflection Coatings

Mashige et al 2009 studied the effect of anti-reflection coating on glare threshold and recovery under scotopic conditions. Their study was done on 37 subjects and concluded that glare vision and recovery times reduced with the usage of lenses with ARC under scotopic conditions. Their study recommends usage of lenses coated with ARC for shortening the period of discomfort and poor vision, thus allowing safe and efficient driving in the night. This study was further strengthened by another study done by Bachman and Weaver 1999 on presbyopic highway patrol troopers. Their study concluded that there was a significant reduction in glare and reflections during the day, night, indoors, and in the patrol vehicle as well as a highly significant improvement in the night vision. The subjects in the study chose the coated lenses as their overall preferred lens. Ross and Bradley 1997 in their study examined the visual and patient preference of anti-reflection coated and uncoated spectacle lenses and concluded that a significantly better contrast sensitivity was observed with anti-reflection coated lenses under dim illumination and also proposed that drivers should wear anti-reflection coated spectacle lenses for better visual comfort and performance.

3.1.4 Methods and Methodology

It was a cross-sectional questionnaire based study conducted in Delhi-NCR region.

3.1.5 Survey Questionnaire

The survey questionnaire was designed and distributed to 5 experts in the field of dispensing optics in India. Comments and suggestions that were received from the experts and which were deemed appropriate to the purpose of study were incorporated into the questionnaire. The study questionnaires were electronically distributed and telephonically confirmed as well as collected in person from the study participants.

The survey questionnaire consisted of three parts. The first part contained the demographic data, namely their age, mode of practice, educational qualification, years of experience in the field of dispensing, and location of practice. The second part consisted of practice details which included the percentage of prescription glasses with anti-reflection coating (ARC) dispensed in their respective practice on an average in a month. The third part of the questionnaire collected in depth information about the factors which helped the dispenser in deciding the type of ARC and also the dispensing habits of the optical dispensers/optometrists in the region. Trained personnel were employed to collect data from the study participants in the region as well as contact them over the telephone. Information on the purpose of the study was explained, and a signed consent was obtained from the participants in the study. The participants were also made aware to remain anonymous in the study. Survey questionnaires were filled and collected at the same time and day of issuance as well.

3.1.6 Statistical Analysis

The means and standard deviation for the linear variables and proportions for categorical variables were estimated. The difference in the mean was tested using unpaired t-test, and difference in the proportions was tested using the chi—square test or Fisher's exact test (for low expected cell counts). The data entered in MS Excel were converted to Stata for analysis using Stata Version 11.

3.1.7 Results

Out of 245 dispensers reached, responses from 104 dispensers were obtained from optical outlets across Delhi-NCR region. The majority of the respondents were males (89%). The dispensers were based in optical eye care set-ups which were either stand-alone practices (52%) or retail chains (48%). The dispensers were segregated based on their educational qualification as those having formal education on dispensing (diploma 27%, B.Optom 39% & M.Optom 10%) that constituted 76% against those who did not have any formal qualification in dispensing constituting 24% as indicated in Graph 3.1. The dispensing patterns and habits amongst dispensers in the two eye care set-ups were studied and correlated. Dispensing is a skill which requires knowledge, practice, and experience, and hence, a cross tab analysis between the dispensers based upon their years of experience in the field of dispensing and educational qualification was performed. The number of years in the field of dispensing amongst the dispensers ranged from 4 months to 45 years (12yrs $\pm$ 10.58). On an average 78% of prescription eyewear dispensed in the Delhi-NCR region were reported to have ARC.

3.1.8 Care and Maintenance of ARC Lenses

The verbal form of communication (87%) was the most preferred mode of explaining care and maintenance to the customers followed by Tab/Ipad/Laptops (21%), written (17%) by the dispensers; while 5% respondents thought newer coatings require no explanation (Graph 3.2).

Graph 3.1 Educational qualification of dispensers in Delhi-NCR region

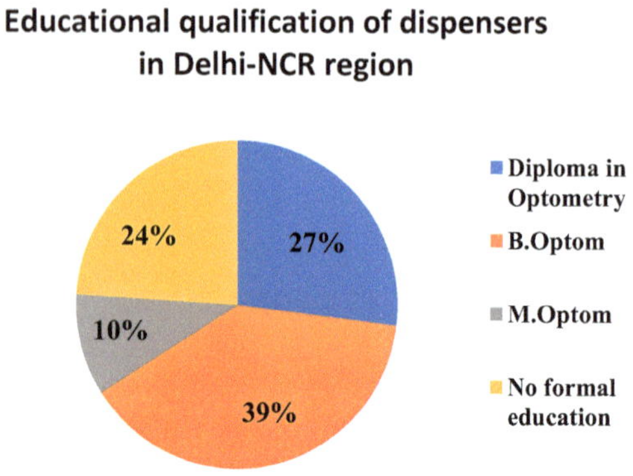

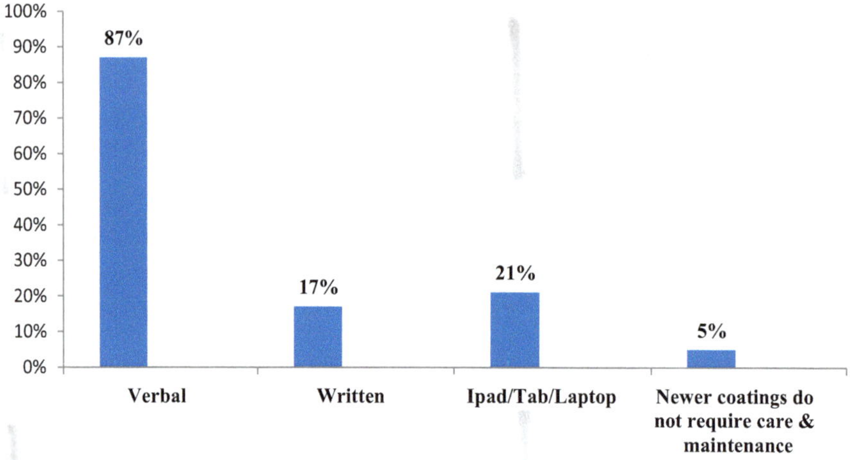

Graph 3.2 Preferred means of explaining care and maintenance of ARC lenses to consumers

3.1.9 Illustration of Benefits of ARC

The verbal form of communication (86%) was again the most popular means of illustrating the benefits of ARC to the customers followed by demonstration kit (40%), printed brochures (37%), and videos (11%) as indicated by the study in the Graph 3.3.

Chapter 1
The Multifaceted Role of Pine Needles in Disaster Management: A Technical Perspective

Vimal Mohan

V. Mohan (✉)
School of Advance Engineering, UPES, Dehradun, Uttarakhand, India
e-mail: vimaluniyal91@gmail.com

N. A. Siddiqui et al. (eds.), *Advances in Disaster Management, Volume 2*, Springer Transactions in Civil and Environmental Engineering,
https://doi.org/10.1007/978-981-96-4047-8_1

1.1 Introduction

Disasters, whether caused by natural occurrences or human activity, pose substantial hazards to the environment and infrastructure, needing meticulous disaster management measures. Pine needles have long been recognized for their usefulness in forest fire prevention due to their capacity to act as a natural mulch, but their utility extends beyond this single function. This study aims to provide a complete evaluation of the multidimensional function that pine needles play in disaster management, with a focus on their applications in environmental preservation and infrastructure development.

Pine needles are essential assets in the field of environmental preservation, helping to mitigate the impact of disasters on ecosystems. Their dense matting on forest floors conserves soil moisture, prevents erosion, and preserves forest biological integrity, lowering vulnerability to natural disasters like landslides and floods. Furthermore, pine forests play an important role in carbon sequestration, which helps to regulate climate and increase resistance to climate-related disasters. Understanding and utilizing the ecological benefits of pine needles are thus critical components of comprehensive disaster management methods aimed at protecting natural habitats and wildlife.

Pine needles also present encouraging opportunities for the building of infrastructure in areas vulnerable to natural disasters. Because of their toughness and fiber composition, they can be used as building materials for a variety of tasks, such as stabilizing roads, eroding barriers, and even serving as parts of sustainable building materials. Incorporating pine needles into infrastructure projects encourages environmentally beneficial behaviors that are consistent with disaster risk reduction and sustainable development concepts, while also enhancing the resilience and durability of the structures.

In order to highlight the value of pine needles as adaptable resources for building infrastructure and environmental resilience, this investigation will examine the various ways in which they are used in disaster management. Pine needles have a lot to offer, and realizing their full potential can help create more sustainable plans for disaster management that put community resilience and the environment first when dealing with changing hazards.

1.2 Pine Needles'Adaptable Function in All-Inclusive Disaster Management

1.2.1 Soil Erosion Management

Soil erosion is an important concern in disaster management because it leads to land deterioration and the loss of valuable soil. Pine needles make an excellent mulch, forming a protective layer on the soil surface. This layer reduces the effect

About the Editors

Prof. (Dr.) Nihal Anwar Siddiqui is Director of Centre of Excellence, Occupational Health, Safety, Fire, and Environment, at GD Goenka University. Dr. Siddiqui specializes in the area of safety management, occupational health and safety, environmental pollution, and env. monitoring and control techniques. In his 28 years of career, Dr. Siddiqui has worked on different organization like Institute of Environment & Management, Lucknow, Paryacon Engineers, Environmental Engineering Division of Central Pulp & Paper Research Institute as Scientist. Dr. Siddiqui did some of the pioneer works in the area of odor monitoring and providing control technology along with Finland scientist. Dr. Siddiqui was also associated with Health, Safety, and Environment Dept. of ICEM College, Muscat, Oman, which is affiliated to University of Central Lancashire, UK. Dr. Siddiqui is Approved Tutor to teach international certificate courses like IOSH, NEBOSH UK. Dr. Siddiqui has received several awards during his 28 years of career.

Dr. Mohammad Moizuddin is Health and Safety Professional and has been working in this field for the last 29 years to manage the hazard and risk of the industry and providing the cost-effective engineering and practical solution to the company along with disaster management. He was Ex-HSE Head of Burj Khalifa-Dubai (world's tallest manmade structure) during construction phase and successfully handed over the iconic project. During this period, he has coordinated with many clients, consultants, contractors, such as Qatar Petroleum, EMAAR (Dubai), BOEING (US), ADCO (Abu Dhabi), Victory Height (Dubai), BFH (Bahrain), and DLF-India. He is Fellow Member of International Institute of Risk and Safety Management (IIRSM-UK), Graduate Member of Institution of Occupational Safety and Health (Grade IOSH-UK), and Member of American Society of Safety Professional (USA). He has achieved the national and internal safety award and guided the company to achieve the safety award.

Prof. Anil Kumar Gupta is working as CEO in IIT Roorkee. Prof. Anil Kumar Gupta is an illustrious figure in DRR internationally, with exemplary contributions to disaster preparedness and resilience through policy-planning at national, sectoral, and international agendas. His vision and striving for human resource capacities and institutional development of training including DMI, NIDM, and ICARS and guidance and support to states/UTs in disaster management and climate actions are widely known. He pioneered CCA-DRR convergence, ECODRR & NBS, DM Plans of Ministries, City DRR Scorecard, DIA, and mitigation analysis in advancing DRR. Besides natural and climatic disasters, he was a driving force in promoting industrial, chemical, and health-related disasters and emergency management in India. He also drives a global move "Accelerating Transformation (ACCTRA)" of pole-vaulting improvements along India's national vision on DRR leadership and cooperation.

Dr. Akbar Ziauddin is Distinguished Professional in Health, Safety, and Environment (HSE) with an impressive academic and professional background. He holds a doctorate in HSE Engineering, with his thesis focusing on "A Study on Exposure Assessment and Evaluation of Occupational Health Hazards of Carbon Nanotubes," alongside a master's in technology in HSE, a master's in environmental studies, and a degree in environmental management. His career spans roles in oil and gas exploration services, where he specialized in HSE advisory, accident investigations, and training senior staff, as well as positions such as Chief Examiner at a leading health and safety institute in India and HSE Engineer and Advisor at Mitchell Drilling International. He has authored and edited more than 13 books, published more than 29 papers, and developed e-learning courses on HSE, fire safety, and environmental management. He is Member of esteemed professional bodies like IOSH, IIRSM, and IEMA.

Dr. Prasenjit Mondal is Associate Professor at Centre of Excellence in Occupational Health, Safety, Fire, and Environment of GD Goenka University, Gurgaon. His area of interest is drinking water and industrial effluent treatment strategies, conversion of waste to wealth, ambient and indoor air quality improvement techniques, industrial hazard, and risk analysis. During his Ph.D., he developed a low budget and highly efficient drinking water filtration system using organic adsorbent. He has worked in several internal and external funded projects as Investigator. He has published more than 50 international journal papers and edited book chapters. His major patents include Cancer Risk Assessment tool using Python language, indoor air pollution control by application of IoT, and water filtration system using bio-adsorbents.

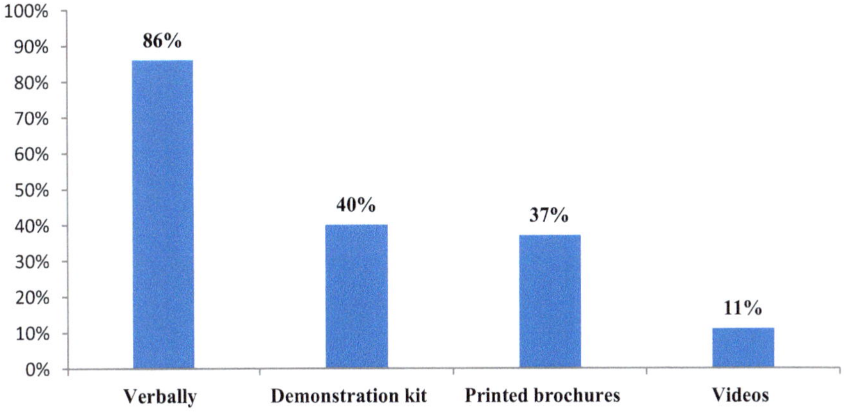

Graph 3.3 Preferred means of explaining benefits of ARC to customers

3.1.10 Cleaning of ARC Lenses

When enquired about the recommended method of cleaning ARC lenses to the consumer, the dispensers gave the following answers which are given in the table Graph 3.4

A Ask to clean circularly with any soft cloth.
B Ask to clean with microfiber cloth circularly.
C Ask to clean by sliding any soft cloth only in one direction.

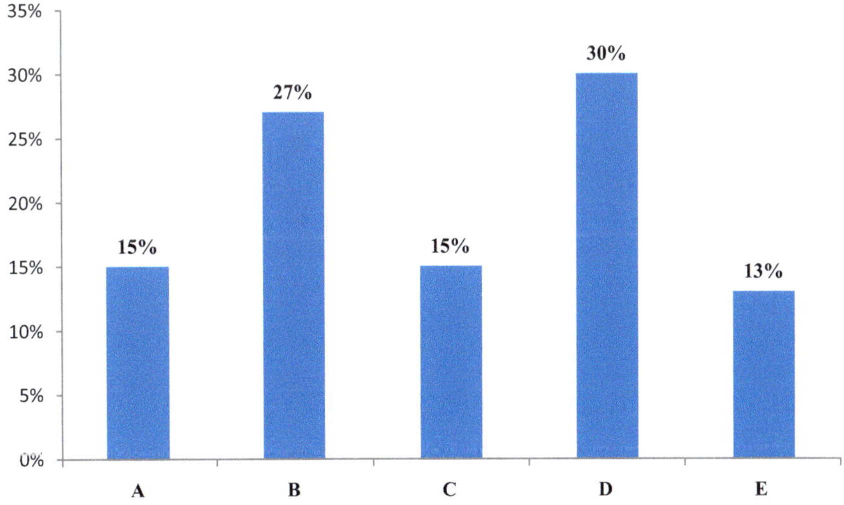

Graph 3.4 Recommended method of cleaning ARC lenses by the dispensers in the study

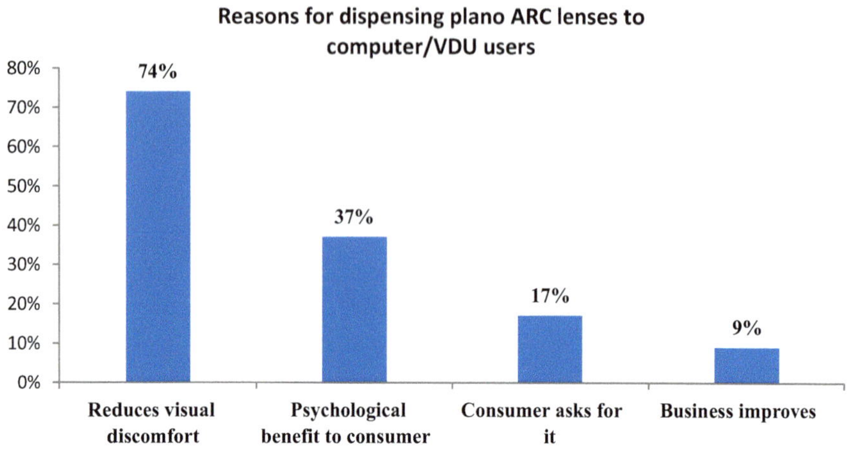

Graph 3.5 Reasons for dispensing plano ARC lenses to computer/VDU users

D Ask to clean by sliding microfiber cloth only in one direction.
E Dip/wet lenses in water or cleaning solution and wipe with microfiber cloth.

3.1.11 Plano ARC Lenses

A majority (95%) of the dispensers in Delhi –NCR region reported that they dispense plano ARC lenses to computer/VDU users because it reduces visual discomfort (74%), provides psychologically benefit to the patient (37%), the consumer asks for it (17%) and because they feel their business improves (9%) (Graph 3.5).

3.1.12 Routine Recommendation of ARC on Every Prescription Lens

The majority (71%) of dispensers in Delhi-NCR region routinely recommend ARC on every prescription lens.

3.1.13 Indications for Recommending ARC Lenses

The indications for recommending ARC lenses are illustrated in the Graph 3.6.

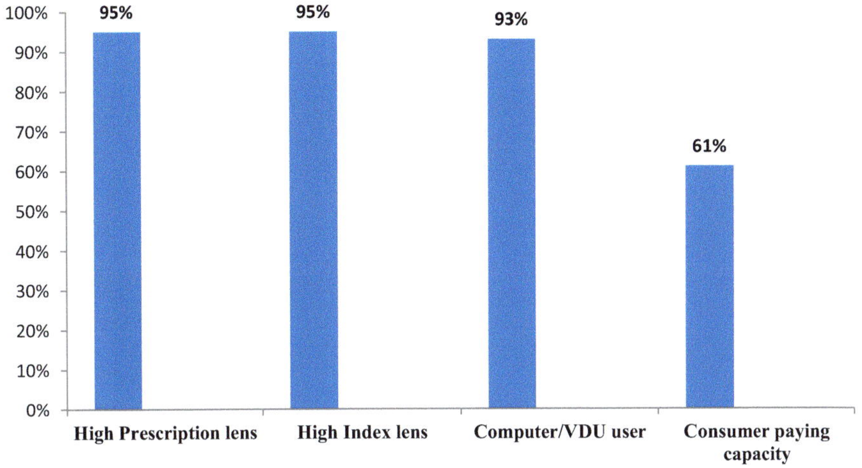

Graph 3.6 Indications for recommending ARC lenses to consumer

3.1.14 Benefits of ARC to Convince Consumers

On enquired about the benefits of ARC that are used to convince a consumer a majority of the dispensers used visual clarity (96%) followed by reduced eye strain (86%), cosmetically better (72%), and reduced glare (47%) as indicated by the Graph 3.7 in the study.

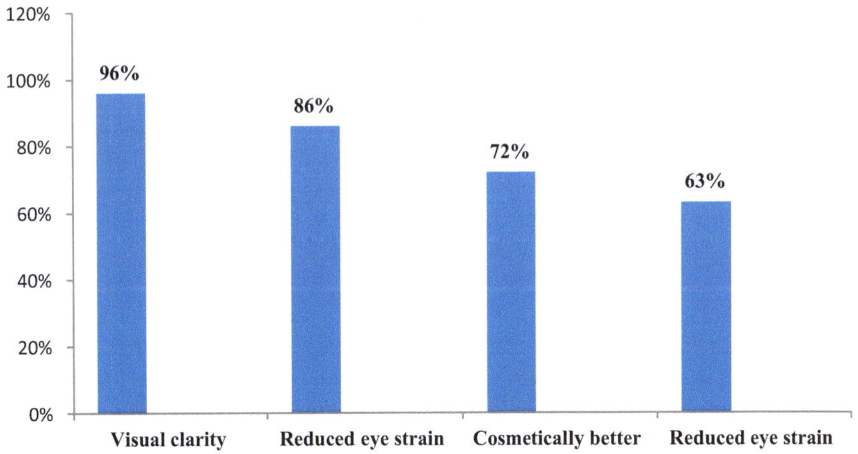

Graph 3.7 Benefits of ARC to convince consumer for ARC lenses

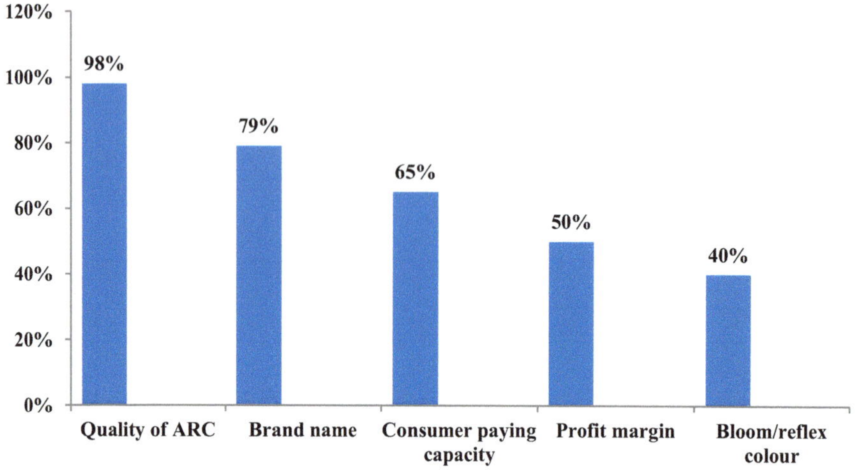

Graph 3.8 Factors contributing to the decision-making of dispensing ARC lenses

3.1.15 Brand/type of ARC

The factors that are used by the dispensers in Delhi-NCR to decide the brand/type of ARC to be dispensed are summarised in the Graph 3.8.

3.1.16 Advanced Coatings

A majority (98%) of the dispensers reported that they recommend advanced coatings such as hydrophobic top coat, oleophobic coat, anti-fog coat, etc., over the ARC to their consumers. The reasons stated have been summarised in the Graph 3.9.

The respondents did not show any discrepancy based on the eye care set-up in which they were dispensing except that the dispensers based in retail chains favoured I pad/tab/lap tops to explain care and maintenance to their customers as compared to the ones based in stand-alone practices ($p = 0.05$).

The dispensers who had no formal education in dispensing felt that the new coatings did not require any care and maintenance; hence, they do not talk about care and maintenance to the consumer as against the dispensers with formal education ($p = 0.011$). The respondents with no formal education do not illustrate the benefits of ARC to consumers with the help of printed brochures as compared to the other group ($p = 0.002$) in the study.

The dispensers were also classified based on the years of experience in the field of dispensing, and a cross tab analysis was done. The most experienced group (> 15 years of experience in dispensing) used reduced eye strain as one of the significant benefits of ARC to convince a consumer for ARC as against the respondents

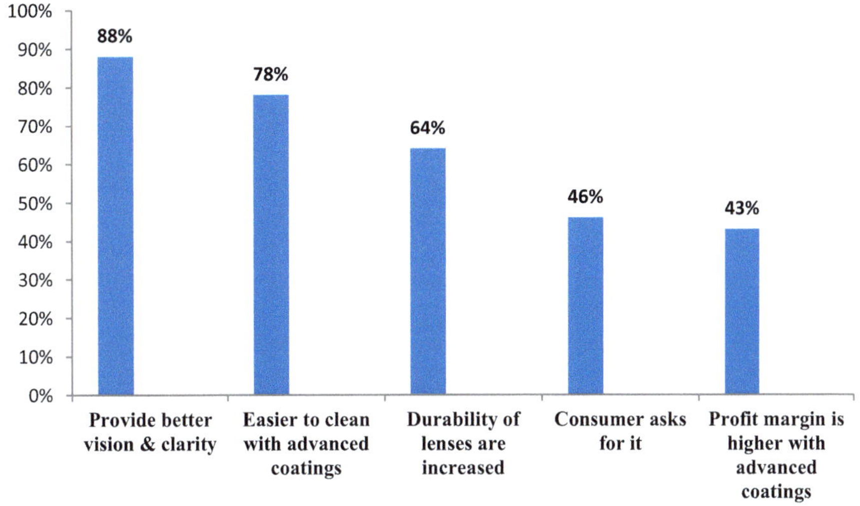

Graph 3.9 Reasons for recommending advanced coatings over ARC to consumer

with lesser years of experience ($p = 0.024$). The respondents with the least years of experience (up to 5 years of experience) felt that the profit margin in advanced coatings(hydrophobic top coat, oleophobic, optifog, etc.) were more; hence, they would strongly recommend advanced coatings over ARC to their consumers as compared to the respondents with more than 5 years of experience ($p = 0.014$).

3.2 Discussion

Anti-reflection coatings enjoy a huge popularity and growth in the optical industry both in the developing and developed countries. There is a high degree of market penetration for spectacles globally due to a large section of the population suffering from refractive errors. This has caused an increase in the usage of prescription eyewear and its products which serves as a driving factor for growth in this sector. Aggressive promotion for prescription eyewear products by the major players in the eyewear industry has created awareness amongst the consumers; hence, the demand and popularity for products such as ARC is on the rise. Spectacle wearing population (Sarath et al 2011) is huge in India, and there is a growing demand for spectacle lenses and its products.

The consumers need to be made aware on the benefits of the anti-reflection coatings, and for this the optical dispensers play a major role. The optical dispensers themselves need to be well-educated and informed on the aspects pertaining to benefits and care and maintenance of anti-reflection coated lenses. Hence, it is important to understand the dispensing pattern and decision-making habits of dispensers with respect to ARC and thereby help identify areas where there are misconceptions in the recommendation and dispensing of ARC lenses.

This study was conducted at the Delhi-NCR region one of the most developed metropolitan cities of India. The average number of ARC lenses dispensed in this region was at par with its global counterparts.

The verbal form of communication was the most preferred method of explaining care and maintenance of ARC lenses by the respondents. Care and maintenance is an important aspect (Bruneni 2002) of any product and more so with ARC lenses. Visual demonstration Lin et al 2012, (http://www.packworld.com/package-design/graphic/visual-trumps-verbal-creating-memorable-brand-identity) would make the consumers understand better along with a verbal approach. The respondents based in retails chains, however, use I pad/tab/laptops to illustrate this aspect indicating that the retail chains are better equipped to make demonstrations easier. Also the respondents with no formal education in dispensing optics in the study reported that the advanced coatings required no care and maintenance and hence felt any need to explain care and maintenance to the consumer. It is important that the dispenser encourages consumers on proper maintenance of ARC-coated lenses and also supply and recommend products to help them care for their lenses. Dispenser education level plays a major role as misconceptions of dispensers would result in incorrect information and recommendation being passed on to the consumer.

It is interesting to note that the dispensers with no formal education reported that they do not use printed brochures to illustrate the benefits of ARC to the consumers. A printed form of information on any lens product like ARC would provide more information about the product and help answer certain frequently asked questions by the consumers.

Premium ARC lenses are being marketed as easy to clean and maintain and are being supplied with cleaning cloth and solutions, and hence, the dispensers do not pay much attention towards illustrating or recommending the right method of cleaning.

Plano ARC lenses do nothing to alleviate the problems of computer/VDU users without any refractive errors. Yet most respondents reported that they dispensed plano ARC lenses to alleviate visual discomfort as well as to provide psychological benefit to the consumer. Part of the reason could be that ARC lenses are being promoted as anti-glare lenses, and hence, the consumers come asking for it. Hence, the dispensers see it as a psychological benefit to the consumer. Another reason being the dispensers see it as a business opportunity to increase their sales.

A majority of the respondents routinely recommend ARC lenses on every prescription lens. High prescription and high index lenses are the major indicators for recommending ARC followed by computer/VDU users. The respondents used visual clarity as the most significant benefit of ARC followed by reduced eye strain, cosmetic benefit, and reduced glare to convince the consumer. This shows that adequate and correct information about the benefits of ARC is passed on to the consumer.

It is reported that the consumer-paying capacity is one of the indicators for recommending ARC to consumers by the most experienced respondents. The respondents with the least years of experience (up to 5 years of experience) felt that the profit margin in advanced coatings (hydrophobic top coat, oleophobic, optifog, etc.) were more; hence, they would recommend advanced coatings over ARC to their consumers as compared to the respondents with more than 5 years of experience. In addition to knowledge and skill, years of experience in the field of dispensing also accounts to dispensing patterns and decision-making habits amongst the dispensers as indicated by the study.

The Vision Council which serves as the global voice for vision care products and services represents the manufacturers and suppliers of the optical industry. The council had announced on 7 May 2009 (http://www.thevisioncouncil.org/blog/anti-reflective-lens-curriculum-introduced-++optometry-schools) an educational grant to the Association of Practice Management Educators (APME) to design and deliver learning modules to increase the understanding and awareness of prescribing anti-reflective technologies. A learning module on this subject was inculcated as part of practice management curriculum in the 17 optometry colleges in the United States as there is no mandatory education on anti-reflection technology in the optometry colleges. This approach should be copied in other schools and colleges of optometry globally. This way the future leaders in eye care are better equipped in providing the right guidance to the consumers and help choose and dispense the best eyewear.

This study reflects the dispensing pattern and decision-making habits of dispensers in Delhi-NCR region and cannot be extrapolated to other regions of the country. Similar studies could be done in the urban and semi-urban regions of the country, thus reaching out to more dispensers to understand the dispensing patterns and decision-making habits of dispensers better. This information could help in clearing the misconceptions at the level of the dispenser and help promoting the benefits of ARC to the consumers.

3.3 Conclusion

In Delhi NCR region, the proportion of ARC lenses dispensed was reported to be at par with global average. Myths pertaining to the benefits of plano ARC need to be cleared as a high number of plano ARC lenses were reported to be dispensed. However, the majority of the dispensers in the study had clarity on indications and benefits of ARC on prescription lenses. The quality of ARC was a major decision-making factor for dispensing ARC lenses by the majority of the dispensers in the study. Most dispensers preferred using verbal instructions to explain care and maintenance and illustrate benefits of ARC. They should use demonstration tools and visual aids more often to create a better impact and help the consumers understand the benefits better. Educational level of the dispenser did not pose as a major barrier in dispensing ARC lenses; however, education on aspects pertaining to functions and benefits does play a role in understanding and communicating the benefits of ARC to the consumers.

References

Bachman WG, Weaver JL (1999) Comparison between anti-reflection-coated and uncoated spectacle lenses for presbyopic highway patrol troopers. J Am Optom Assoc 70(2):103–109
Bruneni, Joseph L. “AR: How Coatings Increase Lens Efficiency.” N.p., Jan2002. Web. 30 Apr. 2014
“Company History.” *Company History*. N.p.,n.d. Web. 29 Apr. 2014.
Neilsen De Souza, Yu Cui, Stephanie Looi, Prakash Paudel, Lakshmi Shinde, Krishna Kumar, Rajbir Berwal, Rajesh Wadhwa, Vinod Daniel, Judith Flanagan, Brien Holden, The role of Optometrists in India: An integral part of an eye health team. Indian J Ophthalmol. 2012 Sep-Oct; 60(5): 401–405.
India Infoline News Service/ 14:30, Mar 22, 2012. www.indiainfoline.com
Tom M.Y Lin, Kuan-Yi Lu, Jia-Jhou Wu, (2012) “The effects of visual information in eWOM communication”, Journal of Research in Interactive Marketing, Vol. 6 Issn: 1, pp.7 – 26
Mashige KP, Thathane NP, Kader F, Nyandoro GD, Sultan AA (2008) The effect of anti-reflection coating on glare threshold and recovery under scotopic conditions. S Afr Optom 67(2):68–76
McQuaid RD (1997) Reflections on antireflection films. J Am Optom Assoc 68(3):191–201
N.p.,n.d. Web http://www.thevisioncouncil.org/blog/anti-reflective-lens-curriculum-introduced-++ optometry-schools
N.p.,n.d. Web http://www.harisingh.com/newsOpticalARCoating.htm
“Optical Market Overview.” N.p., n.d. Web. http://www.thevisioncouncil.org/sites/default/files/The VisionCouncil_2013_AnnualReport2.pdf
Ross J, Bradley A (1997) Visual performance and patient preference: a comparison of anti-reflection coated and uncoated spectacle lenses. J Am Optom Assoc 68(6):361–366
Ravipati Sarath, Jannitha PA and Pavan KV. Preferential and Non- Preferential reasons for Opting Contact Lens. Indian journal of Fundamental and Applied Life Sciences. 2011 Vol. 1 (3) July-September, pp. 221–225

Chapter 4
Favipiravir: A Promising Nucleotide Analog for Treatment of COVID-19

Kirti Sharma, Asim Kumar, Atul Sharma, and Mohit Sanduja

4.1 Introduction

The increasing cross-species transmission due to the increasing interactions of viral outbreaks with humans and animals have been creating threat to health of humans as well as of animals from the times of nineteenth century (Sharma et al. 2020). With rapidly growing globalization and anthropogenic activities, pathogenic transmission across the world has intensified leading to several pandemics, especially viral pandemics (Liu et al. 2016; Qu et al. 2020). Over the last two decades, many new pathogenic viruses have been identified such as Middle East respiratory syndrome coronavirus (MERS-CoV) in Saudi Arabia (Ghosh et al. xxxx), haemorrhagic fever viruses (Lassa, Ebola) in West Africa (Goba et al. 2016), and novel coronaviruses including severe acute respiratory syndrome coronavirus (SARS-CoV) (Ghosh et al. 2020) and influenza (avian influenza A H7N9, pandemic H1N1) (Kaye and Pringle 2005) All these pandemics caused significant number of deaths. Such situations result in drastic mortality and economic losses all across the world (Sharma et al. 2020). In recent days again, the world is witnessing a pandemic as COVID-19. The

A. Kumar
Amity Institute of Pharmacy, Amity University, Gurugram, Haryana, India

A. Sharma
Department of Pharmaceutical Chemistry, SGT College of Pharmacy, SGT University, Gurugram, Haryana, India

K. Sharma
Saraswati Institute of Pharmaceutical Education and Research, Saraswati Group of Colleges, Mohali, India

M. Sanduja (✉)
School of Pharmaceutical Sciences and Technology, Sardar Bhagwan Singh University, Dehradun, India
e-mail: mohitsanduja@hotmail.com

N. A. Siddiqui et al. (eds.), *Advances in Disaster Management, Volume 2*, Springer Transactions in Civil and Environmental Engineering,
https://doi.org/10.1007/978-981-96-4047-8_4

human coronavirus or COVID-19 was 1st reported or recognized in December 2019 in Wuhan, China (Bogoch and Thomas 2020; Hui et al. 2020).

COVID-19 is a respiratory malady showing symptoms as fever, fatigue, dry cough, muscle aches, shortness of breath, and in severe cases lead to pneumonia (Hussin and Siddappa 2020; Cynthia et al. 2020). The disease spectrum of COVID-19 is diverse in severity ranging from asymptomatic to respiratory infection and ultimately to pneumonia (Ko et al. 2020). In cases of severity, it causes ARDS, i.e. acute respiratory distress syndrome which is a known lung inflammation so drastic that the fluid gets up around and also in the lungs leading to a situation of septic shock because of fall in the blood pressure causing the bodily organs to starve for oxygen (Shah et al. 2020). As per the reports of WHO, World Health Organization in January 2020, any person who travelled to Wuhan, Hubei Province in China, two weeks before the onset of the symptoms got suspected to be a COVID -19 patient (Bogoch and Thomas 2020). COVID-19 has been declared as a pandemic by WHO on 12 march 2020 (Zhu et al. 2020). Coronavirus is found to be a zoonotic which means of animal origin (mainly marine) and gets transmitted to humans from the animals and reports suggests that it transmits from animals to humans therefore, it is known to be an animal-oriented virus (Cascella et al. 2020). The incubation period of this virus stands for 1–14 days approximately, and the symptoms vary from patient to patient. Children below the age group of 6, senior citizens, person having past medical history or person having compromised immune system are more susceptible to this virus and are at stake (Shah et al. 2020). On 20 January 2020, National Health Commission of China confirmed the human to human transmission of this virus which made it very difficult to regulate (Elfiky 2020). As of now 187 countries have been severely affected by this virus with over 38,186,663 confirmed cases out of which 1,087,315 deaths have been reported as on 13 October 2020. (https://www.worldometers.info/). USA has become the worst affected country with 8,046,993 cases. The clinical and the epidemiological features of COVID-19 patients show that the CoV-2 infection can lead to intensive care unit (ICU) admission and high mortality. Around 16%–21% of people in China have acquire severe illness, having a mortality rate of 2–3% (Wang et al. 2020; Chen et al. 2020). Therefore, it is a need of the hour to identify some of the effective antiviral agents to combat and explore the clinical manifestations of antiviral drugs. Our approach will be to discover effective drugs and test whether the existing antiviral drugs are effective or not in treating other related viral infections. Many of the antiviral drugs, such as ribavirin, interferon (IFN), favipiravir (FPV), and lopinavir (LPV)/ritonavir (RTV), have already been used in patients with SARS or MERS with good results, but still the efficacy of some of the drugs remains controversial (Qingxian et al. 2019). Recently, it has been reported that, as a prodrug, FPV (half-maximal effective concentration (EC50) = 61.88 $\mu mol.L^{-1}$, half-maximal cytotoxic concentration (CC50) > 400 $\mu mol.L^{-1}$, and selectivity index (SI) > 6.46) efficiently inhibits the SARS-CoV-2 infection in Vero E6 cells (ATCC-1586) (Wang et al. 2020). It has also been made clear that FPV is effective in protecting mice against Ebola virus, although its EC50 value in Vero E6 cells was as high as 67 $\mu mol.L^{-1}$ (Oestereich et al. 2014). In this study, we are going for a review on the efficacy, the anti-viral properties, and the clinical status of FPV as a drug of choice for fighting the SARS-CoV-19.

4.2 Causing Agent

The causing agent of COVID-19 is severe acute respiratory syndrome coronavirus-2 (SARS-CoV-2) which belongs to the same class as of Middle East respiratory syndrome (MERS) virus and SARS–CoV (Elfiky 2020; Wrapp et al. 2020). SARS-CoV was first detected in Guandong province, China, in November 2002, (Ghosh et al. xxxx) and MERS was first reported in Saudi Arabia in 2012 (Ghosh et al. 2020). HCoVs are generally positive sense and very long single-stranded RNA viruses. Two groups of protein basically involved in the characterization of HCoVs one being the structural such as spike (S), nucleocapsid (N), matrix (M), envelope (E), and the other being non-structural part which includes proteins such as RNA-dependent RNA polymerase (RdRp) (nsp12) which is a very crucial enzyme in RNA virus life cycle (Elfiky et al. 2017). The SARS-CoV-2 is a new virus which is spherical in shape with a spike protein similar or mushroom-shaped proteins known as spikes that resemble to the shape of a "CROWN" (Pant et al. 2021). SARS virus family consists of 14 binding residues from which 8 amino acids are specifically reserved for SARS-CoV-2, and importantly, the binding residues of this family have strong interaction with ACE-2 (Angiotensin converting enzyme-2) (Cynthia et al. 2020; Abdul et al. 2020). The SARS-CoV-2 virus mainly spreads via the droplets, saliva, or nose discharge of an infected person after he sneezes or coughs (Pant et al. 2021). SARS-CoV-2 is a zoonotic virus. According to the virologist, human and the zoonotic coronavirus belong to the Coronaviridae family in the order Nidovirales. As per now, 4 genera in the subfamily Coronavirinae of family Coronaviridae which includes: alphacoronavirus, betacoronavirus, deltacoronavirus, and gammacoronavirus are known (Ko et al. 2020). Earlier, 6 human respiratory coronavirus were recognized which include HCoV-229E (from alphacoronavirus), HCoV-0C43 (Betacoronavirus), HCoV-NL63 (Alphacoronavirus) and HKU1 (betacoronavirus), SARS and MERS-CoV, and now with the emergence of SARS-CoV-2(Coronavirus-2), the number has increased to 7 (Chen et al. 2020). COVID-19 is also a member of the betacoronavirus genera like the SARS and MERS (Pant et al. 2021; Chan et al. 2015).

For the effective treatment, any traditional drug discovery program or any novel discovery in such a short period of time is almost not feasible as it is a very time-consuming and costly process which sometimes takes decades to complete (Romano and Tatonetti 2019). Many of the already existing antiviral drugs have been restructured for the effective treatment, and some of them have shown good results including FPV, ramdesivir, chloroquine, ritonavir, olsetamivir, many protease inhibitors, etc., (Shah et al. 2020; Yang et al. 2020) but the FPV found to be most effective with minimal side effects.

4.3 Favipiravir

Favipiravir (FPV) (T-705; 6-fluoro-3-hydroxy-2-pyrazinecarboxamide) is an antiviral agent which has the capacity to selectively and potently inhibit the RNA-dependent RNA polymerase (RdRp) of RNA viruses (Furuta et al. 2002). As the catalytic domain of RdRp is conserved among various types of RNA viruses, the discovered mechanism of action (Furuta et al. 2005) shows a broad spectrum of antiviral activities of FPV. FPV has been found effective for a wide range of influenza viruses (Furuta et al. 2002), including resistant strains to existing anti-influenza drugs and the RNA viruses. An unique and different antiviral profile makes FPV a potentially and promising drug candidate for specifically untreatable RNA viral infections. FPV has proved to be a selective and potent inhibitor of influenza viral RNA polymerase and also against the various subtypes and different strains of influenza viruses along with the one which have been found resistant against marketed neuraminidase and M2 inhibitors (Furuta et al. 2017). This indicates that FPV is a promising drug and a suitable candidate to be studied and explored for the treatment of infections by not only influenza virus but also a wide range of RNA viruses including COVID-19.

FPV (T705) is a purine nucleic acid-based analog and is one of the antiviral candidates for COVID-19. It is a derivative of pyrazine carboxamide (6-fluoro-3-hydroxy-2-pyrazinecarboxamide) and is a broad-spectrum antiviral drug which was used and approved in 2014 in Japan for the effective treatment of influenza (Scavone et al. 2020). It is also a prodrug which is ribosylated and phosphorylated intracellularly to make the active metabolite FPV ibofuranosyl-5'-triphosphate (T-705-RTP) (Furuta et al. 2005; Sangawa et al. 2013). T-705-RTP competes with the purine nucleosides of the virus and interferes with the viral replication by getting incorporated in the virus RNA and thus potentially inhibits the RNA-dependent RNA polymerase (RdRp) enzyme of RNA virus. As recently shown by (Qingxian et al. 2019), favipiravir was found effective in alleviating the SARS-CoV-2 infection *in vitro.*

In March 2020, FPV has been registered by the National Medical Products Administration of China as the 1st anti-COVID-19 drug during the 1st wave of COVID-19 in China, because of the efficacy shown and the minimal side effects shown by the drug (Du and Chen 2019; Yanai 2020).

4.4 Mechanism of Action

FPV is a prodrug which gets converted into its active form T-705-RTP by ribosylation and phosphorylation (Fig. 4.1). The activity of FPV gets decreased in the presence of purine nucleotides, i.e. ATP and GTP, showing that FPV-RTP can be identified as a pseudopurine by the viral RNA-dependent RNA polymerase (RdRp) enzyme (Furuta et al. 2005; Delang et al. 2018). It has been hypothesized that the FPV-RTP can be wrongly incorporated in a growing viral RNA chain, or it might act by binding to the conserved polymerase domains, thereby preventing the viral RNA replication.

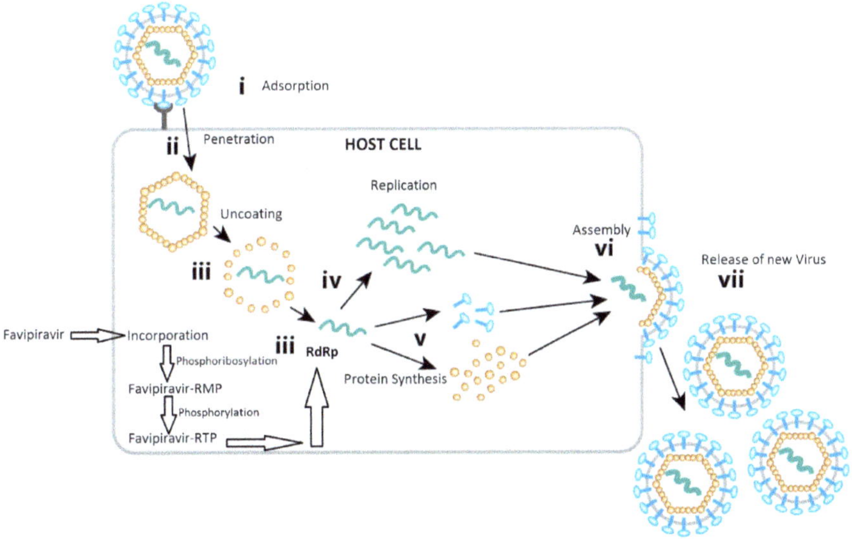

Fig. 4.1 Mechanism of action of Favipiravir

Addition of FPV-RTP in the nascent viral RNA may result in lethal mutagenesis by either ambiguous base-pairing or by chain termination. For lethal mutagenesis, FPV-RTP has to be incorporated into the RNA without going for immediate chain termination (Furuta et al. 2017; Sangawa et al. 2013; Baranovich et al. 2013). Several studies have supported the hypothesis of lethal mutagenesis as a mechanism of action of FPV examples being: influenza virus (Baranovich et al. 2013), West Nile virus, (Morrey et al. 2008) etc.

Figure 4.1 Mechanism of action of Favipiravir: FPV is converted to active triphosphate form and selectively inhibits the activity of the virus. The numerals indicate:

i. Adsorption
ii. Penetration
iii. Uncoating
iv. Replication
v. Protein synthesis
vi. Assembly
vii. Release of new virus.

4.5 Anti Viral Potential of Favipiravir

FPV has been found to be effective against a spectrum of viruses, especially virus subtypes (Furuta et al. 2013), arenavirus, bunyavirus (Gowen et al. 2007), filovirus (Furuta et al. 2017), West Nile virus (Morrey et al. 2008), Western equine encephalitis virus (WEEV) (Julander et al. 2009) and more recently it has also shown in vitro inhibition of norovirus replication (Rocha et al. 2012). Moreover, FPV is also found to be effective for the treatment of patients infected with Ebola and Lassa viruses (Goba et al. 2016). It has been already found effective for the treatment of patients infected with Ebola and Lassa viruses (Goba et al. 2016). However, clear conclusions about the efficacy profile of the drug could not be drawn (Delang et al. 2018). Patients found with an initial diagnosis of Ct $\geq$ 20 (Cycle threshold value, which is mainly considered a surrogate measure of viral load, higher viral load in a patient's acute specimen (with lower Ct value) is associated with increased rates of transmission (Reichler et al. 2020) for EBOV RNA and on-trial mortality was found to be 20.0% (95% confidence interval 11.6–32.4), which was 33% lower than target value (30%). A retrospective analysis of patients with (EVD) concluded that compared to the patients who received the WHO-recommended supportive therapy, the patients who went for additional FPV treatment showed a higher survival rate and a longer average survival time, and also a higher percentage of patients with a > 100-fold viral load reduction (Bai 2016).

4.6 Antiviral Activity Against Covid-19

As per a report by Watanabe (Watanabe et al. 2020), FPV was administered as drug during a clinical trial to 200 patients infected from COVID-19 at different hospitals in Wuhan and Shenzhen. The results showed that patients who have been receiving the drug tested negative in a relatively short time (4 days as compared to 11 days in the control group), while the symptoms of pneumonia also significantly reduced. Chen (Chen et al. 2020) published the results of a random clinical trial for the treatment of 240 patients with COVID-19 and hospitalized in three different hospitals from 20 February 2020 to 12 March 2020, and the results showed that the 7-day clinical recovery rate was 71.43% in the FPV group. The Japanese Association for Infectious Diseases reported an elderly patient (http://www.kansensho.or.jp/uploads/files/topics/2019ncov/covid19_casereport_en_200408_2.pdf.) or a haemodialysis patient (Eri and Saito 2019), whose COVID-19 pneumonia got alleviated by using FPV. The Japanese Association for Infectious Diseases also reported an improvement of 90%, 85%, and 61%, in mild, moderate, and severe COVID-19 cases, respectively, after 14 days from the start of administration of FPV. Cai (Qingxian et al. 2019) performed an open-label control study for evaluation of treatment with FPV for COVID-19.

Patients were assigned oral FPV plus interferon-α aerosol inhalation (n = 35), or LPV/RTV which is a fixed-dose protease inhibitor plus interferon-α aerosol inhalation (n = 45). The results showed that a viral clearance time was significantly shorter in the FPV group than the LPV/RTV group by 7 days (median value). Further studies through multivariable Cox regression revealed that FPV was independently also associated with faster viral clearance. The FPV group also showed significant improvement in chest imaging compared with the LPV/RTV group, having an improvement rate of 91.43% versus 62.22% after 4 days of treatment.

4.7 Structural Analysis of Favipiravir

Structural analysis of FPV was done by Kun Harismah, Mahmoud Mirzaei (Harismah and Mirzaei 2020) by means of exploring tautomer formation (Fig. 4.2), and they found that four tautomeric structures can be considered as ligands obtained by density functional theory (DFT) calculations at B3LYP/6-31G* theoretical level (Frisch et al. 2013). Tautomeric conformational change is a very common phenomenon in the cyclic organic compounds (Yaraghi et al. 2015; Kouchaki et al. 2016). FPV also undergo tautomerism process. Hence, the structures of all tautomeric forms of FPV have been studied based upon quantum chemical computations. The different numerical values including total energy (E), relative energy of stabilization (delta-E), energy levels of the highest occupied and the lowest unoccupied molecular orbitals (HOMO and LUMO), and the energy gap (EG) were obtained for all of the optimized structures (Table 4.1). As per the study, four tautomers were found possible for FPV, (Fig. 4.3) among which the stability could vary regarding the values of total energy (E) (Harismah and Mirzaei 2020). These tautomers can be made available if the energy required for tautomer formation is supplied by some other resources like that of intermolecular interactions and bonding. Each of the four investigated tautomeric ligands have the different interacting properties for each protease and polymerase targets. They found F1 of the tautomers as stable compounds based on the results of structural analysis, and one of the tautomers F3 is found to be the most active ligand against macromolecules based on the molecular docking simulations. So as per the above study done by Kun Harismah, Mahmoud Mirzaei, it can be concluded that FPV in its tautomeric form also exhibited the antiviral activity which provides us a scope for synthesizing other forms of FPV also along with the basic molecule.

Fig. 4.2 Chemical structure of Favipiravir

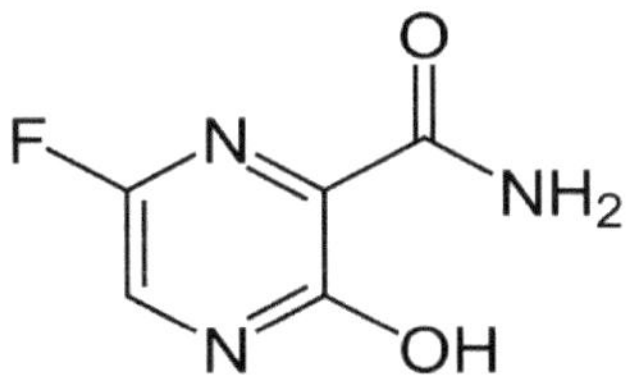

Table 4.1 Molecular properties (As per data collected by Kun Harismah, Mahmoud Mirzaei) (Harismah and Mirzaei 2020)

Property	F1	F2	F3	F4
E (kcal/mol)	−380,886	−380,871	−380,891	−380,876
ΔE (kcal/mol)	5	20	0	15
LUMO (eV)	−2.24	−2.23	−2.66	−2.69
HOMO (eV)	−6.98	−7.05	−6.71	−6.62
EG (eV)	4.74	4.82	4.05	3.93

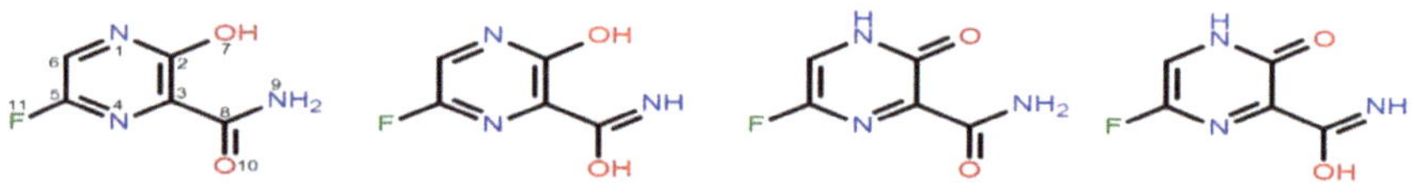

F1 F2 F3 F4

Fig. 4.3 Tautomers of Favipiravir

4.8 Clinical Trials

To find out the suitability of FPV for repurposing in COVID-19, clinical trials were done on FPV as an investigational drug, and the different clinical trials done with their responses or results are listed (https://clinicaltrials.gov/ct2/who_table).

Registration no.	Public title	Primary sponsor	Regd. date	Target size	Study type	Phase	Outcome
ChiCTR2000029600	Clinical study for safety and efficacy of favipiravir in the treatment of novel coronavirus pneumonia (COVID-19)	The Third People's Hospital of Shenzhen	6/2/2020	Group A:30; Group B:30; Group C:30;	Non randomized, interventional study	0	1.Declining speed of SARS-CoV-2 by PCR 2. Incidence rate of chest imaging and liver enzyme
ChiCTR2000029544	A randomized controlled trial for the efficacy and safety of baloxavir, favipiravir, and marboxil tablets in COVID-19	The First Hospital Affiliated to Zhejiang University's Medical School	3/2/2020	Experimental group 1:10; Experimental group 2:10 ;Control group:10;	Parallel, randomized, interventional study	0	1.Time to clinical improvement
ChiCTR2000029548	Randomized, open-label, controlled trial for evaluating of the efficacy and safety of baloxavir marboxil, favipiravir, and lopinavir-ritonavir against COVID-19	The First Affiliated Hospital, Zhejiang University School of Medicine	4/2/2020	A:10;B:10;C:10;	Randomized, open label, parallel, interventional study	0	1. Time to clinical improvement 2. Time to viral negativity by RT-PCR
ChiCTR2000030113	Trial for safety and efficacy of favipiravir in the treatment of novel coronavirus pneumonia (COVID-19) with poorly responsive ritonavir/ritonavir	The Third People's Hospital of Shenzhen	23/2/2020	Control group:15; Experimental group:15;	Randomized, parallel, interventional study	0	1. Blood routine test 2. Liver function

(continued)

(continued)

Registration no.	Public title	Primary sponsor	Regd. date	Target size	Study type	Phase	Outcome
ChiCTR2000029996	A randomized, open-label, controlled trial for the efficacy and safety of favipiravir tablets in the treatment of patients with COVID-19	Beijing Chaoyang Hospital, Capital Medical University	20/2/2020	Low-dose group:20;Middle-dose group:20;High-dose group:20;	Randomized, open label, parallel, interventional study	2	1. Time to clinical recovery
ChiCTR2000030894	Favipiravir combined with tocilizumab in the treatment of novel coronavirus pneumonia (COVID-19)	Peking University First Hospital	16/3/2020	Group 1:90;Group 2:30;Group 3:30;	Randomized, controlled, parallel, interventional study	4	1. Clinical cure rate
ChiCTR2000030987	A randomized controlled trial for favipiravir tablets combine with chloroquine phosphate in the treatment of novel coronavirus pneumonia (COVID-19)	Beijing Chao-yang Hospital, Capital Medical University	20/3/2020	Experimental group 1:50; experimental group 2:50; Control group:50;	Randomized, controlled, parallel, interventional study	N/A	1. Improvement or recovery of respiratory system 2. Viral nucleic acid shedding

4.9 Conclusion

Based upon the previous antiviral studies of favipiravir, it can be concluded that favipiravir is an effective broad-spectrum antiviral, and from the study of its clinical trials and mode of action against COVID-19, it is interfered that favipiravir can be used as a possible efficacious therapeutic agent against COVID-19 due to its minimal side effects as also shown in the studies that during the trials or study of the drug on COVID-19 patients, there were very less safety concerns. And its structural analysis also showed its tautomers to be good ligands for viral protease and polymerase enzymes inhibiting the viral activity. Recently DGCA (India) has also approved the sale of favipiravir based on its clinical efficiency, low toxicity levels, and minimal side effects as it has led to alleviation of severity of COVID-19 patients. From the complete study, we can conclude that there are also other antiviral, ACE-2 inhibitors, and other traditional medicines which have been found effective for the treatment of COVID-19 but despite the fact, favipiravir has also proven itself as a competitor to the more potent and efficacious drugs. Hence, favipiravir could be a suitable candidate for repurposing as a drug against COVID-19.

4.10 Future Perspectives

As per the details and the data collected from various literature, it has been justified that favipiravir has proved to be an effective drug in combating the COVID-19 and can be used as a possible alternative to other antiviral drugs such as remdesivir, ribavirin, chloroquine, etc. Due to the found activity of favipiravir against COVID-19 and based on the clinical trials, it has been manufactured and marketed worldwide as it has been made the first drug to be introduced in China for COVID-19. In India also, where the impact of COVID-19 has been the most accounting for nearly a third of the world's COVID-19 cases, the use of favipiravir has been approved by the Drug Controller General of India (DCGI) for restricted emergency use, and favipiravir has been marketed by Indian pharma major Glenmark as FabiFlu and was launched in the Indian market on 21 June 2020. Patients of COVID-19 were treated with oral tablets of favipiravir for two weeks at the most. Up to 3600 mg or up to 18 tablets in dosage of 200 mg each were given to the patients on the first day. "From the second day, a total of eight tablets, four in the morning and four in the evening, forming a total dosage of 1600 mg were given, most patients gradually recovered in a week, while other patients took almost two weeks to recover (Hindustan Times report 2020). Generic drugmaker Dr Reddy's Laboratories launched antiviral drug Avigan (favipiravir) tablets, currently being manufactured by innovator Japanese drug giant Fujifilm Toyama Chemical Co, for the potential treatment of mild to moderate COVID-19 patients (The Hindu Report). Fusion Health Care Pvt Ltd, a leading innovative pharmaceutical company based at Hyderabad, has announced the launch of its Favipiravir drug under the brand name XARAVIR, for the treatment of mild to moderate COVID-19 cases (EP News Bureau). Drugmaker Sun Pharmaceutical Industries Ltd. has launched favipiravir for the treatment of mild to moderate cases of COVID-19 as Fluguard (Economic Times report). Based on the above facts and the results of the clinical trials, we can say that favipiravir finds a promising future as antiviral drug for treating different strains of virus, and its complete clinical efficacy can be known only after its administration in the patients till now which has proven to be a not complete but partial treatment at least going for the recovery of the patients with COVID-19.

Acknowlegements We would like to appreciate and thanks Amity University for the honourable contribution.

Conflict of Interest All of authors declare that they have no conflicts of interest to this work.

References

Abdul MB, Areeba K, Usman A, Hira S (2020) Evidence of the COVID-19 virus targeting the CNS: tissue distribution, host−virus interaction, and proposed neurotropic mechanisms. ACS Chem Neurosci. https://doi.org/10.1021/acschemneuro.0c00122

Bai CQ (2016) Clinical and virological characteristics of Ebola virus disease patients treated with favipiravir (T-705). Clin Infect Dis 63:1288–1294

Baranovich T, Wong SS, Armstrong J, Marjuki H, Webby RJ, Webster RG, Govorkova EA (2013) T-705 (favipiravir) induces lethal mutagenesis in influenza A H1N1 viruses in vitro. J Virol 87:3741–3751

Bogoch W, Thomas B (2020) Pneumonia of unknown etiology in Wuhan, China: potential for international spread via commercial air travel. J Travel Med. https://doi.org/10.1093/jtm/taaa008

Cascella M, Rajnik M, Cuomo A, Dulebohn SC, Napoli RD (2020) Features, evaluation and treatment Coronavirus (COVID-19). StatPearls 2020

Chan JF, Lau SK, Cheng VC, Woo PC, Yuen KY (2015) Middle East respiratory syndrome coronavirus: another zoonotic betacoronavirus causing SARS like disease. Clin Microbiol Rev 28:465–522

Chen N, Zhou M, Dong X, Qu J, Gong F, Han Y (2020) Epidemiological and clinical characteristics of 99 cases of 2019 novel coronavirus pneumonia in Wuhan, China: a descriptive study. Lancet 395(10223):507–513

Chen Y, Liu Q, Guo D (2020) Emerging coronaviruses: genome structure, replication and pathogenesis. J Med Virol 92:418–423

Chen C, Huang J, Yin P, Zhang Y, Cheng Z, Wu J, Yin P, Favipiravir versus arbidol for COVID-19: a randomized clinical trial 2020.https://doi.org/10.1101/2020.03.17. 20037432

Cynthia L, Qiongqiong Z, Yingzhu L (2020) Research and Development on therapeutic agents and vaccines for COVID-19 and related human coronavirus diseases. ACS Cent Sci. https://doi.org/10.1021/acscentsci.0c00272

Delang L, Abdelnabi R, Neyts J (2018) Favipiravir as a potential counter measure against neglected and emerging RNA viruses. Antiviral Res 153:85–94

Du YX, Chen XP (2019) Favipiravir: Pharmacokinetics and concerns about clinical trials for 2019-nCoV infection clinical pharmacology & therapeutics. www.cpt-journal.com

Elfiky A (2020) Anti-HCV, nucleotide inhibitors, repurposing against COVID-19. Life Sci 248:117477

Elfiky AA, Mahdy SM, Elshemey WM (2017) Quantitative structure-activity relationship and molecular docking revealed a potency of anti-hepatitis C virus drugs against human corona viruses. J Med Virol 89:1040–1047

Eri K, Saito S (2019) Efficacy of favipiravir for an end stage renal disease patient on maintenance hemodialysis infected with novel coronavirus disease. CEN Case Report 2020:1–6. https://doi.org/10.1007/s13730-020-00534-1

Frisch MJ, Trucks GW, Schlegel HB, Scuseria GE, Robb MA, Cheeseman JR (2013) Gaussian 09, Revision D.01.Gaussian, Inc., Wallingford CT

Furuta Y, Takahashi K, Fukuda Y, Kuno M, Kamiyama T, Kozaki K, Nomura N, Egawa H, Minami S, Watanabe Y, Narita H, Shiraki K (2002) In vitro and in vivo activities of anti-influenza virus compound T-705. Antimicrob Agents Chemother 46:977–981

Furuta Y, Takahashi K, Kuno M, Sangawa H, Uehara S, Kozaki K, Nomura N, Egawa H, Shiraki K (2005) Mechanism of action of T-705 against influenza virus. Antimicrob Agents Chemother 49:981–986

Furuta Y, Gowen BB, Takahashi K, Shiraki K, Smee DF, Barnard DL (2013) Favipiravir (T-705), a novel viral RNA polymerase inhibitor. Antiviral Res 100:446–454

Furuta Y, Takashi K, Takaaki N (2017) Favipiravir (T-705), a broad spectrum inhibitor of viral RNA polymerase. Proceedings of the Japan Academy, Ser. B Physical and Biological Sciences 2017, 93(7):449–463

Ghosh AK, Brindisi M, Shahabi D, Chapman ME, Mesecar AD, Drug development and medicinal chemistry efforts toward SARS-coronavirus and covid-19 therapeutics. ChemMedChem https://doi.org/10.1002/cmdc.202000223

Goba A, Khan S. H, Fonnie M, Fullah M, Moigboi A, Kovoma A (2016) An outbreak of ebola virus disease in the lassa fever zone. Int J Infect Dis 214:S110–S121

Gowen BB, Wong MH, Jung KH, Sanders AB, Mendenhall M, Bailey KW, Furuta Y, Sidwell RW (2007) In vitro and in vivo activities of T-705 against arenavirus and bunyavirus infections. Antimicrob Agents Chemother 51:3168–3176

Harismah K, Mirzaei M (2020) Favipiravir: structural analysis and activity against COVID-19. Adv J Chem, Sect B 2(2):55–60

http://www.kansensho.or.jp/uploads/files/topics/2019ncov/covid19_casereport_en_200408_2.pdf Accessed on 22 June 2020. xxxxhttps://clinicaltrials.gov/ct2/who_table) Accessed on 22 June 2020

https://economictimes.indiatimes.com/industry/healthcare/biotech/pharmaceuticals/sun-pharma-latest-to-sell-covid-19-drug-favipiravir-in india/articleshow/77349409.cms?from=mdr Last Updated: Aug 04, 2020, 04:10 PM IST Accessed on: 20 September 2020

https://www.expresspharma.in/covid19-updates/fusion-health-care-launches-favipiravir-for-covid-19-treatment/ By EP News Bureau on Aug 18, 2020 Accessed on: 20 September 2020

https://www.hindustantimes.com/india-news/glenmark-pharma-drops-price-of-covid-19-drug-favipiravir-to-rs-75-tablet/story-VNVG3UWN0strJyqXkY1gTM.html. Updated: Jul 13, 2020 21:28 IST Accessed on: 18 August 2020

https://www.thehindu.com/news/national/dr-reddys-launches-avigan-favipiravir-in-india-for-covid-19-treatment/article32394457.ece. Updated: August 19, 2020 19:01IST. Accessed on: 2 September 2020

https://www.thehindubusinessline.com/companies/early-clinical-trial-results-of-favipiravir-for-covid-19-promising-glenmark/article31876374.ece Updated on June 20, 2020 Published on June 20, 2020 accessed on: 18 August 2020

Hui DS, Azhar IE, Madani TA, Ntoumi F, Kock R, Dar O (2020) The continuing 2019-nCoV epidemic threat of novel coronaviruses to global health—the latest 2019 novel coronavirus outbreak in Wuhan, China. Int J Infect Dis 91:264–266

Hussin R, Siddappa NB (2020) The epidemiology and pathogenesis of coronavirus disease (COVID-19) outbreak. J Autoimmune Dis 102433. https://doi.org/10.1016/j.jaut.2020.102433

Julander JG, Smee DF, Morrey JD, Furuta Y (2009) Effect of T-705 treatment on western equine encephalitis in a mouse model. Antiviral Res 82:169–171

Kaye D, Pringle CR (2005) Avian influenza viruses and their implication for human health. Clin Infect Dis 40:108–112

Ko WC, Rolain JM, Lee NY, Chen PL, Huang CT, Lee PI, Hsueh PR (2020) Arguments in favour of remdesivir for treating SARS-CoV-2 infections. Int J Antimicrob Agents 55(4):105933

Kouchaki A, Gülseren O, Hadipour N, Mirzaei M (2016) Relaxations of fluorouracil tautomers by decorations of fullerene-like SiCs: DFT studies. Phys Lett A 380:2160–2166

Liu J, Dai S, Wang M, Hu Z, Wang H, Deng F (2016) Virus like particle-based vaccines against emerging infectious disease viruses. Virol Sin 31:279–287

Morrey JD, Taro BS, Siddharthan V, Wang H, Smee DF, Christensen AJ, Furuta Y (2008) Efficacy of orally administered T-705 pyrazine analog on lethal West Nile virus infection in rodents. Antiviral Res 80:377–379

Oestereich L, Lüdtke A, Wurr S, Rieger T, Muñoz-Fontela C, Günther S (2014) Successful treatment of advanced Ebola virus infection with T-705 (favipiravir) in a small animal model. Antiviral Res 105:17–21

Pant S, Singh M, Ravichandiran V, Murty USN, Srivastava H, Peptide-like and small-molecule inhibitors against Covid-19. J Biomol Struct Dyn. https://doi.org/10.1080/07391102.2020.1757510

Qingxian C, Minghui Y, Dongjing L, Jun C, Dan S, Junxia X, Xuejiao L, Yuanbo G, Qiue C, Yang Y, Chenguang S, Xiaohe L, Ling P, Deliang H, Jing Z, Shurong Z, Fuxiang W, Jiaye L, Li C, Shuyan C, Zhaoqin W, Zheng Z, Ruiyuan C, Wu Z, Yingxia L, Lei L, Experimental treatment with favipiravir for COVID-19: an open-label control study journal. www.elsevier.com/locate/eng

Qu G, Li X, Hu L, Jiang G (2020) An imperative need for research on the role of environmental factors in transmission of novel coronavirus (COVID-19). Environ Sci Technol 54:3730–3732

Reichler MR, Bruden D, Harold T, Bobbie E, Barbara K, Nadia D, Klena J, Hennessy T (2020) Ebola patient virus cycle threshold and risk of household transmission of Ebola virus. J Infect Dis 221:707–714

Rocha PJ, Jochman D, Dallmeier K, Leyssen P, Nascimento MS, Neyts J (2012) Favipiravir (T-705) inhibits in vitro norovirus replication. Biochem Biophys Res Commun 424:777–780

Romano JD, Tatonetti NP (2019) Informatics and computational methods in natural product drug discovery: a review and perspectives. Front Genet 10:368

Sangawa H, Komeno T, Nishikawa H, Yoshida A, Takahashi K, Nomura N, Furuta Y (2013) Mechanism of action of T-705 ribosyl triphosphate against influenza virus RNA polymerase. Antimicrob Agents Chemother 57:5202–5208

Scavone C, Brusco S, Bertini M, Sportiello L, Rafaniello C, Zoccoli A, Berrino L, Racagni G, Rossi F, Capuano A (2020) Current pharmacological treatments for COVID-19: What's next? Br J Pharmacol, 1–12

Shah B, Modia P, Sagara SR (2020) In silico studies on therapeutic agents for COVID-19: Drug repurposing Approach. Life Sci 252:117652

Sharma A, Tiwari S, Deb MK, Marty JL (2020) Severe acute respiratory syndrome coronavirus-2 (SARS-CoV-2): a global pandemic and treatment strategies. Int J Antimicrob Agents 56:106054

Wang M, Cao R, Zhang L, Yang X, Liu J, Xu M (2020) Remdesivir and chloroquine effectively inhibit the recently emerged novel coronavirus (2019-nCoV) *in vitro*. Cell Res 30:269–271

Wang D, Hu B, Hu C, Zhu F, Liu X, Zhang J (2020) Clinical characteristics of 138 hospitalized patients with 2019 novel coronavirus-infected pneumonia in Wuhan, China. J Am Med Assoc

Watanabe S, Chan M, Suzuki W (2020) China says Japan-developed drug Avigan works against coronavirus. Positive reception by Chinese government contrasts with reservations in Japan. https://asia.nikkei.com/Spotlight/Coronavirus/China-says-Japan-developed-drug-Avigan-works-against-coronavirus

Wrapp D, NianShuang W, Kizzmekia SC (2020) Cryo-EM structure of the 2019- nCoV spike in the prefusion conformation. Science. https://doi.org/10.1126/science.abb2507

Yanai H (2020) Favipiravir: a possible pharmaceutical treatment for COVID-19. J Endocrinol Metabolism 10(2):33–34

Yang LC, Qiu X, Zeng YK, Jiang M, Fan HR, Zhang ZM (2020) Coronavirus disease 2019: a clinical review. Eur Rev Med Pharmacol Sci 24:4585–4596
Yaraghi A, Ozkendir OM, Mirzaei M (2015) DFT studies of 5-fluorouracil tautomers on a silicon graphene nanosheet. Superlattices Microstructres 85:784–788
Zhu N, Zhang D, Wang W, Li X, Yang B, Song J (2020) A novel coronavirus from patients with pneumonia in China, 2019. N Engl J Med 382:727–733

Chapter 5
Managing Deep Venous Thrombosis with the Synergistic Approach of Physical Therapy and Medication in a Young Female: A Case Study

Priyanshi Aggarwal, Priyanka Rishi, and Sonia Pawaria

5.1 Introduction

Deep vein thrombosis (DVT) is a medical condition characterized by the formation of blood clots, or thrombi, in deep veins, typically in the legs. These clots can obstruct blood flow in the affected veins and may lead to serious complications if they dislodge and travel to the lungs, causing a pulmonary embolism (Office of the Surgeon General 2008). Venous thromboembolism (VTE), encompassing disorders such as deep vein thrombosis (DVT) and pulmonary embolism (PE), is closely linked to arterial thrombotic conditions. It poses a rising challenge to public health due to increased morbidity, premature mortality, hospital admissions, and elevated healthcare expenses (Kunutsor et al. 2020). Its incidence is 17.46 cases per 10,000 hospital admissions in India. The leading contributing factor was malignancy, accounting for 31% of cases, followed by postoperative status at 30%. Specifically related to surgeries, the incidence of VTE was five cases per 10,000 operations (Lee et al. 2009). Clinical risk factors associated with venous thromboembolism include factors such as advancing age, paralysis, prior occurrences of venous thromboembolism, surgical procedures involving the abdomen, pelvis, hip, leg, heart, or lungs, heart-related problems, trauma, and extended periods of immobility (Kwon et al. 2003). Post-thrombotic syndrome (PTS) is a long-term outcome that results from deep vein thrombosis (DVT). It is marked by ongoing symptoms of chronic venous insufficiency, which may include leg pain, swelling, a sensation of heaviness, and in severe cases, the development of venous ulcers (Jasionowska et al. 2022). Pharmacological

P. Aggarwal (✉) · P. Rishi · S. Pawaria
Faculty of Physiotherapy, SGT University, Gurugram, India
e-mail: Priyanshiaggarwal28@gmail.com

N. A. Siddiqui et al. (eds.), *Advances in Disaster Management, Volume 2*, Springer Transactions in Civil and Environmental Engineering,
https://doi.org/10.1007/978-981-96-4047-8_5

interventions for the management of DVT often include the use of low molecular-weight heparin and oral anticoagulants (Kp et al. 2017). In addition to medication, non-pharmacological strategies like gentle physical activity and physiotherapy have been recommended as preventive measures for DVT (Hashemzadeh et al. 2021).

5.2 Case Report

A 19-year-old female presented to the ortho clinic with chief complaints of sudden onset of pain in the left inguinal region radiating down to the thigh with swelling in the left lower limb, inability to bear weight on the left leg, and inability to walk along with fever and chills at night for a week. On examination there was swelling of the thigh and ankle, discoloration of the leg, and an increase in local temperature. Homan's sign was also positive. A peripheral left lower extremity venous Doppler ultrasound (USG) and coagulation profile were requested to rule out a vascular disorder. The USG investigation demonstrated DVT involving the left side common iliac and external iliac (Fig. 5.1), common femoral, and superficial femoral veins. Great saphenous veins were also thrombosed in the thigh. mild diffuse cutaneous edema was seen in the leg and foot. (Fig. 5.2).

The activated partial thromboplastin time (APTT) is valued at 31.20 s (normal 27-40 s), prothrombin time (PT) is valued at 15.50 s (11-16 s), and international normalization ratio (INR) is valued at 1.30 (normal < 1.2) with elevated ESR and normal ECG findings. The vital signs were as follows: Blood pressure showed a slight elevation, measuring 148/88 mmHg, the heart rate was recorded at 110 beats per minute, and oxygen saturation was at 100%. The patient was instructed to begin anticoagulation therapy on the first day following the diagnosis for 1 month. She was

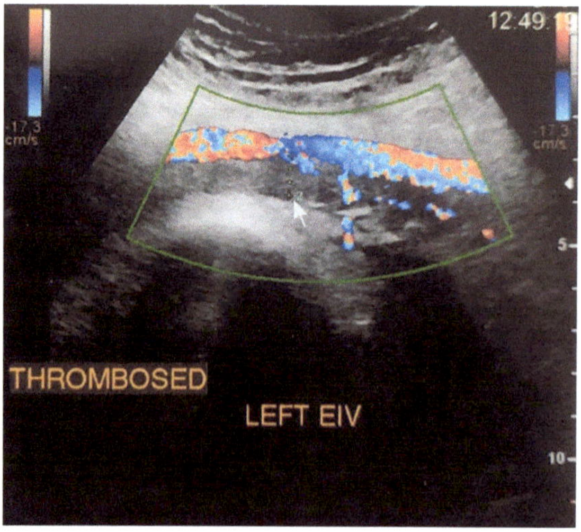

Fig. 5.1 Thrombosed left external iliac vein on venous Doppler ultrasound

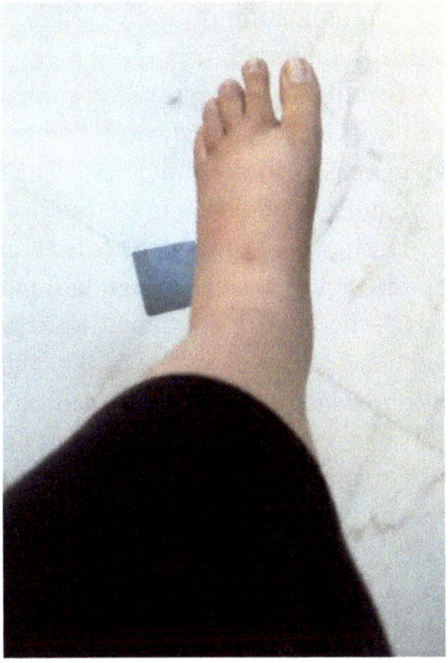

Fig. 5.2 Cutaneous edema of foot

advised to take 5 mg of warfarin per day, and with restrictions in physical activity and movement she was also instructed to wear compression stockings. Simultaneously low-molecular-weight heparin (60 mg) was injected subcutaneously twice a day. PT/INR was ordered after 3 doses. One week later, the dosage was adjusted to 7.5 mg and 5 mg of warfarin, on alternate days for the rest of the treatment time. The patient was also advised to follow up with her primary physician for regular monitoring every week for 1 month and to refrain from participating in activities involving contact or collision while undergoing anticoagulation. The potential risks associated with warfarin therapy and the likelihood of developing a pulmonary embolism were thoroughly and comprehensively discussed.

The patient was engaged in gyming activities 5 months before the diagnosis and recalled a traumatic "high-impact" injury to the posterior area of her left lower extremity (calf). After 1-month post-injury, the patient noted mild pain in her ankle and calf with ambulation. The patient visited her primary doctor one-month post injury and was diagnosed with a minor soft tissue injury with slightly raised blood Pressure. The patient's past medical and family history did not provide any significant information related to a history of thrombophilia or other major thrombotic risk factors.

The patient visited the hospital for a follow-up appointment 1.5 months after being diagnosed with deep venous thrombosis. During this follow-up examination, there was swelling with a reduction in pain, a color Doppler study was performed, which revealed reduced flow velocity in the great saphenous veins and confirmed the presence of deep venous thrombosis (DVT) in the common iliac, external iliac, and common femoral veins. Vit D3 and B12 levels were checked. D3 level is valued at 23 (normal 30–100), and B12 level is valued at 203 (normal 211–911). The physician saw no indication of an IVC filter and decided to proceed conservatively. Warfarin was discontinued, and the patient was prescribed Pradaxa at a dosage of 150 mg twice daily and ecosprin 75 mg once daily for 6 months. Additionally, limb elevation and walking with thigh-high compression stockings were recommended, along with an increase in oral fluid intake. The frequency of monitoring was adjusted to be once a month for the duration of the treatment period.

During the subsequent follow-up appointment, which occurred 2.5 years after commencing pharmacological treatment following a deep venous thrombosis diagnosis, the patient continued to exhibit complete blockage of the left external iliac vein and partial blockage of the common femoral vein, with the presence of chronic thrombus with swelling on exertion and decreased quality of life. The use of Pradaxa was discontinued, and the patient was recommended to take a daily dose of 20 mg of Rivaroxaban. The monitoring schedule was modified to occur once every 6 months for the duration of the treatment period.

The patient was referred to the physical therapy department as a follow-up case for inadequately recanalized deep venous thrombosis (DVT). The recommended physical therapy regimen, in conjunction with pharmacological therapy, extends over six months and involves a combination of components that are implemented five days a week. It includes a daily 3-km walk while wearing compression stockings, 15 min of low-resistance static cycling, ankle-toe pumps consisting of 3 sets with 30 repetitions each day, stair-climbing exercises, a low-intensity treadmill workout without incline, and a 20-min daily session of mechanical compression using an intermittent pneumatic compressor. (Fig. 5.3). The patient's oxygen saturation was assessed using a probe attached to her hand while exercising, and her blood pressure was consistently monitored. Following the physical therapy program, the patient continued to wear compression stockings, and she was recommended to elevate the limb at night with the support of a pillow in case of swelling. Another color Doppler study was requested six months later, revealing the presence of only a minor chronic thrombus in the left common femoral vein. The next follow-up date was set for 1 month. The concluding Doppler examination displayed unobstructed blood flow in all the veins (Fig. 5.4).

The patient successfully resumed her regular activities with a comprehensive awareness of the heightened risk of recurrent DVT and potential future complications. A follow-up appointment was scheduled for one year from that point to monitor the possibility of recurrence.

5.3 Discussion

This case illustrates the complex nature of DVT in young individuals with multiple risk factors. While anticoagulation therapy is crucial in DVT management, the persistence of chronic thrombus necessitates a multifaceted approach.

The majority of deep vein thrombi originates in the calf region, often in areas like the soleal sinusoids and behind the venous valves (Cronan 1993). Virchow's triad

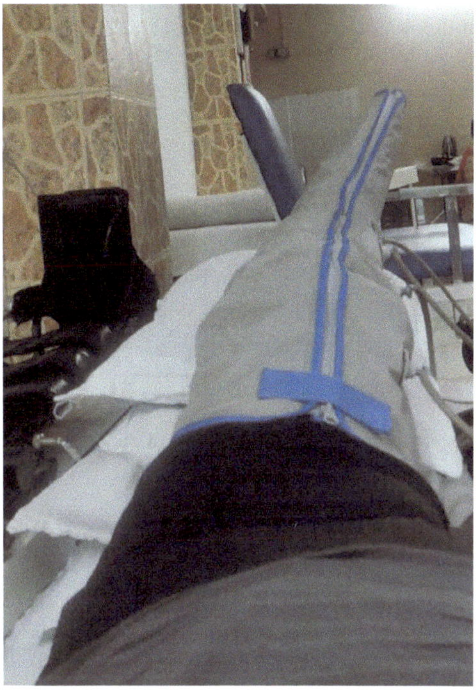

Fig. 5.3 Pneumatic compression applied for 20 min

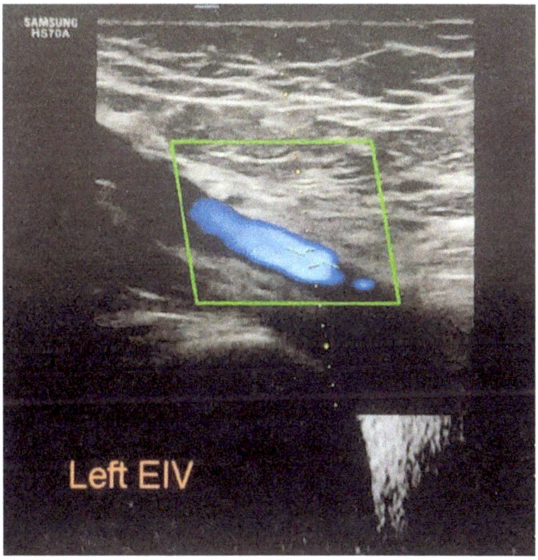

Fig. 5.4 Unobstructed flow in external iliac vein after treatment on a Doppler ultrasound

outlines the three contributing factors to thrombosis: alterations in the integrity of the blood vessel's inner lining, modifications in blood flow characterized by stasis or reduced flow, and modifications in the composition of blood, leading to a state of increased coagulability. These elements promote the formation of clots by disturbing the equilibrium between the coagulation and fibrinolytic systems (Virchow xxxx). The importance of DVT is associated with the potential occurrence of pulmonary thromboembolism (Sudo et al. 2003). In addition, deep venous thromboembolism has the potential to result in nonfatal issues like post-thrombotic syndrome and chronic thromboembolic pulmonary hypertension (Kakkar et al. 2011). Post-thrombotic syndrome leads to leg-related symptoms like a sensation of heaviness, pain, itching, cramps, and paresthesia.

Although the normal endothelium is non-thrombogenic, damage or injury to the endothelium can trigger the activation of platelets and coagulation. Direct trauma has the potential to harm the vascular endothelium (Prins and Hirsh 1991). Vascular injury triggers the expression of tissue factors, which can occur either directly from endothelial cells or from monocytes that migrate to the damaged area. Additionally, when blood is exposed to the subendothelial layer, it results in platelet adhesion and aggregation. Apart from initiating the coagulation process, tissue damage can also hinder fibrinolysis because inflammatory cytokines induce endothelial cells to produce plasminogen activator inhibitor-1. Endothelial cell activation occurs in two phases. In the first phase, which takes seconds to minutes, endothelial cells retract from each other, revealing the subendothelial layer. This is accompanied by the translocation of P-selectin, the release of von Willebrand's factor, and the secretion of platelet-activating factor. In the second phase, which takes place within 4 to 6 h, endothelial cells initiate the transcription of several genes, including cytokines such as IL-1, IL-6, and IL-8, along with tissue factor, which is the primary initiator of the coagulation cascade (Line 2001).Engaging in physical activity is recognized for its positive impact on factors associated with the pathogenesis of venous thromboembolism. Active exercises involving both lower limbs can effectively decrease venous stasis and the formation of blood clots (thrombus) (Kp et al. 2017). Physical activity can enhance insulin sensitivity, alleviate abnormal lipid profiles in the bloodstream, normalize elevated blood pressure, reduce blood thickness, stimulate endothelial nitric oxide production, and enhance leptin sensitivity, all of which collectively contribute to safeguarding the health of the heart and blood vessels (Xu et al. 2021). Mechanical techniques using physical activity to improve venous thromboembolism include ankle exercises, leg elevation, the utilization of compression stockings, intermittent pressure, and continuous passive range of motion.

The most effective method is ankle-toe pumps and leg elevation, which contributes to enhancing cardiac output by augmenting venous return when muscles contract, thereby compressing the veins and pushing blood back toward the heart (Kwon et al. 2003). According to the findings of Fleming et al. (2000), the velocity of venous blood flow in the common femoral vein demonstrated a substantial increase of 102.8% when individuals performed foot pump exercises in the reverse-Trendelenburg position (with the feet lowered) (Fleming, et al. 2000). Additionally, as demonstrated by Sochart and Hardinge (1999), the simultaneous active movement involving ankle flexion and rotation resulted in the highest velocity of blood flow, showing a 38% boost in the average venous blood flow velocity and a 58% increase in peak flow when the leg was elevated at a 30° angle (Sochart and Hardinge 1999).

Compression therapy can be frequently employed for all individuals with deep vein thrombosis (DVT) due to its capacity to diminish tissue swelling and improve the contractile function of the calf muscles (Xu et al. 2021).According to Kahn SR et al. 2003, a 30-min treadmill exercise session did not exacerbate venous symptoms like heaviness, swelling, and aching. In individuals with post-thrombotic syndrome, this exercise regimen resulted in noteworthy improvements in calf muscle flexibility. The use of elastic compression stockings during exercise had an impact on leg swelling or joint flexibility (Kahn et al. 2003).Anticoagulation is successful in halting the expansion of deep vein thrombosis (DVT) and preventing pulmonary embolism (PE), but it lacks direct fibrinolytic properties. This form of therapy permits the body's natural fibrinolysis processes to take place. Radiologically observable clot dissolution is observed in only 50% of individuals undergoing anticoagulation, and complete resolution occurs in less than 5% of cases (Baldwin et al. 2004). Exercise is believed to function as a safeguard against thrombosis by maintaining a controlled equilibrium between the coagulation and fibrinolytic pathways activated during physical activity (Smith 2003). The overall findings of this case support a positive role for exercise in patients with deep venous thrombosis. The majority of patients can successfully participate in training programs that encompass stretching, strength training, mechanical exercises, and aerobic exercises. The exercise intensity typically falls within the range of 60% to 85% of the maximum heart rate and should be tailored according to the different states and needs of the patients under the supervision of a physiotherapist and close monitoring of the condition with the physician.

The role of exercise in the management of deep venous thrombosis (DVT) lacks sufficient evidence. This case report underscores the use of a combined approach involving anticoagulation and physiotherapy to facilitate the early lysis of the thrombus as there is growing evidence that the early lysis is more likely to preserve valve function, decreasing the likelihood of DVT recurrence, and the occurrence of PTS.

The potential limitations of this case study include the lack of testing for prothrombin mutation, fibrinolytic parameters (level of tPA, PAI-1 or PAI-1 polymorphism 4G/5G), Lupant anticoagulant, Factor V Leiden, Antithrombin, ANA, Protein C, Protein S, and RPR.

5.4 Conclusion

Deep vein thrombosis (DVT) poses a significant risk leading to potentially life-threatening complications such as pulmonary thromboembolism. While anticoagulation remains a primary treatment for DVT, compression therapy and physiotherapy play crucial roles in mitigating its impact. Exercise not only enhances overall cardiovascular health but also contributes to reducing venous stasis and blood clot formation. The combination of anticoagulation and physiotherapy shows promise in early resolution and minimizing the risk of DVT recurrence and post-thrombotic syndrome. Understanding the multifaceted nature of DVT and its management strategies is vital in preventing its debilitating consequences and improving patient outcomes.

References

Baldwin ZK, Comerota AJ, Schwartz LB (2004) Catheter-directed thrombolysis for deep venous thrombosis. Vasc Endovasc Surg 38(1):1–9. https://doi.org/10.1177/153857440403800101

Cronan JJ (1993) Venous thromboembolic disease: the role of US. Radiology 186(3):619–630. https://doi.org/10.1148/radiology.186.3.8430164

Fleming P et al (2000) The effect of the position of the limb on venous impulse foot pumps. J Bone Jt Surg 82(3). https://doi.org/10.1302/0301-620x.82b3.9834

Hashemzadeh K, Dehdilani M, Gol MK (2021) Study of the effects of simple exercise with or without physiotherapy on prevention of deep vein thrombosis among postmenopausal women requiring coronary artery bypass graft surgery. Int J Womens Health Reprod Sci 9(1):69–74

Jasionowska S et al (2022) Systematic review of exercise therapy in the management of post-thrombotic syndrome. Phlebology 37(10):695–700. https://doi.org/10.1177/02683555221129738

Kahn SR et al (2003) Acute effects of exercise in patients with previous deep venous thrombosis. Chest 123(2):399–405. https://doi.org/10.1378/chest.123.2.399

Kakkar AK et al (2011) Low-molecular-weight heparin and mortality in acutely ill medical patients. N Engl J Med 365(26):2463–2472. https://doi.org/10.1056/nejmoa1111288

Kp DR et al (2017) Comparison between physiotherapeutic alone and combined physiotherapeutic and pharmacological measures in deep venous thrombosis prophylaxis in joint replacement and spinal surgeries. Int J Orthop Sci 3(4l):845–848. https://doi.org/10.22271/ortho.2017.v3.i4l.115

Kunutsor SK et al (2020) Physical activity and risk of venous thromboembolism: systematic review and meta-analysis of prospective cohort studies. Eur J Epidemiol 35(5):431–442. https://doi.org/10.1007/s10654-019-00579-2

Kwon O-Y et al (2003) Effects of ankle exercise combined with deep breathing on the blood flow velocity in the femoral vein. Aust J Physiother 49(4):253–258. https://doi.org/10.1016/s0004-9514(14)60141-0

Lee AD et al (2009) Venous thrombo-embolism in India. Eur J Vasc Endovasc Surg: off J Eur Soc Vasc Surg 37(4):482–485. https://doi.org/10.1016/j.ejvs.2008.11.031

Line BR (2001) Pathophysiology and diagnosis of deep venous thrombosis. Semin Nucl Med 31(2):90–101. https://doi.org/10.1053/snuc.2001.21406

Office of the Surgeon General (US) and National Heart, Lung, and Blood Institute (US) (2008) INTRODUCTION: Definitions of deep vein thrombosis and pulmonary embolism. Office of the Surgeon General

Prins MH, Hirsh J (1991) A critical review of the evidence supporting a relationship between impaired fibrinolytic activity and venous thromboembolism. Arch Intern Med, 151(9). Available at: https://pubmed.ncbi.nlm.nih.gov/1888237/ Accessed: 31 Oct 2023

Smith JE (2003) Effects of strenuous exercise on hemostasis. Br J Sports Med 37(5):433–435. https://doi.org/10.1136/bjsm.37.5.433

Sochart DH, Hardinge K (1999) The relationship of foot and ankle movements to venous return in the lower limb. J Bone Jt Surg. 81(4). https://doi.org/10.1302/0301-620x.81b4.8909

Sudo A et al (2003) The incidence of deep vein thrombosis after hip and knee arthroplasties in Japanese patients: a prospective study. J Orthop Surg (Hong Kong) 11(2):174–177. https://doi.org/10.1177/230949900301100212

Virchow R: Die verstopfund den lungenarteries und ihre folgen. Betir Exp Pathol Physiol 21, 1846

Xu L et al (2021) The effectiveness of exercise training in treating venous thromboembolism: a systematic review. Phys Sportsmed 49(1):1–11. https://doi.org/10.1080/00913847.2020.1786746

Chapter 6
Integrating Flaxseed as a Nutraceutical and Lifestyle Modifications for Preventing PCOS: A Comprehensive Review

Harmanjeet Kaur and Santosh Kumar Verma

6.1 Introduction

During the reproductive years, polycystic ovary syndrome is a prevalent issue that leads to dysfunction in the reproductive system such as high levels of androgen, infertility, and irregularity in menstrual cycle. Prevalence of this disease is very high because it affected 18 –20% of women in their child bearing age. The unknown etiology of this disease is linked to numerous clinical manifestations such as missed or irregular menstruation, ovarian cyst, excessive hair growth on face and body (hirsutism), and tint skin tone among patients. Furthermore, this illness presents numerous other indications related to mental health like depression and confusion in excessiveness. Endocrine interruption and high level of androgens directly associated with PCOS, and it could have an adverse effect on development of follicular growth and ovarian function; therefore, this disease can point to infertility and hyperplasia and cancer. In older age there is some more points helping to increase chances of developing this disease, and they are IR (insulin resistance), overweight, high blood pressure, type 2 diabetes, and weaken lipid profile or uterine cancer. Treatment aims in women with PCOS include modified healthy diet, lifestyle such as good sleep pattern, supplements with functional food, weight loss, and regular exercise. Weight loss is one of the key point of effective plan to regulate maturational cycle and handle other symptoms related to this disease (Saadati, et al. 2021).

Moreover, a control diet to keep up the management of insulin in natural way is the most important treatment for PCOS; uses of functional foods would therefore be a good regard to the treatment, such as aloe vera, green tea, cinnamon and chamomile,

H. Kaur (✉) · S. K. Verma
Quantum School of Health Sciences, Quantum University, Roorkee, India
e-mail: santosh.hs@quantumeducation.in

N. A. Siddiqui et al. (eds.), *Advances in Disaster Management, Volume 2*, Springer Transactions in Civil and Environmental Engineering,
https://doi.org/10.1007/978-981-96-4047-8_6

and white mulberry which effect on lipid and glucose profile of blood (Szczuko, et al. 2021). Because properties of these herbs as a functional food manage lipid profile and metabolism of carbohydrate, they can be used by all stereotypes of PCOS women. Another group of herbs indicated biochemical evidence for PCOS women: green mint and licorice root give antiandrogen effect, (O'Connor et al. 2010) (Sabbadin, et al. 2019). However, licorice is not a perfect solution having the potential to bring on hypertension, hypokalaemia, and metabolic alkalosis; therefore, patients having high level of cortisol avoid this (Szczuko, et al. 2021). According to last studies as a functional food, the flaxseed lignans are best found. In this review, we discuss about all beneficial effect of these flaxseeds on PCOS.

Modified lifestyle, comprising diet and consistent exercise, is regarded the first-line treatment for the management of PCOS and complications linked with it (Moran, et al. 2006) (Wild, et al. 2010). However, reduction in weight through moderate restriction in energy intake and consistent exercise put beneficial effect on insulin resistance and high androgen level and cardiovascular disease risk in PCOS women with obesity (Moran, et al. 2003) (Moran, et al. 2013). To maintain appropriate body weight, regular aerobic exercise following healthy dietary intervention and avoiding uses of tobacco and alcohol play important role in prevention and treatment of metabolic disorders and are included in clinical guidelines for various conditions. Moreover, it is personal choice of everyone for overall well-being, and it is very first step toward a healthier life but it not an instant fix (Szczuko, et al. 2021).

At the reproductive stage, polycystic ovary syndrome (PCOS) stands as a prevalent concern leading to disruptions in the reproductive system, characterized due to heightened androgen levels, irregularities in the menstrual cycle, and difficulties with fertility. This multifaceted condition affects a significant proportion, around 18–20%, of women in their fertile age. The elusive origin of PCOS presents through various clinical manifestations, including inconsistent menstrual cycles, cysts in the ovaries, hirsutism (excessive hair growth on the face and body), and alterations in skin pigmentation among affected individuals. Additionally, mental health-related symptoms such as depression and cognitive disturbances are observed. The etiology of PCOS, underscored by endocrine imbalances and heightened androgen levels, significantly impacts follicular development and ovarian function. This association with infertility, hyperplasia, and potential risks of ovarian cancer indicates the seriousness of the condition. Advancing age further compounds the likelihood of developing PCOS, particularly the existence of elements like insulin resistance, obesity, hypertension, type 2 diabetes, and compromised lipid profiles, which could even lead to uterine cancer. The treatment focus for women with PCOS encompasses a multifaceted approach. This entails dietary modifications, lifestyle adjustments including proper sleep patterns, the incorporation of functional food supplements, weight management, and regular physical activity. Weight reduction assumes a pivotal role in regulating menstrual cycles and alleviating associated symptoms of PCOS (Saadati, et al. 2021).

Nutraceutical interventions stand out as a pivotal element in the handling of PCOS. Certain functional foods like aloe vera, green tea, cinnamon, chamomile, and white mulberry exhibit positive influences on blood lipid and glucose profiles

(Szczuko, et al. 2021).These elements aid in managing lipid profiles and carbohydrate metabolism, proving beneficial for all subtypes of PCOS. Specific herbs like green mint and licorice root show antiandrogenic effects (Wild, et al. 2010), (Sabbadin, et al. 2019), although caution is advised with licorice by reason of its potential negative impacts on individuals with heightened cortisol levels which can lead to hypertension, hypokalemia, and metabolic alkalosis (Szczuko, et al. 2021).

Recent studies highlight flaxseed lignans as particularly beneficial in managing PCOS. A modified lifestyle, which includes dietary changes and regular exercise, remains the frontline strategy in addressing PCOS and its associated complications (Moran et al., 2006) (Wild, et al. 2010).Weight reduction through controlled energy intake and exercise has shown positive impacts on insulin resistance, androgen levels, and cardiovascular risks in obese women with PCOS (Moran, et al. 2003) (Moran, et al. 2013).

In summary, the use of nutraceuticals, in combination with lifestyle modifications, plays a crucial role in managing PCOS and its associated complications. The integration of specific functional foods and herbs into dietary plans, along with exercise regimens, represents an essential approach for the comprehensive treatment of PCOS. Moreover, it is a personal choice of everyone for overall well-being, and it is a very first step towards a healthier life but it not an instant fix (Szczuko, et al. 2021).

6.2 Flaxseed as a Nutraceutical Value

The oil flax seed is composed of approximately 30–45% fat, primarily a drying oil rich in polyunsaturated acids such as linoleic and linolenic acids. Additionally, it comprises around 20–30% protein, 8% water, and can contain up to 28% of both soluble and insoluble fiber. Notably, the seed coat harbors soluble fiber, capable of forming a gel that can expand to several times its original size. Within the seed, approximately 1% of cyanide glycosides are present, including well-known compounds like lignans, accompanied by a notably high content of vitamin E. (Sirotkin AV 2023).

Flaxseed contains a special compound called secoisolariciresinol diglucoside. When animals eat flaxseed, their stomachs (rumen) change this compound into helpful substances called enterodiol and enterolactone. Scientists have found that the tiny organisms living in the stomach (microbiota) play a big role in making this change happen. While no adverse side effects have been identified in the consumption of flaxseed products at therapeutic doses, it is essential to note that the ingestion of large doses could potentially lead to intoxication by hydrogen cyanide. This toxic release may occur during the milling process of the seeds. (Sirotkin AV 2023).

It is strongly advised that long-term use of flaxseed at therapeutic doses be undertaken under medical supervision, especially in cases involving intestinal blockage, acute appendicitis, pancreatitis, peritonitis, and acute painful hernia. For pregnant women, opting for flaxseed oil rather than milled flaxseed is recommended. Flaxseed oil lacks lignans with a phytoestrogenic effect, which could potentially trigger complications during pregnancy. It is important to acknowledge that these potential dangers have not been definitively validated by experiments, although flaxseed products are generally considered safe.

The U.S. Food and Drug Administration (FDA) regards flaxseed products as safe for nursing women and for the nutritional needs of children (Sirotkin AV 2023).

6.3 Flaxseed Impact on Outcomes

6.3.1 Anthropometric Improvements

Various phytoestrogens, including flavonoids and lignans, were identified in plants, exhibiting natural antiestrogenic properties. Previous data suggested that flaxseed lignans were yielding more favorable outcomes (Szczuko, et al. 2021)). Flaxseed powder was utilized as a supplementation to address PCOS, leading to a substantial decrease in body weight after 12 weeks ($p > 0.05$), while waist circumferences decreased notably ($p = 0.001$) (Haidari, et al. 2020). In a study, a flaxseed dose of 30 g per day was administered to PCOS patients for 93 days, demonstrating a more pronounced effect on body mass index *(*Haidari, et al. 2020).

6.3.2 Hormonal Regulation and Metabolism

Plant contains various phytoestrogens, natural compounds with antiestrogenic properties. It includes flaxseed (Linum usitatissimum), which contains lignans that might have reduced the activity of key enzymes involved in estrogen synthesis (e.g., aromatase) to modulate the relative levels of circulating sex hormones, (Hassanzadeh et al. 2013). In a recent study, it was found that flaxseed oil, rich in α-linolenic acid, produced favorable impacts on polycystic ovary syndrome through the sex steroid hormones–microbiota–inflammation axis in rats. However, other sources of α-linolenic acid will likely produce an equally positive effect (Wang et al. 2020).

In a study, after treatment, women had their regular menstrual cycle restored to eumenorrhea (12 out of 21 women), with a cycle length of 28 ± 5 days (Hassanzadeh et al. 2013)). In another study, a combination of flaxseed with spearmint significantly improved progesterone levels and decreased testosterone and estradiol, (Mehraban et al. 2020).

The addition of flaxseed powder resulted in a reduction in serum insulin levels ($p = 0.003$), HOMA-IR ($p = 0.002$), and triglycerides ($p = 0.003$), LEPTIN ($p = 0.016$), and significantly increased HDL ($p = 0.008$) and adiponectin ($p = 0.03$) (Haidari, et al. 2020).

6.4 Lifestyle Modification

6.4.1 *Physical Activity*

Among professional's exercise is more acceptable to manage PCOS. According to a recent study the different outcomes improved by regular exercise than only diet or dose. Its result analysis supports the regular exercise, and it gives a significant impact on insulin resistance and body composition (Patten et al. 2020). Insulin resistance, measured using the HOMA-IR and BMI, demonstrated a notable reduction with moderate and high certainty (MD–0.57; 95% confidence interval (CI), – 0.98 to – 0.16, and $p = 0.01$; MD–1.90, 95% CI – 3.37, – 0.42, and $p = 0.01$), respectively (Santos, et al. 2021). In a recent review study, authors concluded that intense aerobic activity and resistance training enhance insulin sensitivity, and level of androgen is validated for women's fight with PCOS (Shele et al. 2020). Aerobic activities should be 120 min per week minimum required to get better results (Patten et al. 2020).

6.4.2 *Sleep*

A large data analysis show that huge proportion of psychological disorder with PCOS is sleep disorder. It impacts directly to the development of the PCOS. So, management or modification of sleep is considered as valuable part of lifestyle modification in PCOS patients. Moreover, sleep deprivation is linked with increased risk if insulin resistance and obesity which further related to type 2 diabetes (Chen, et al. 2020).

Some studies reported that women with PCOS and diabetes highly disturbed having less sleep length not more than 5 h per night. This situation leads to low quality of life (Gu, et al. 2022). So, it is ensuring that adequate and good sleep pattern maintained and decreased the danger of mental disturbance in PCOS obese and diabetic women.

6.4.3 *Diet*

Diet nutrition is advised as the initial treatment for polycystic ovary syndrome (PCOS) because it is an important environmental factor related to lifestyle that affects it. Some studies analysis shows that energy intake from any macronutrients showed no significant effect on the levels of the any analyzed parameters. But lowering in the level of calorie content of the diet with low GI put significant effect on different parameters such as insulin resistance and body mass index (Moran, et al. 2003; Szczuko, et al. 2021). Moreover, a recent comparative study examining the effects of high glycemic index (GI) and low GI diets found that the low GI diet was more effective in improving insulin sensitivity, as evidenced by better outcomes in homeostaatics model assessment for insulin resistance (HOMA-IR), fasting insulin, total cholesterol, and triglyceride level (Kazemi, et al. 2021). Appetite-regulating hormones such as ghrelin and glucagon influenced by low glycemic diet, and it increased glucagon and decreased ghrelin in women with PCOS (Hoover, et al. 2021). Different studies show that the low glycemic diet is an effective and supportable diet pattern to regulate insulin resistance. So, this dietary pattern advice should be offered to all patients by professionals to those who suffered from metabolic dysfunctions (Shang, et al. 2020).

Another modified diet, keto diet patterns for patients improve blood glucose, weight, and liver functions and regulate menstrual cycle. Moreover, keto diet use with combination of Mediterranean diet give great result on body weight (−9.43 kg), improve body mass index (-3.35) and significant effect on fat body mass (8.29 kg), and also beneficial for management of metabolic dysfunction (Paoli, et al. 2020). Furthermore, a comparative study found that a very low calorie ketogenic diet consisting of animal, vegetable, and whey protien sources, administrered over 45 days, significantly improved lipid metabolism and glycemic control, while also resulting in substantial body weight reduction in patients with obesity and insulin resistance. Moreover, this whey and vegetable protein have a secure profile and resulted in improving gut microbes (Basciani, et al. 2020). But in complexity, this modified diet must be recommended and reconcile by qualified professionals in the field and also necessary to individualize the treatment and evaluate the contradictions and unfavorable effects (Che, et al. 2021).

6.5 Conclusion

Researches show that PCOS is associated with a low to moderate quality of life, and proper management can help improve its outcomes. Lifestyle modification is regarded as cornerstone of all interventions to manage anthropometric disbalance and metabolic dysfunctions. This is the most applicable intervention: modified diet with functional food and increased physical activity, exercise, and strategies to prevent PCOS. There is more need to focus on quality life with some natural functional food supplementation (e.g., flaxseed).

The addition of flaxseed supplementation to lifestyle modifications has shown enhanced effectiveness in influencing biochemical and anthropometric variables among patients with polycystic ovary syndrome (PCOS), compared to relying solely on lifestyle modifications. This combined approach appears to yield more promising outcomes in managing various parameters linked to PCOS. The incorporation of flaxseed supplementation alongside lifestyle adjustments exhibits a synergistic effect, contributing to improved alterations in biochemical markers and physical measurements within individuals diagnosed with PCOS.

Studies suggest that the incorporation of flaxseed, rich in omega-3 fatty acids and lignans, as a supplementary component to lifestyle modifications, plays a contributory role in positively impacting key markers associated with PCOS, such as hormonal imbalances, lipid profiles, and weight management. The collective impact of flaxseed supplementation in tandem with lifestyle adjustments appears to produce more significant changes in these variables when compared to solely implementing lifestyle modifications.

These findings imply that the combination of flaxseed supplementation and lifestyle adjustments may offer a more comprehensive and potentially more effective approach to managing PCOS-related indicators. The potential synergy between lifestyle modifications and the bioactive components found in flaxseed could be further explored and utilized in clinical settings for more tailored and efficient management strategies for patients with PCOS.

References

Basciani S et al (2020) Very-low-calorie ketogenic diets with whey, vegetable, or animal protein in patients with obesity: a randomized pilot study. J Clin Endocrinol Metabolism, 105(9):2939–2949. Available at: https://doi.org/10.1210/clinem/dgaa336

Che X et al (2021) dietary interventions: a promising treatment for polycystic ovary syndrome. Ann Nutr Metablsm, 77(6):313–323. Available at: https://doi.org/10.1159/000519302

Chen X. et al (2020) Association of polycystic ovary syndrome or anovulatory infertility with offspring psychiatric and mild neurodevelopmental disorders: a Finnish population-based cohort study. Hum Reprod, 35(10):2336–2347. Available at: https://doi.org/10.1093/humrep/deaa192

Gu Y et al (2022) Life modifications and PCOS: old story but new tales. Front Endocrinol, 13:1–7. Available at: https://doi.org/10.3389/fendo.2022.808898

Haidari F et al (2020) The effects of flaxseed supplementation on metabolic status in women with polycystic ovary syndrome: a randomized open-labeled controlled clinical trial. Nutr J, 19(1):1–11. Available at: https://doi.org/10.1186/s12937-020-0524-5

Hassanzadeh M, Ahmad S, Mousavifar N (2013) Ijpr-12–475.Pdf. 12:475–481

Hoover SE et al (2021) Changes in ghrelin and glucagon following a low glycemic load diet in women with PCOS. J Clin Endocrinol Metabolism, 106(5):E2151–E2161. Available at: https://doi.org/10.1210/clinem/dgab028

Kazemi M et al (2021) Effects of dietary glycemic index and glycemic load on cardiometabolic and reproductive profiles in women with polycystic ovary syndrome: a systematic review and meta-analysis of randomized controlled trials. Adv Nutr, 12(1):161–178. Available at: https://doi.org/10.1093/advances/nmaa092

Mehraban M, Jelodar G, Rahmanifar F (2020) A combination of spearmint and flaxseed extract improved endocrine and histomorphology of ovary in experimental PCOS. J Ovarian Res, 13(1):1–8. Available at: https://doi.org/10.1186/s13048-020-00633-8

Moran LJ et al. (2003) Dietary composition in restoring reproductive and metabolic physiology in overweight women with polycystic ovary syndrome. J Clin Endocrinol Metabolism, 88(2):812–819. Available at: https://doi.org/10.1210/jc.2002-020815

Moran LJ et al (2006) Effects of lifestyle modification in polycystic ovarian syndrome. Reprod BioMedicine Online, 12(5):569–578. Available at: https://doi.org/10.1016/S1472-6483(10)61182-0

Moran LJ et al (2013) The contribution of diet, physical activity and sedentary behaviour to body mass index in women with and without polycystic ovary syndrome. Human Reproduction, 28(8):2276–2283. Available at: https://doi.org/10.1093/humrep/det256

O'Connor A, Gibney J, Roche HM (2010) Metabolic and hormonal aspects of polycystic ovary syndrome: the impact of diet. Proceedings of the Nutrition Society, 69(4):628–635. Available at: https://doi.org/10.1017/S0029665110002016

Paoli A et al. (2020) Effects of a ketogenic diet in overweight women with polycystic ovary syndrome. Journal of Translational Medicine, 18(1):1–11. Available at: https://doi.org/10.1186/s12967-020-02277-0
Patten RK et al (2020) exercise interventions in polycystic ovary syndrome: a systematic review and meta-analysis. Front Physiol, 11. Available at: https://doi.org/10.3389/fphys.2020.00606
Saadati N et al (2021) The effect of low glycemic index diet on the reproductive and clinical profile in women with polycystic ovarian syndrome: A systematic review and meta-analysis. Heliyon, 7(11):e08338. Available at: https://doi.org/10.1016/j.heliyon.2021.e08338
Sabbadin C et al (2019) Licorice: From pseudo hyperaldosteronism to therapeutic uses. Front Endocrinol, 10:1–6. Available at: https://doi.org/10.3389/fendo.2019.00484
Santos IK dos et al (2021) Effect of high-intensity interval training on metabolic parameters in women with polycystic ovary syndrome: a systematic review and meta-analysis of randomized controlled trials. PLoS ONE, 16:1–18. Available at: https://doi.org/10.1371/journal.pone.0245023
Shang Y et al (2020) Effect of diet on insulin resistance in polycystic ovary syndrome. J Clin Endocrinol Metabolism, 105(10):1–15. Available at: https://doi.org/10.1210/clinem/dgaa425
Shele G, Genkil J, Speelman D (2020) A systematic review of the effects of exercise on hormones in women with polycystic ovary syndrome. J Funct Morphol Kinesiol, 5(2):14–33. Available at: https://doi.org/10.3390/jfmk5020035
(Sirotkin AV 2023)'Influence of Flaxseed (Linum usitatissimum) on Female Reproduction Planta Med 2023; 89: 608–615|© 2023
Szczuko M et al (2021) Nutrition strategy and life style in polycystic ovary syndrome—narrative review. Nutrients, 13(7):1–18. Available at: https://doi.org/10.3390/nu13072452
Wang T et al. (2020) Dietary α-linolenic acid-rich flaxseed oil exerts beneficial effects on polycystic ovary syndrome through sex steroid hormones—microbiota—inflammation axis in rats. Front Endocrinol, 11. Available at: https://doi.org/10.3389/fendo.2020.00284
Wild RA et al (2010) Assessment of cardiovascular risk and prevention of cardiovascular disease in women with the polycystic ovary syndrome: a consensus statement by the androgen excess and polycystic ovary syndrome (AE-PCOS) society. J Clin Endocrinol Metabolism, 95(5):2038–2049. Available at: https://doi.org/10.1210/jc.2009-2724

Chapter 7
Study of Association Between Dry Eyes and Non-strabismic Binocular Vision Disorders in Young Adults

Krishna Kumar Gupta, Kumari Diksha, Renu Thakur, and Sachitanand Singh

7.1 Introduction

Dry eye and binocular vision disorder are two common eye conditions that can affect individuals of all ages. Dry eye and binocular vision anomalies can share some similar symptoms, such as eye strain, eye fatigue, and headaches. They can have overlapping symptoms, making it difficult to differentiate between the two conditions. Some of the shared symptoms include eye strain, headache, blurred vision, and discomfort while reading or using digital screens. Dry eye, a common condition affecting the eye, affects millions of people worldwide, also termed as keratoconjunctivitis sicca. "Dry eye is a multifactorial disease characterized by a persistently unstable and/or deficient tear film (TF) causing discomfort and/or visual impairment, accompanied by variable degrees of ocular surface epitheliopathy, inflammation, and neurosensory abnormalities (Nelson et al. 2017)." It happens when the tear film that covers the front of the eye is disrupted due to inadequate tear production or tears of low quality. Depending on the population being examined, the diagnostic standards being followed, and the geography, different populations have different prevalence of dry eye. Dry eye prevalence is 32% in north India, according to Tityal et al. (2018) The prevalence of dry eye illness ranges from 5% to 50% in various populations, with an estimated global prevalence of 11%; according to the TFOS DEWS II epidemiological report, the risk factors of dry eye include age, gender (women are more likely to be affected),particular prescription drugs, specific medical conditions, and environmental influences including dry climates or continuous use of digital screens Stapleton et al. (2017). According to a study conducted in the year 2019, the overall

K. K. Gupta (✉) · K. Diksha · R. Thakur · S. Singh
Department of Optometry, Chitkara School of Health Science, Chitkara University, Rajpura, Punjab, India
e-mail: Krishnakumar.gupta@gdgu.org

N. A. Siddiqui et al. (eds.), *Advances in Disaster Management, Volume 2*, Springer Transactions in Civil and Environmental Engineering,
https://doi.org/10.1007/978-981-96-4047-8_8

incidence of dry eye disorder in India ranges from 5.9% to 33.5% (Tandon et al. 2020). According to the study done by Gomes et al. (2019) dry eye was more common among women than men and in urban areas compared to rural ones. Additional risk factors for dry eye in India include age, contact lens use, and a few systemic diseases such thyroid problems and diabetes.

It is important to note that because many people do not seek medical attention for their symptoms, the prevalence of dry eye may be underestimated. Yang et al. (2021) conducted a cross-sectional study to find out the prevalence of dry eye using the OSDI questionnaire and they found that 59.64% of students suffered from dry eye. Another study by Supriyaphun et al. (2021), found that history of contact lens use, male gender, and screen time are associated with dry eye. A study among university students by Gracia Ayuso et al. (2022), using an OSDI questionnaire found that the prevalence of symptomatic dry eye was 51.8%. Based on a study by Majunder and Toh (2020), 76.5% of college students exhibit binocular vision disorders. 40% of students at an independent school located in Malaysia had non-strabismic binocular vision defects, according to another study on this subject. A prospective cross-sectional study among high school children found that 56% of the population suffers from accommodative anomalies (Alkabbani et al. 2021). In another study on ocular surface health at Shanghai University, it was found that subjects with DED had more symptoms of anxiety and there is a correlation between eye strain for a long time and DED (Li et al. 2018). In research on the impact of lifestyle intervention on dry eye in-office employees, Kawashima et al. (2018) concluded that the lifestyle intervention improved office workers' dry eye disease status. Another study titled "Prevalence of Ocular Surface Disease Symptoms and Risk Factors in Group of University Students in Montreal" found that 70.4% of students had dry eyes, contact lens use, hours spent in front of computers, or a history of refractive surgery has a negligible impact on the likelihood of presenting an ocular disease (Garza-León et al. 2016). The prevalence of binocular vision anomalies was found to be 31.5% and 29.6%, respectively, in a study by Hussaindeen et al. (2017) There were no statistically significant differences in the prevalence between the rural and urban areas, nor were there any gender preferences. According to Rao (2014) research, 78.02% of college students have binocular vision anomalies, which is higher than the national average. When using the OSDI and CISS questionnaires to compare the symptoms of binocular vision disorders and dry eye, Rueff et al. (2014) discovered a positive correlation. The findings of this study suggested that there may be considerable overlap between the symptoms of binocular vision disorders and those of dry eye. The cornea is covered with a layer of tears, which aids in keeping the surface of the eye wet (Shaikh and Bandamwar 2022) Dry eye disease (DED) diagnosis and treatment typically involve assessment of the ocular and systemic health, way of life, and environmental factors (Sah et al. 2021). To evaluate tear film stability, optometrists employ a variety of tests. These tests may include tear film breakup time (TBUT) measurement, Schirmer's test, corneal staining, and meibomian gland evaluation. Overall, a thorough analysis of the tear film is necessary for identifying and treating dry eye disease, and several tests can offer insightful data to help with treatment selection. Patients' responses on dry eye questionnaires, such as the OSDI, DEQ-5, McMonnies questionnaire, SPEED, and

IDEEL, are frequently used to learn more about the symptoms of the condition and how it affects their everyday lives. To help with diagnosis and therapy planning, these surveys offer insightful information on the patient's subjective experience of dry eye illness. One of the most popular and well-validated questionnaires for assessing dry eye illness is the OSDI.

7.2 Binocular Vision Anomalies

The ability to see can also be considerably affected by binocular vision problems such as convergence insufficiency and accommodative dysfunction. While accommodative dysfunction affects the ability to focus at various distances, convergence insufficiency affects the coordinated inward turning of the eyes while focusing on close objects. These disorders are more common in certain populations than others; according to some research, rates in various age groups and geographical areas range from 6.5% to 22%. The most prevalent form of convergence abnormality, convergence insufficiency (CI), impairs the eyes' ability to come together when focusing on close objects. In rural India, it affects 5.5% of children aged 5–15 years, while in the US, it affects 5–10% of people, with greater incidence in children and young adults. 9.9% of school-age children in New Delhi, India, had the prevalence, with males having greater rates than girls. Studies have shown that CI prevalence rates vary by nation, with rates of 4.2% in US adults aged 20–39 and 12.6% in Mexican schoolchildren aged 7 to 11. The reverse of convergence inhibition (CI), convergence excess (CE), occurs when the eyes over converge, resulting in impaired vision and eye fatigue when focused on distant objects. In India, it affects 8% of computer professionals. When focused on distant things, basic exophoria (BE) causes the eyes to drift outward from their normal position, whereas basic esophoria (BO) causes the eyes to drift inward (Stein et al. 2017). Research on the incidence of BE and BO in India is scarce. An uncommon illness known as convergence paralysis prevents the eyes from converging because of nerve issues, whereas divergence insufficiency prevents the eyes from moving outward when focused on distant things (Shah and Patel 2015). Another uncommon illness that causes the eyes to over converge uncontrollably is convergence spasm (Tamhankar 2019). To identify vision issues that may be the source of symptoms including headaches, eye strain, double vision, and difficulties reading, a binocular vision evaluation is crucial. Binocular vision abilities are crucial for both academic and athletic success.

There isn't much epidemiological research on the relationship between dry eyes and abnormal binocular vision among Indian university students. The majority of research has employed questionnaires to evaluate binocular vision and dry eye, although clinical examination of both disorders is necessary. The goals of this study are to determine the prevalence of dry eye and binocular vision issues among college students and to assess how these two conditions are related.

7.3 Methods

This study was approved by the institutional ethics committee chitkara university, punjab (Ethics Committee Registration No. EC/NEW/INST/2022/531/132, issued under Biomedical & Health Research (NECRBHR), Department of Health Research (DHR), Government of India, New Delhi 110001) with approval number ECNEWINST/2022/531/132. The participants were college students between the ages of 18 and 25. A convenient sampling method was used to draw in a total of 325 students. Each participant in the study provided informed consent after the risks and benefits were explained. An eye exam was conducted, including vision assessment, refraction, ocular health examination, dry eye measurement, and binocular vision assessment. The approval for the research conduct was taken from the dean of various departments from where the students were participating in this study. Initially, the OSDI questionnaire was administered to the participant then a detailed history was taken from each participant including their ocular and systemic condition, dry eye medication, undergoing certain vision therapy procedures, history of refractive surgery, contact lens usage, strabismus, and amblyopic condition. This study comprised participants without any ocular or systemic conditions and with a best corrected visual acuity of 6/6 and those subjects who were on some dry eye medication, undergoing vision therapy, had strabismus, amblyopia, was using contact lens and had refractive surgery in past was excluded from the study. After that refraction was done followed by sensory evaluation was done. By using the cover test, Maddox rod test, Thorington test, and ac/a ratio, the motor function of the eye was evaluated. By performing accommodative tests such the near point of accommodation, negative relative accommodation, positive relative accommodation, monocular estimation method (MEM), monocular accommodative facility test, and binocular accommodative facility test, the accommodative state was assessed. The convergence test was done by performing the near point of convergence test, negative fusional vergence, positive fusional vergence, and vergence facility test. The TBUT and the Schirmer test method were used to accomplish the clinical evaluation of the dry eye test.

Sample size calculation: On the basis of an expected prevalence of dry eye and binocular vision abnormalities of 30%, the sample size for the current study was determined to be 322 students.

$$\text{Sample Size Formula } (n) = Z^2 * p * (1 - p)/d2$$

Where,

z is the z-score for 95% is 1.96
d is the margin of error which is 5%
p is the prevalence of 30

$$
\begin{aligned}
&Z^2 * p * (1 - p)/d2 \\
&= 1.962 * 0.30 * (1 - 0.30)/0.052 \\
&= (3.8416 * 0.30 * 0.70)/0.0025 \\
&= 0.806736/0.0025 \\
&= 322.4
\end{aligned}
$$

By calculating the sample size for 95% CI, 5% margin of error, and 30% population proportion. 322.4 sample size was optioned.

The subject used for the study joined from the Chitkara School of Health Science, and all subjects were selected based on Systematic sampling. Best Corrected Visual Acuity should be 6/6, there were no individuals in the research with systemic or ocular disease taking dry eye medication. Vision Therapy, Strabismus, and Amblyopia were excluded from the study. The null hypothesis stated that there will be no association between dry eye and binocular vision anomalies, and the alternative hypothesis states that there will be an association between dry eye and binocular vision anomalies.

Data Analysis: This study aimed to investigate the prevalence of non-strabismic binocular vision anomalies and dry eye, as well as their potential relationships. The analysis was conducted using statistical software SPSS. Dry eye severity was assessed based on the OSDI score, categorized as normal, mild, moderate, or severe. Clinical tests for dry eye included the Schirmer test and tear breakup time, with specific criteria for interpreting the results. The study utilized the Pearson correlation test to examine the connection between dry eye and non-strabismic binocular vision disorders.

7.4 Results

The study involved 325 students in all, 116 were men and 209 were women. The subjects ages ranged from 17 to 25, with an average of 20.44 +/− 1.361. The female participation with 64.3% outnumbered the males with 35.7% of the participation. According to the OSDI score, 198 students that are 60.9% were having normal ocular surface conditions, while 74 students that are 22.8% were found to have mild dry eyes., 23 students were found to be suffering from moderate dry eye that is 7.1%. A total of 30 students were found to have severe dry eyes, which account for 9.2% of the population and hence the prevalence of dry eye according to OSDI was found to be 39.1% (Fig. 7.1).

In the right eye, the normal reading of Schirmer is > 15 mm which was found in 141 students that are 43.3%, the low normal reading (10–15 mm) was found in 79 students 24.3%, the borderline reading (6–10 mm) was found in 73 individuals that are 22..5%, and abnormal reading(< 6 mm) was found in 32 individual that is 9.8%. Hence, in the right eye, the prevalence of DED was found to be 32.3%. In the left eye, the normal reading of the Schirmer test that is > 15 mm was found in 145 students is 44.6%, the low normal reading (10–15 mm) was found in 83 students that is 25.5%, the borderline reading (6–10 mm) was found in 73 individuals that are 22.5%, and

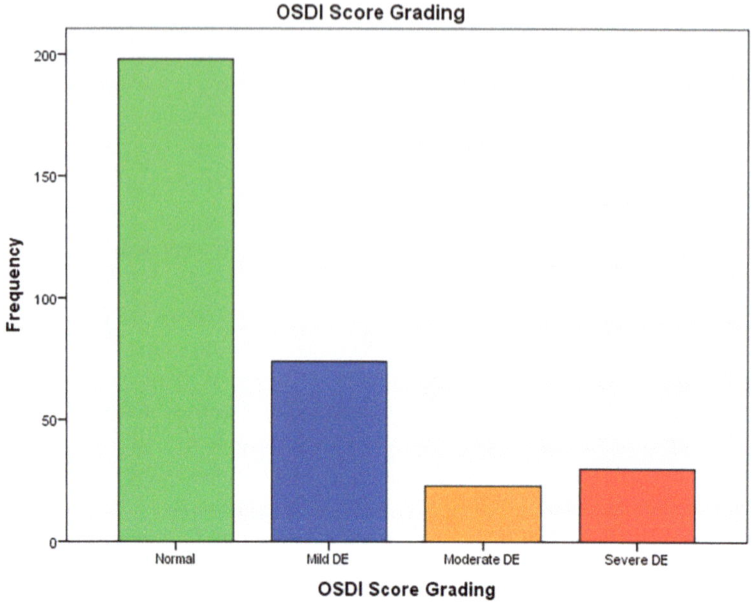

Fig. 7.1 *X* axis represents the OSDI score grading and *Y* axis represents the number of students. Graphical representation of the dry eye condition according to the OSDI scoring, normal (0–12), mild (13–22), moderate (23–32), and severe (33–100) which shows that most students suffer from mild dry eyes

abnormal reading (< 6 mm) was found in 24 individual that is 7.4%. Hence, in the left eye, the prevalence of DED was found to be 29.8%. In the right eye, the tear breakup time test found dry eye (< 5 s) in 15 individuals are 4.6%, marginal dry eye was found in 127 students that are 39.1% (5–10 s), normal tear film (> 10 s) was found in 183 students that are 56.3%. In the left eye, the tear breakup time test found dry eye (< 5 s) in 13 individuals is 4%, marginal dry eye was found in 132 students that is 40.61% (5–10 s), normal tear film (> 10 s) found in 180 students that is 55.38% students. The most common NSBVD has been identified to be convergence insufficiency at 18.2%, followed by accommodative insufficiency at 9.2%, convergence excess at 7.1%, accommodative infacility at 7.1%, accommodative excess at 4.3% and fusional vergence dysfunction with 1.2% (Fig. 7.2) (Table 7.1).

According to the study, university students have a significant prevalence of dry eyes and non-strabismic binocular vision problem. According to the Ocular Surface Disease Index (OSDI) score, 198 students (60.9%) had normal ocular surface conditions, whereas 74 students (22.8%) had mild dry eye, 23 students (7.1%) had moderate dry eye, and 30 students (9.2%) had severe dry eye. This gave the OSDI score a dry eye prevalence of 39.1%. The Schirmer test revealed a dry eye prevalence in the right eye of 32.3% and a dry eye prevalence in the left eye of 29.8%. The tear breakup time test in the right eye found dry eye (< 5 s) in 15 individuals (4.6%), and marginal

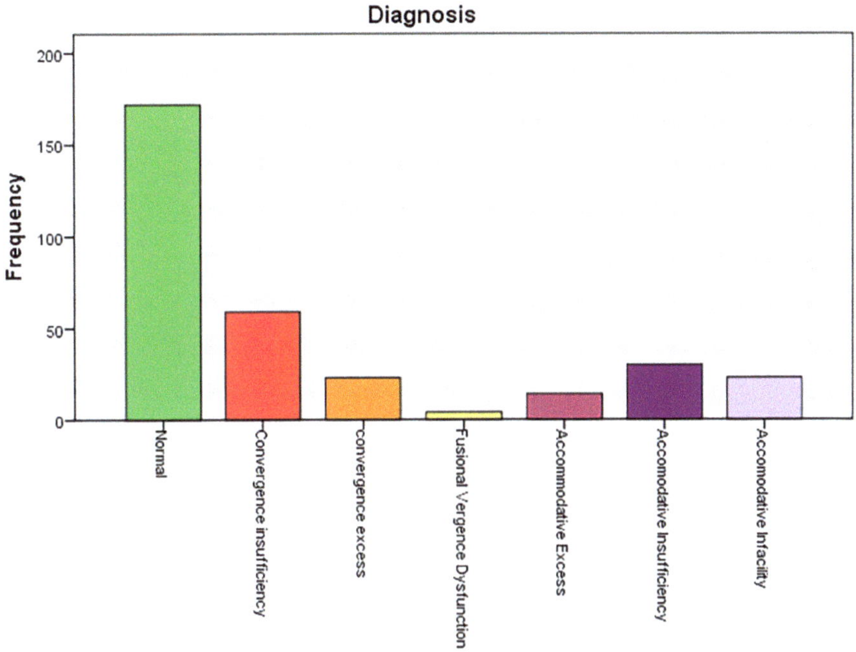

Fig. 7.2 *X* axis represents the types of NSBVA and *Y* axis represents the frequency of the students. Graphical representation of the prevalence of different types of non-strabismic binocular vision anomalies. The figure shows that convergence insufficiency is the most prevalent binocular vision anomaly, followed by accommodative insufficiency, convergence excess, accommodative infacility, and accommodative excess and very few of the students had fusional vergence dysfunction

dry eye (5–10 s) in 127 students (39.1%). Convergence insufficiency, with a prevalence of 18.2%, was the most common non-strabismic binocular vision problem, followed by accommodative insufficiency (prevalence of 9.2%), convergence excess (prevalence of 7.1%), accommodative infacility (prevalence of 7.1%), accommodative excess (prevalence of 4.3%), and fusional vergence dysfunction (prevalence of 1.2%). There is a strong positive correlation between the Schirmer test value and the tear breakup time value.

7.5 Discussion

The study found a high prevalence of dry eyes among university students, with a rate of 39.1% based on the OSDI score. This prevalence is notably higher than a previous study in Brazil by Yang et al. (2021) where only 23.5% of students had severe dry eye. Another study in Bangkok by C. Supiyaphun et al. Thailand, identified risk factors for dry eye among university students, including wearing contact lenses, being male, and excessive screen time. Similarly, the current study found that using contact

Table 7.1 Correlations between different tests of dry eye and binocular vision anomalies

		STOD Schirmer test right eye	STOS Schirmer test left eye	TBOD Tear breakup test on the right eye	TBOS Tear breakup test on the left eye	OSDISCOR ES Ocular surface disease index score	Non-strabismic binocular vision anomalies
STOD Schirmer test right eye	Pearson correlation	1	0.908**	0.610**	0.814*	– 0.543	– 0.062
	Sig. (2-tailed)		0.000	0.000	0.000	0.000	0.264
	N	325	325	325	325	325	325
STOS Schirmer test left eye	Pearson correlation	0.908**	1	0.610**	0.829*	– 0.526	0.011**
	Sig. (2-tailed)	0.000		0.000	0.000	0.000	0.840
	N	325	325	325	325	325	325
TBOD Tear Breakup test on the right eye	Pearson correlation	0.610**	0.610**	1	0.750	– 0.556	0.014**
	Sig. (2-tailed)	0.000	0.000		0.000	0.000	0.804
	N	325	325	325	325	325	325
TBOS Tear Breakup test on the left eye	Pearson correlation	0.814**	0.829**	0.750**	1	– 0.570	– 0.012**
	Sig. (2-tailed)	0.000	0.000	0.000		0.000	0.000
	N	325	325	325	325	325	325
OSDISCO RES Ocular Surface disease index score	Pearson correlation	– 0.543**	– 0.526**	– 0.556**	– 0.570**	1	0.006**

(continued)

Table 7.1 (continued)

		STOD Schirmer test right eye	STOS Schirmer test left eye	TBOD Tear breakup test on the right eye	TBOS Tear breakup test on the left eye	OSDISCOR ES Ocular surface disease index score	Non-strabismic binocular vision anomalies
	Sig. (2-tailed)	0.000	0.000	0.000	0.000		0.916
	N	325	325	325	325	325	325
Non-Strabismic Binocular Vision anomalies	Pearson correlation	− 0.062	0.011	0.014	− 0.012	0.006	1
	Sig. (2-tailed)	0.264	0.840	0.804	0.000	0.916	
	N	325	325	325	325	325	325

** Correlation is significant at the 0.01 level (2-tailed)
STOD Schirmer test right eye
STOS Schirmer test left eye
OSDI Ocular surface disease index
TBOD Tear breakup time test right eye
TBOS Tear breakup time test right eye

lenses, being female, and taking more online classes were associated with a higher frequency of symptomatic dry eye, consistent with previous research findings. The investigation on the similarities between dry eye symptoms and those of binocular vision problems was conducted by Rueff EM et al. They used OSDI and CISS and discovered a positive correlation between the two, and present study had also positive but weak correlation between the OSDI and CISS. Since the university students has huge visual demand due to which the screen time is also very high. The near visual demand might be the reason for the accommodative stress and more the screen time more will be the evaporation. Hence, there is weak positive correlation observed. The findings revealed that symptoms of binocular vision abnormalities and symptoms of dry eye may be extremely similar. It is crucial to remember that the prevalence of dry eye illness in India may be underreported since many people may choose not to seek medical treatment for their symptoms or may have their eyes misdiagnosed with other eye diseases. Therefore, it is essential to spread knowledge about the value of routine eye exams as well as the effective identification and management of dry eye disease.

7.6 Conclusion

In this particular study, our focus was on assessing the ocular health of students and determining the prevalence of dry eye among them. Our analysis revealed some interesting statistics. Out of all the participants, 60.9% exhibited a normal ocular surface condition and overall prevalence of dry eye among the students was 39.1%. Additionally, non-strabismic binocular vision abnormalities in students were investigated in the current study. At 18.2%, convergence insufficiency was determined to be the most common. To delve further into the matter, we conducted Schirmer tests to measure tear production. In the right eye, a dry eye prevalence of 32.3% was found. In the left eye, a prevalence of 29.8% was found. Present study observed a significant positive correlation between the Schirmer test results and the tear breakup time value. The correlation between the OSDI score and non-strabismic binocular vision anomalies, however, was only marginally positive. These discoveries provide valuable insights into the ocular health of students, the prevalence of dry eye, and related ocular conditions.

References

Alkabbani S, Jeyaseelan L, Rao AP, Thakur SP, Warhekar PT (2021) The prevalence, severity, and risk factors for dry eye disease in Dubai–a cross sectional study. BMC Ophthalmol 21:1–7

García-Ayuso D, Di Pierdomenico J, Moya-Rodríguez E, Valiente-Soriano FJ, Galindo-Romero C, Sobrado-Calvo P (2022) Assessment of dry eye symptoms among university students during the COVID-19 pandemic. Clin Exp Optom 105(5):507–513

Garza-León M, Valencia-Garza M, Martínez-Leal B, Villarreal-Peña P, Marcos-Abdala HG, Cortéz-Guajardo AL, Jasso-Banda A (2016) Prevalence of ocular surface disease symptoms and risk factors in group of university students in Monterrey, Mexico. J Ophthalmic Inflamm Infect 6(1):1–7
Gomes JA, Santo RM (2019) The impact of dry eye disease treatment on patient satisfaction and quality of life: a review. Ocul Surf 17(1):9–19
Hussaindeen JR, Rakshit A, Singh NK, George R, Swaminathan M, Kapur S, Scheiman M, Ramani KK (2017) Prevalence of non-strabismic anomalies of binocular vision in Tamil Nadu: report 2 of BAND study. Clin Exp Optom 100(6):642–648
Kawashima M, Sano K, Takechi S, Tsubota K (2018) Impact of lifestyle intervention on dry eye disease in office workers: a randomized controlled trial. J Occup Health 60(4):281–288
Li S, He J, Chen Q, Zhu J, Zou H, Xu X (2018) Ocular surface health in Shanghai University students: a cross-sectional study. BMC Ophthalmol 18:1–11
Majunder C, Toh CL (2020) Non-strabismic binocular visión anomalies among students of a Malaysian private university uses visual display unit. Guoji Yanke Zazhi (Int Eye Sci) 20(6):940–5
Nelson JD, Craig JP, Akpek EK, Azar DT, Belmonte C, Bron AJ, Clayton JA, Dogru M, Dua HS, Foulks GN, Gomes JA (2017) TFOS DEWS II introduction. Ocul Surf
Rao D (2014) Prevalence of non strabismic binocular vision disorders in patients with asthenopia. J Multidiscip Res Healthc 1(1):33–41
Rueff EM, Sinnott LT, Bailey MD, King-Smith PE (2014) The similarity between symptoms of binocular vision disorders and dry eye. Invest Ophthalmol vis Sci 55(13):1990–1990
Sah SK, Thakur R, Adhikari PR (2021) Vision screening among hearing-impaired school children in Biratnagar Nepal. Med Hypothesis, Discov Innov Optom 2(1):36–40
Sah KK, Optom M (2020) Prevalence of accommodative anomaly in high school children age from 13 to 17 year in clinical population
Shah J, Patel S (2015) Strabismus:-symptoms, pathophysiology, management & precautions. Int J Sci Res 4(7):1510–1514
Shaikh S, Bandamwar K (2022) Assessment of digital eye strain with the digital eye strain scoring index (Dessi). J Pharm Negat Results 4567–4575
Stapleton F, Alves M, Bunya VY, Jalbert I, Lekhanont K, Malet F, Na KS, Schaumberg D, Uchino M, Vehof J, Viso E (2017) Tfos dews II epidemiology report. Ocul Surf 15(3):334–365
Stein HA, Stein RM, Freeman MI (2017) The ophthalmic assistant e-book: a text for allied and associated ophthalmic personnel. Elsevier Health Sciences
Supiyaphun C, Jongkhajornpong P, Rattanasiri S, Lekhanont K (2021) Prevalence and risk factors of dry eye disease among University students in Bangkok. Thailand. Plos One 16(10):e0258217
Tamhankar MA (2019) Eye movement disorders: third, fourth, and sixth nerve palsies and other causes of diplopia and ocular misalignment. In: Liu, volpe, and Galetta's neuro-ophthalmology. Elsevier, pp 489–547
Tandon R, Vashist P, Gupta N, Gupta V, Sahay P, Deka D, Singh S, Vishwanath K, Murthy GVS (2020) Association of dry eye disease and sun exposure in geographically diverse adult ($\geq$ 40 years) populations of India: the SEED (sun exposure, environment and dry eye disease) study-second report of the ICMR-EYE SEE study group. Ocul Surf 18(4):718–730
Titiyal JS, Falera RC, Kaur M, Sharma V, Sharma N (2018) Prevalence and risk factors of dry eye disease in North India: ocular surface disease index-based cross-sectional hospital study. Indian J Ophthalmol 66(2):207
Yang I, Wakamatsu T, Sacho IBI, Fazzi JH, de Aquino AC, Ayub G, Rebello PA, Gomes JÁP, Alves M (2021) Prevalence and associated risk factors for dry eye disease among Brazilian undergraduate students. PLoS ONE 16(11):e0259399

Chapter 8
Effectiveness of Manual Therapy on Acute and Subacute Non-specific Low Back Pain: A Literature Review

Kumari Muskan, Anushree Rai, and Priyanka Rishi

8.1 Introduction

People with pain in lower back can be found in all age groups and socioeconomic backgrounds. Acute, subacute, and chronic low back pain affected 577 million people globally in 2017; the point prevalence was 7.83%. One of the most prevalent and expensive musculoskeletal issues in contemporary society is low back pain (LBP). Between 70 and 80% of adults have dealt with it at some point in their lives. Despite the fact that LBP is a benign and self-limiting illness, many patients seek therapy in an effort to enhance their function and reduce their symptoms. In industrialized nations, pain in lower back remains the primary reason of incapacity and lost productivity. Global burden of disease study indicates that LBP is the most common Bhat et al. (2021).

Due mostly to population growth and aging, low back pain accounted for approximately 65 million years lived with disability in 2017.

Over the last 20 years, there has been a significant ~ 50% increase in all of these measures. Women bear a slightly higher burden of all LBP cases than men do. It increases gradually from birth to a peak between 40 and 50 years of age, after which it gradually declines. An examination of each age group reveals that the prevalence of LBP increases along with age for all human diseases, with an especially noticeable

K. Muskan (✉)
MPT Student, Faculty of Physiotherapy, SGT University, Gurugram, Haryana 122505, India
e-mail: muskaansingh750@gmail.com

A. Rai
Assistant Professor, Faculty of Physiotherapy, SGT University, Gurugram, Haryana 122505, India
e-mail: anushree.rai.21@gmail.com

P. Rishi
Associate Professor, Faculty of Physiotherapy, SGT University, Gurugram, Haryana 122505, India
e-mail: Priyanka.physio@sgtuniversity.org

N. A. Siddiqui et al. (eds.), *Advances in Disaster Management, Volume 2*, Springer Transactions in Civil and Environmental Engineering,
https://doi.org/10.1007/978-981-96-4047-8_9

increase beyond the age of 80. The risk of lung cancer (BLD) increases in proportion to the sociodemographic index (SDI); in nations with high SDIs, the risk of LBD is more than three times higher than in those with low SDIs. According to a linear fit of data spanning the preceding 20 years, there may have been an additional 1.4-fold increase in LBP incidence, prevalence, and DALYs by 2050 (Mattiuzzi 2020). Pain in the lower spine is one of the public health problems that 50–80% of grown-ups will involvement at some point in their lives. 2010 saw an estimated 21.8 million disability-adjusted life years (DALYs) as a result of ergonomic exposures at work. Of these, men accounted for 13.5 million and women for 8.3 million DALYs (Lehtola et al. 2016).

According to reports, 42% to 83% of Indians are thought to have LBP (4, 5). According to a current study by Ganesan et al., the prevalence of LBP in young Indian adults between the ages of 18 and 35 is 42.4% annually and 22.8% weekly (Bansal et al 2020). According to (Gupta et al. 2015), the lifetime prevalence ranges from 35 to 80%, while the annual prevalence ranges from 5% to as high as 65%. According to cross-sectional data, the first signs of lower back pain typically appear around the age of 30, and they peak in frequency between the ages of 45 and 60. LBP is no longer an age-related condition. Remarkably, 39.8% of teenagers are also reported to have lower back pain. It restricts daily activities for 10–40% of teenagers (Aggarwal et al. 2013).

Rather, it is advised to stay active and start treating acute LBP with basic analgesics. Exercise is frequently prescribed for patients with acute pain in many countries, despite current recommendations. Evidence on the elects of exercise in acute and subacute populations is of moderate quality (Wright et al. 2005).

Currently, manual therapy (MT) in its various forms is employed in physical therapy practices to treat LBP. In addition to a variety of passive techniques like mobilization and manipulation, physical physiotherapists use a variety of therapeutic modalities, such as different workout plans.

A point prevalence estimates of LBP ranging from 1.0 to 58.0% were published by Hoy et al. According to estimates, high-income countries have a higher prevalence of LBP (30.0%) than low-income countries (18.0%). Those with a 1-year incidence of any episode of LBP and those with their first episode ever fell into the 1.5–36.0% and 6.3–15.4% categories, respectively. The prevalence of LBP was 1.0%, 6.0%, and 18.0% for children ages 7 to 16, 10 to 14, and 14 to 16 years old, respectively, according to a nationwide study carried out in Finland. According to a current research on the gender-age environmental correlates of teenage lower backache, girls were more likely than boys to report having LBP (Fatoye et al. 2019).

The effects of the manual therapy approach have not yet been subjected to a literature review. This literature review aims to determine the efficacy of various manual therapy in treating soft tissue constraints, lowering pain, improving tissue flexibility, joint range of motion, reducing disability and improving patient functionality in the treatment of musculoskeletal diseases.

8.2 Methods

8.2.1 Eligibility Criteria

(1) Population: Participants ranged in age from 16 to 85 years old, regardless of gender. Patient experiencing non-specific sudden and intermittent pain in the lower back for less than three to eight weeks. Individuals with a history of spinal trauma, joint dysfunction within the last three months or congenital deformities at the hip or lumbar region, neurological issues or nerve root compression, recent fractures to related joints within the last six months, a history of surgeries within the last three months for the abdomen, hip, or lumbar region, a positive SI joint compression test, a history of amputation, systemic disorders, or refusal to participate in the study were excluded from the study.
(2) Intervention: Interventions include all the manual therapy [Mulligan, Manual thrust manipulation, compression myofascial trigger point therapy, active back program, and spinal manipulation] which is active for decreasing acute and subacute lower back pain.
(3) Comparison: Comparison is done between various manual therapy versus other physiotherapy protocol in relieving various outcomes.
(4) Outcome: Pain, Disability, Flexibility, Functional status, Fear
(5) Study design: Randomized clinical trial, parallel-group study, randomized controlled trial, cross-sectional, and Cohort study.

This review did not include any studies written in languages other than English.

8.2.2 Search Strategy

We looked through the following databases. The following search terms were combined to perform a systematic search across PubMed, Google Scholar, Pedro, and libraries between 2005 and the present: "manual therapy, acute and subacute low back pain, disability, pain." Boolean operators OR and AND were used in conjunction with these search terms to either widen or narrow the search.

8.2.3 Assortment and Analyzing Data

8.2.3.1 Quality Valuation of Trials

The authors employed an 11-point PEDro scale, which is intended to evaluate the quality of RCTs on the Physiotherapy Evidence Database, to evaluate the methodological quality of the evidence they had retrieved. The original list created by

Table 8.1 Sample and study characteristics

Author	Sample size	Age	Duration exercise program	Data collection	Intervention group	Control group
Bhat et al. (2021) (India)	n = 65 MFR (n = 33) or SNAGs (n = 32)	18–60 years	6 sessions	Day-0 and day-6	1. Procedure for SNAGs (Mulligan, 2010) + Strengthening exercises 2. Procedure for MFR + Strengthening exercises	No exercise
Schneider et al. (2015) (Pittsburgh)	n = 107	18 years	4 weeks	Week = 0 and week = 4 follow up	1. Manual Thrust Manipulation (MTM) + Usual Medical Care (UMC) 2. Mechanical-Assisted Manipulation (MAM) + Usual Medical Care (UMC)	No exercise
Takamoto et al. (2015) (Japan)	n = 63 MTrPs (N = 23), non-trigger points (N = 21), superficial massage (N = 19)	16 and 65 years	2 weeks	Week = 0 week = 1 month = 1	1. Compression at MTrPs 2. Compression at non-trigger points	Superficial massage

(continued)

Table 8.1 (continued)

Author	Sample size	Age	Duration exercise program	Data collection	Intervention group	Control group
Hondras et al. (2009) (Davenport)	N = 240 (105 women and 135 men) sss	55 years old, 63.1 ± 6.7	6 week	Week = 0 week = 3 week = 6	1. High-Velocity Low Amplitude Spinal Manipulation 2. Low-Velocity Variable Amplitude Spinal Mobilization 3. Minimal Conservative Medical Care	Home exercise instruction
Lehtola et al. (2016) (Finland)	61 patients SMCE n = 30 and general exercise n = 31	Age between 16 and 65 years	3 month period	Week = 0 month = 3 follow-up = 12	1. Specific movement control exercise	General exercise
Sung et al. (2015) (USA)	66 patients	Thirty-three gender, age (±5 years)	6 months	Month = 6	NA	NA
Wright et al. (2005)	111 N = 56 [normal routine care] N = 55[active back program]	18–65 year	2 month	Week = 0 month = 1 month = 2	1. Active back program	Normal routine care
	239 participants	NA	12 weeks	4 weeks 12 weeks	NA	NA

8.3.2.3 Outcome Measures

Pain, Disability, Flexibility, Functional status, Fear, Visual Analog Scale (VAS) (Khan et al., 2022), Patient-Specific Function Scale (PSFS), Modified Oswestry disability index (MODI), Roland–Morris's questionnaire (RMQ), Fear-Avoidance Beliefs Questionnaire [FABQ].

8.3.3 Result

See Table 8.2.

Quality Evaluation of the Trials

For all of the trials, the average PEDro rating was 6/10 (good quality). Two trial got a rating of 6 /10, two received a 7/10, and one received an 8/10. Furthermore, just a few studies incorporated techniques of randomization and concealed grouping of individuals. In every trial, the reporting of differences between groups after interventions using measures of variability and point estimations was highly praised (Table 8.3).

8.4 Discussion

The most prevalent and fourth-leading cause of musculoskeletal disorders globally is non-specific back pain (NSBP). An annual incidence of 15% to 50% is predicted for neck pain, which affects 70% of people at some point in their lives. The purpose of the literature review was to evaluate different treatment methods for non-specific lower spine pain, both acute and subacute. In 2017, there were 577 million cases of sudden, intermittent, serious lower spine pain worldwide; the point prevalence was 7.83%. Over the previous 20 years of disability, there has been an increase in reduced range of motion, lower functional status, decreased flexibility, and decreased range of motion.

Our review aimed to assess the treatment effectiveness of manual therapy on reducing discomfort, disability and refining functionality, eminence of lifespan. Eight studies were included in our review but varied in their study number, outcomes, treatment protocol, method, thus preventing the direct comparison between studies. But all the eight studies have the same types of population. 3 studies have shown the statistical reduction in pain.

Results with respect to pain, disability, functionality, and quality of life, fear. Study of Bhat et al. (2021) found evidence of effectiveness of mulligan therapy in terms of reducing pain and disability. By activating mechanoreceptors, MFR and SNAGs aid in the correction of anomalies within the components of the movement system. This may enhance the paraspinal muscles' activation pattern, leading to an improvement in pain-free range of motion. But in the current study, lumbar flexion

Table 8.2 Overview of results

No	Author, year	Population	Study type	Outcome	Treatment	Conclusion
1	Bhat et al. 2021 (India)	65 patients	Parallel-group study	Visual analog scale (VAS), patient-specific function scale (PSFS), modified oswestry disability index (MODI)	Procedure for SNAGs (Mulligan, 2010) Procedure for MFR	Pain and restricted function associated with NSLBP can be improved using SNAGs or MFR, along with strengthening exercises. For limited lumbar flexion ROM, Mulligan SNAGs have a better outcome than MFR over the short-term
2	Schneider et al. (2015) (Pittsburgh)	107 adults	Randomized clinical trial	Pain (0–10 scale) and disability (20 on 0–100 scale)	1. Manual thrust manipulation (MTM) 2. Mechanical-assisted manipulation (MAM) 3. Usual medical care (UMC)	MTM provides greater short-term reductions in self-reported disability and pain scores compared to UMC or MAM
3	Takamoto et al. (2015) (Japan)	63 patients	Randomized controlled trial	Visual analog scale (VAS) and Roland–Morris questionnaire (RMQ)	Three groups 1. Compression at myofascial trigger points (MTrP group) 2. Compression at non-trigger points (non-MTrPs group) 3. Effleurage (effleurage group	These results indicate that compression at MTrPs is effective to treat acute low back pain compared with compression at non-MTrPs and superficial massage

(continued)

Table 8.2 (continued)

No	Author, year	Population	Study type	Outcome	Treatment	Conclusion
4	Hondras et al. (2009) (Davenport)	240 participants (105 women and 135 men) ages	Randomized controlled trial	Roland-Morris disability questionnaire	1. High-velocity low amplitude spinal manipulation 2. Low-velocity variable amplitude spinal mobilization 3. Minimal conservative medical care	n. Participants who received either form of SM had improvements on average in functional status ranging from 1 to 2.2 over those who received MCMC. From an evidence-based care perspective, patient preference and clinical experience should drive how clinicians and patients make the SM procedure decision for this patient population
5	Lehtola et al. (2016) (Finland)	61 patients	Randomized controlled trial	Disability Roland-Morris disability questionnaire (RMDQ)	1. General exercise 2. Specific movement control exercise	For subjects with non-specific recurrent subacute LBP and MCI an intervention consisting of SMCE and manual therapy combined may be superior to general exercise combined with manual therapy

(continued)

Table 8.2 (continued)

No	Author, year	Population	Study type	Outcome	Treatment	Conclusion
6	Sung et al. (2015) (USA)	66 patients	Cross-sectional design	Pain, disability, fear	NA	Patients with clinical identification of trunk MCI demonstrated decreased trunk motor 8 control, suggesting impairments in proprioception, motor output, or central processing occur early in 9 the back pain episode. This information may help to guide interventions to address these specific 10 limitations, improving delivery of care

(continued)

Table 8.2 (continued)

No	Author, year	Population	Study type	Outcome	Treatment	Conclusion
7	Wright et al. (2005) (United Kingdom)	111 (aged 18–65)	Randomized controlled trial	1. Rate of return to work, 2. Pain scores and health status, and 3. Cost effectiveness of interventions versus the financial cost to the individual or employer	1. Group one–Back book, advice and normal route of care (n = 56) 2. Group two–back book, advice and back program	The results indicate that the costs of this active back program are more than reimbursed as a consequence of earlier return to work
8)	Hiebert et al. (2012) (USA)	253	Cohort study	(Fear-avoidance beliefs questionnaire [FABQ] and the Tampa scale of Kinesiophobia), distress (the pain catastrophizing scale), clinical depression (The center for epidemiologic studies depression scale), a numeric pain intensity scale, self-perceived disability (oswestry disability index), and general health status (12-item short form health survey)	Two groups, "full duty" (consisting of "full duty" and "medically released to full duty") and "not at full duty" (consisting of "light duty," "SIQ," and "limited duty")	The findings that fear-avoidance beliefs were predictive of subsequent work status among active duty service personnel in this study population (after adjusting for clinical, demographic, and psychological covariates) suggest the clinical utility of addressing these factors during treatment of back pain episodes in the military. These findings reflect the important role that psychological factors may play in the return to work process in an active duty military population

Table 8.3 Quality scoring of studies using PEDro scale

Sr. no.	Authors	Random allocation	Concealed allocation	Group similarity at baseline	Participant blinding	Therapist blinding	Assess or blinding	< 15% dropouts	Intention to treat analysis	Between group differences reported	Point estimates and variability reported	Total score (out of 10)	Quality
1	Schneider et al. (2015)	Yes	Yes	Yes	No	No	Yes	Yes	Yes	No	No	6	Good
2	Takamoto et al. (2015)	Yes	No	Yes	No	No	Yes	Yes	Yes	Yes	Yes	7	Good
3	Hondras et al. (2009) (Davenport)	Yes	Yes	Yes	No	No	Yes	No	Yes	Yes	Yes	7	Good
4	Lehatola et al. (2016)	Yes	Yes	Yes	No	No	Yes	Yes	Yes	Yes	Yes	8	Excellent
5	Wright et al. (2005)	Yes	No	Yes	No	No	No	Yes	Yes	Yes	Yes	6	Good

range of motion improved only for the SNAGs group, whereas both groups' lumbar extension improved. This result was in contrast to the outcome in patients with non-specific neck pain, where neck ROM was improved in all planes by both MFR and SNAGs.

Study of Schneider et al. (2015) in this manual thrust manipulation [MTM] has shown better outcomes in term of reducing pain and disability when compare to other therapy such as medical care with visual aids and mechanical assistance. The characteristics of our trial cohort, which included only patients with recent onset of LBP, palpably palpable localized lumbar/buttock pain, and no pain distal to the knee, may have had an impact on the reported efficacy of MTM. This was intentional since past studies have shown that these features reflect important clinical findings in a subset of individuals with low back pain (LPP) who will likely benefit from spinal manipulation and can inform joint decision-making.

Study of Takamoto et al. (2015) has shown that compression myofascial trigger points have better results than non-compression myofascial trigger point in terms of pain and functional status. Excessive release of acetylcholine is caused by hyper-activation of the neuromuscular junction, which is brought on by either acute or chronic muscle overload. Constant contractions caused by this overabundance of acetylcholine release eventually cause contraction knots to form in the muscle fibers. These tightening bulges then result in the progress of limited hypoxia and ischemia, and the loss of energy resulting from the contraction triggers the release of noxious substances that sensitize the area, increasing pain and tenderness there (Simons, 2004). Moreover, autonomic nervous system activity can influence the release of acetylcholine and support the positive feedback loop. Compression at MTrPs has been shown to diminish muscle spasm by spinal reflex mechanism in one study (Simons et al., 1999); however, in another study (Patrick and Melzack, 1984), it was found to induce reactive hyperemia in the MTrP region. Additionally, one study (Takamoto et al., 2009) discovered that compression at MTrPs markedly enhanced parasympathetic nervous activity and ameliorated fatigue; a different study (Gerwin et al., 2004) suggested that sympathetic nervous activity may enhance the release of acetylcholine from the motor nerve terminals. According to these findings, there are two possible ways that compression at MTrPs reduces pain: (1) by increasing peripheral blood flow and subsequently removing harmful substances, and (2) by raising parasympathetic tone, which prevents excessive acetylcholine release.

Study of Hondras et al. (2009) has shown that the high velocity low amplitude spinal manipulation has shown better outcomes in reducing disability. When compared to MCMC, low-velocity, variable-amplitude SM—a maneuver whose biomechanical properties could be interpreted as a "mobilizing" technique—seemed to have a slight clinical and statistical advantage. While not always statistically significant, HVLA-SM, the more conventional chiropractic "adjustment," seemed to show some minor clinical benefits to MCMC similar to those of LVVA-SM.

Study of Lehatola et al. (2016) found the evidence for specific movement control exercise in terms of reducing disability. According to a study by Sung et al. (2015) patients with clinically diagnosed trunk MCI had reduced motor control of trunk, indicating that initial in the back pain episode, damages in proprioception, motor output, or central processing may have occurred. This could have implications for pain, disability, and fear. According to Wright et al. (2005), the active back program is far more successful at lowering pain and enhancing functional status. A study by Hiebert et al. (2012) found that after adjusting for clinical, demographic, and psychological variables, fear-avoidance beliefs among active duty military members in this study sample were predictive of subsequent job status. In terms of lowering pain, anxiety, and depression during back pain episodes, this suggests the therapeutic value of addressing these factors. However, in comparison with other manual therapy only Mulligan, Manual thrust manipulation, compression myofascial trigger point therapy, active back program and spinal manipulation has showed the better outcomes in relations of lowering pain, incapacity and improving purposeful status in the patient with non-specific sudden and intermittent pain in the lower back.

8.5 Conclusion

This literature review summary offers a thorough synthesis of the data pertaining to the various forms of manual therapy and their effectiveness in mitigating the symptoms of both non-specific sudden and intermittent pain in the lower back. According to Mulligan's evidence from various studies, patients with non-specific sudden and intermittent pain in the lower spine have shown better results in relations of discomfort, incapacity, and functional status when treated with manual thrust manipulation, compression myofascial trigger point therapy, active back programs, and spinal manipulation.

References

Aggarwal N, Anand T, Kishore J, Ingle GK (2013) Low back pain and associated risk factors among undergraduate students of a medical college in Delhi. Education for Health 26(2):103–108

Bansal D, Asrar MM, Pharm M, Ghai B, Pushpendra D (2020) Prevalence and impact of low back pain in a community-based population in northern India. Pain Physician 23(4):E389

Bhat V, Patel VD, Eapen C, Shenoy M, Milanese S (2021) Myofascial release versus Mulligan sustained natural apophyseal glides' immediate and short-term effects on pain, function, and mobility in non-specific low back pain. PeerJ 15(9):e10706

Fatoye F, Gebrye T, Odeyemi I (2019) Real-world incidence and prevalence of low back pain using routinely collected data. Rheumatol Int 10(39):619–626

Gerwin RD, Dommerholt J, Shah JP (2004) An expansion of Simons' integrated hypothesis of trigger point formation. Curr Pain Headache Rep 8:468–475

Gupta G, Nandini N (2015) Prevalence of low back pain in non-working rural housewives of Kanpur, India. Int J Occup Med Environ Health 28(2):313–320

Hiebert R, Campello MA, Weiser S, Ziemke GW, Fox BA, Nordin M (2012) Predictors of short-term work-related disability among active duty US Navy personnel: a cohort study in patients with acute and subacute low back pain. The Spine J 12(9):806–816

Hondras MA, Long CR, Cao Y, Rowell RM, Meeker WC (2009) A randomized controlled trial comparing 2 types of spinal manipulation and minimal conservative medical care for adults 55

years and older with subacute or chronic low back pain. J Manipulative Physiol Ther 32(5):330–343
Khan ZK, Ahmed SI, Baig AA, Farooqui WA (2022) Effect of post-isometric relaxation versus myofascial release therapy on pain, functional disability, rom and qol in the management of non-specific neck pain: a randomized controlled trial. BMC Musculoskelet Disord 23(1):567
Lehtola V, Luomajoki H, Leinonen V, Gibbons S, Airaksinen O (2016) Sub-classification based specific movement control exercises are superior to general exercise in sub-acute low back pain when both are combined with manual therapy: a randomized controlled trial. BMC Musculoskelet Disord 17:1–9
Mattiuzzi C, Lippi G, Bovo C (2020) Current epidemiology of low back pain. J Hosp Manag Health Policy 4(1)
Mulligan B (2010) Manual therapy, NAGS, SNAGS MWMS etc, 5th edn. Plane View Services Limited, Wellington
Patrick DW, Melzack R (1984) Textbook of pain. Churchill Livingstone, New York, p 691
Schneider M, Haas M, Glick R, Stevans J, Landsittel D (2015) A comparison of spinal manipulation methods and usual medical care for acute and Sub-acute low back pain: a randomized clinical trial. Spine 40(4):209
Simons DG (2004) Review of enigmatic MTrPs as a common cause of enigmatic musculoskeletal pain and dysfunction. J Electromyogr Kinesiol 14:95–107
Simons DG, Travell JG, Simons LS (1999) Myofascial pain and dysfunction, the trigger point manual, the upper extremities, vol 1, 2nd edn. Williams and Wilkins, Baltimore
Sung W, Abraham M, Plastaras C, Silfies SP (2015) Trunk motor control deficits in acute and subacute low back pain are not associated with pain or fear of movement. The Spine J 15(8):1772–1782
Takamoto K, Bito I, Urakawa S, Sakai S, Kigawa M, Ono T, Nishijo H (2015) Effects of compression at myofascial trigger points in patients with acute low back pain: a randomized controlled trial. Eur J Pain 19(8):1186–1196
Takamoto K, Sakai S, Hori E, Urakawa S, Umeno K, Ono T, Nishijo H (2009) Compression on trigger points in the leg muscle increases parasympathetic nervous activity based on heart rate variability. J Physiol Sci 59:191–197
Verhagen AP, de Vet HCW, de Bie RA, et al (1998) The Delphi List: a criteria list for quality assessment of randomized clinical trials for conducting systematic reviews developed by Delphi consensus. J Clin Epidemiol 51:1235–1241
Wright A, Lloyd-Davies A, Williams S, Ellis R, Strike P (2005) Individual active treatment combined with group exercise for acute and subacute low back pain

Chapter 9
Impact of Healthcare Digitalization on Medical Device Trade in India

Manvi Aggarwal, Ketki Sahore, Vinod Kumar, and Jyoti Sinha

9.1 Introduction

The medical device and healthcare industries have expanded dramatically over the last ten years. India is among the top 20 nations for healthcare supplies worldwide. Despite that, there is a large gap between the supply and demand for medical devices in the Indian market, thus offering an enormous potential for Indian manufacturers of medical devices (IBEF 2023).

To continually enhance the standard of treatment at every stage, medical devices' role is now expanded with technological advancements (Davey et al. 2011). Medical devices help maintain lifestyle by regularly monitoring health indicators to prevent diseases and screening, diagnosing, and treating patients (Deloitte India) (Fig. 9.1).

In recent years, the Government of India put into action diverse regulations to stand the demanding situations of the medical devices industry. One includes the indigenous initiative 'Make in India' for promoting domestic manufacturing. It is beneficial to start manufacturing domestically and achieve the goals of accessibility and affordability (Deloitte India).

As per Worldwide Health Organization (WHO) definitions, a medical device includes any instrument, apparatus, implement, machine, appliance, implant, reagent for in vitro use, software, material, or other different or related article intended by

M. Aggarwal (✉) · J. Sinha
Department of Pharmacy, School of Health Sciences, Sushant University, Gurugram, Haryana, India
e-mail: manvi0502@gmail.com

K. Sahore
GNA School of Pharmacy, GNA University, Phagwara, India

V. Kumar
Department of Pharmacy, GD Goenka University, Sohna, Gurugram, India

N. A. Siddiqui et al. (eds.), *Advances in Disaster Management, Volume 2*, Springer Transactions in Civil and Environmental Engineering,
https://doi.org/10.1007/978-981-96-4047-8_10

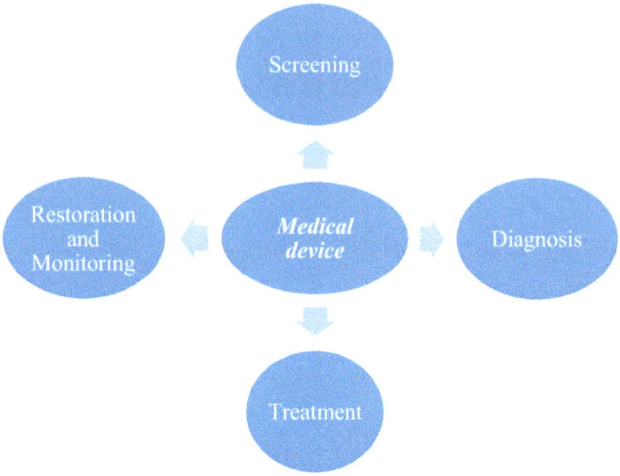

Fig. 9.1 Depicts the various role of medical devices throughout the healthcare spectrum (Deloitte India)

the manufacturer to be used, alone or in combination, for humans for one or more of the following specific medical purposes:

- The study, replacement, modification, or reinforcement of the anatomy or a physiological procedure
- Identifying, preventing, treating, monitoring, and preventing illness or injury. preserve or maintain life
- Sterilizing medical equipment
- Information is provided utilizing in vitro analysis of human body specimens
- Support for people with disabilities
- Devices made with human or animal tissues
- Instruments used in in vitro fertilization or fertility treatments
- Decontamination agents (Dang and Sharma 2019)

In India, medical devices were regulated under the Drug and Cosmetic Act of 1940 as drugs (pharmaceutical products). Later, WHO introduced Medical Device Rules 2017, which became operative on January 1, 2018. As per WHO guidelines, Medical Devices are categorized as given in Table 9.1 (Kalaiselvan et al. 2020).

Table 9.1 Classification of medical device as per Medical Device Rule 2017 (Hoda et al. 2020; Jarrow et al. 2015)

Class of medical device risk	Risk level	Examples
Class A	Low	Nasopharyngeal catheter, bolster suture, alcohol swaps
Class B	Low-moderate	Intravenous catheters, disinfectants, rectal catheters
Class C	Moderate-high	Biliary stents, imaging catheter, bone cement
Class D	High	Heart valve, coronary stent, copper-T

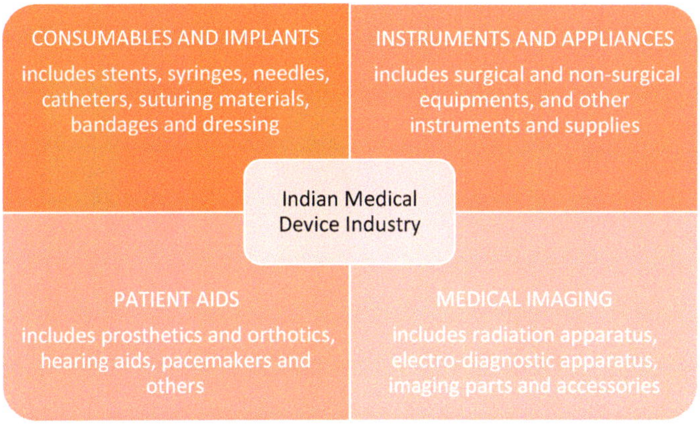

Fig. 9.2 Categorization of Indian medical devices (Deloitte India)

The classifications are based on the severity of the patient's medical device, and the level of risk is regulated and controlled by the FDA, which is required to distribute the device legally in the market. The threat to the patient and the extent of FDA regulatory control rise with class level. Indian medical devices industry is broadly divided into four categories. Figure 9.2 depicts the types of Indian medical devices. (Deloitte India).

9.2 Current Status of the Indian Medical Device Trade Industry

In India, the healthcare supply industry is an emerging sector expanding rapidly. India's medical device market is projected from US$ 11 billion in 2020 to US$ 65 billion in 2024 (IBEF, 2023). Future trends in nanotechnology, stem cell research, micromechanical systems, and far-off medical technology can dramatically alternate the accepted clinical patient care pathways (Davey et al. 2011).

India's experience with the COVID-19 epidemic reminds us of the advantages of investing in the healthcare sector. Industry analysts predict that India's health care spending will increase dramatically over the next several years, partly due to the lessons learned from the pandemic (Van Eck et al. 2009; Kaye et al. 2021). With increased health care spending in India, the market for medical equipment is anticipated to grow in lockstep. Market potential is limited by the increasing price controls on pharmaceuticals and medical equipment, even though entry barriers in the healthcare sector are lower than in other industries (Anon 2023). In addition, enforcing intellectual property laws hinders the biopharmaceutical industry's exports to India. Even though India's economy has grown significantly over the past 25 years, the market still needs to be more challenging. India tends to be a market with fierce

- Middle-Class Growth: The Indian middle class is predicted to grow by 73 million households over the next ten years, bringing with them more purchasing power, in particular when it comes to medical devices.
- 8% of the population will have an increase in discretionary income, and 20% of the population will have health insurance by 2026. This figure is anticipated to increase with increasing earnings and accelerating urban growth.
- Increase in Medical Tourism—India's comparatively affordable healthcare, medical tourism is expanding by 22–25%. Over $2 billion is added to the Indian healthcare sector as a result. A boom in medical tourism has increased interest in medical services and healthcare equipment, leading to the development of medical device parks in four states. Himachal Pradesh, Madhya Pradesh, Uttar Pradesh, and Tamil Nadu.
- 100% Foreign Direct Investment (FDI) is permitted in Greenfield and Brownfield projects. An active and diverse start-up ecosystem exists, with comparatively fewer entrance obstacles as compared to other industries (Precedence research 2023; Invest India, 2018)

9.5 Major Achievements

a
- The SSI-Mantra, a medical technology firm called SS Innovation, will be delivered to the Rajiv Gandhi Cancer Institute and Research Center (RGCI) in New Delhi in July 2022 as the nation's first-ever surgical robotic system made in India.

b
- The Indian Institute of Technology (IIT) Delhi has established a national center for medical technology development to help medical device startups manufacture their products in a facility that has obtained ISO accreditation and obtained the necessary certificates (WHO, 2017 & IIT- Delhi ,2022).

c
- The Ultrahuman Ring is a real-time physiological characteristic tracker that can measure movement, sleep, and metabolism, was introduced by Ultrahuman in July 2022 (The Indian express, 2022)

d
- India's first handheld, wireless spirometer based on a pneumotach was introduced by Cipla in November 2021 under the name "Spirofy" (Bureau,2021).

e
- The first of its type in South Asia, the Surgical Robot Experience Center (SREC) by Medtronic has opened in Gurugram, Haryana. The SREC's primary objectives will be to educate and train surgeons in robotic surgery (Cipla, 2023).

9.6 Opportunity and Challenges

9.6.1 Opportunities

The dynamics of supply and demand offer exceptional potential in India for the production of medical equipment.

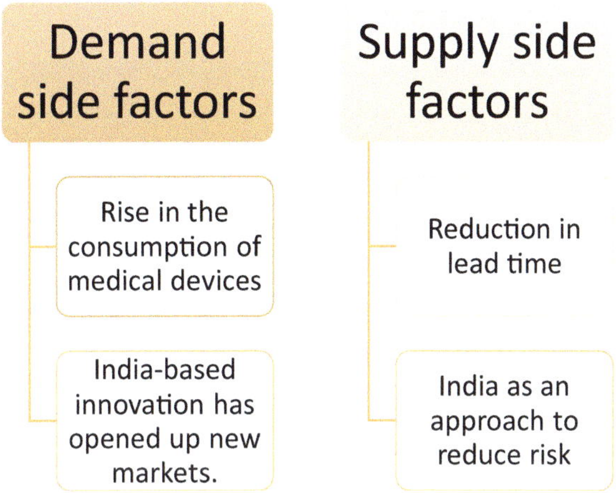

9.6.2 Factors Related to Demand

(a) An increase in the use of medical equipment per person. In the next ten years, the Indian medical device industry is anticipated to expand quickly at a pace of 15%, driven by government initiatives to achieve universal health insurance coverage, initiatives like Make in India, and expansion by private healthcare firms. Despite a double-digit growth rate of the medical devices market at 10% in the past decade, the Indian per capita consumption of medical devices remains significantly low at ~ USD 3.0. It is much lower than the global average per capita consumption of USD 47 and the per capita consumption of developed nations like the USA and Germany (USD 415 and USD 313, respectively, in 2015). One of the primary causes for low per capita consumption is the challenges of affordability for high-end technology devices among large portions of the Indian population. As a result, just like many other sectors of the Indian manufacturing sector, the manufacturing of medical devices primarily concentrates on frugal engineering to create low-cost items tailored to India's lower and middle-income segments. As demand rises and the per capita use of medical equipment increases, it is anticipated that this market's potential will be achieved (IBEF 2023; Marucheck et al. 2011).

(b) Newer Market access as a result of innovation from India. Even though manufacturing is primarily focused on creating low-tech products, a small number of domestic businesses and MNCs with manufacturing operations in India have successfully generated low-cost goods with quality equivalent to those currently on the market but need sophisticated technological expertise to produce. Consequently, these products created a niche market in many world regions. It is not unexpected that medical device exports have experienced rapid growth over the past ten years, given these success stories. At USD 1.17 billion (2014), the Consumables and Implants segment has significantly contributed to India's exports. In terms of value, the USA (USD 171 million), Singapore (USD 81 million), and China (USD 72 million) are India's top three export destinations.

9.6.3 Factors Related to Supply

(a) Reduction in lead time and enhanced suitability- India is poised to become a top location for manufacturing operations due to its possibility for high domestic demand and additional beneficial elements. This is especially true for multinational corporations seeking to adjust their global manufacturing footprint to reflect changing consumer trends. Reduction in lead time and the opportunity to considerably improve service standards are encouraging signs for India's expanding healthcare market (Martin et al. 2008).
(b) India as a risk-reduction strategy for the region- India provides multinational corporations with an excellent opportunity to mitigate risks related to the expanding domestic market and local and global threats to their business. India has great potential to develop into an export-focused nation and is poised to become a significant hub for consumption. China offers low-cost manufacturing and a large domestic market that has grown into an important export center—where exports are much higher than domestic sales. Some examples are Ireland, Singapore, and Puerto Rico (Deloitte, India)

9.7 Challenges

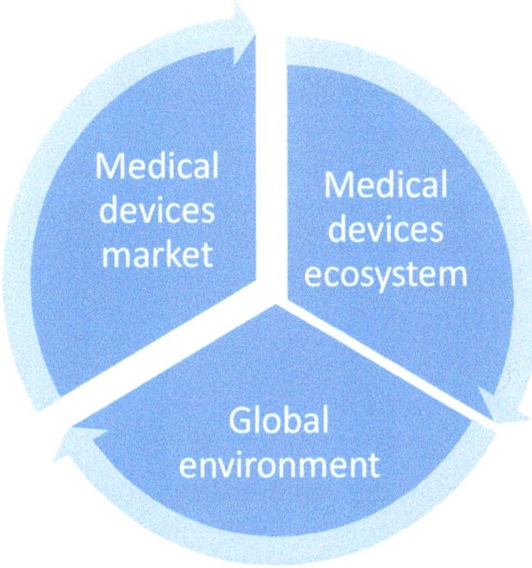

9.7.1 Medical Devices Market

- Unfavorable work framework: Imports are less expensive than Indian manufacturing in many segments or sub-segments due to a detrimental duty structure. This further diminishes the possibility of local value addition, particularly in areas that are currently helpful to manufacturing.
- Need for domestic demand for specific market niches or product categories: Most medical device industry's segments and sub-segments have small markets, restricting investment because any investment needs scale to be profitable. An inverted duty structure combined with low demand makes the case for domestic manufacturing less intense.
- Absence of comprehensive legislation and weak enforcement practices for intellectual property protection: Intellectual property protection for the latest innovations is essential to the profitability of multinational medical device companies that invest outside their home markets. Global players would not feel confident enough to invest in manufacturing assets if there were no enforceable IP laws, thus limiting Indian manufacturing.
- The absence of an indigenous "quality certification" authority: As many countries demand a certificate from the land of creation, the lack of a licensing authority limits export opportunities. To serve a portion of the Indian and global markets, domestic manufacturers must obtain an FDA/CE certification. The time to market may be impacted by the approval process, which can take a long time (Thomas).

9.7.2 Medical Devices Ecosystem

- Inadequate ecosystem support: Establishing a manufacturing facility involves more than just setting it up; it also requires having a functional component supply chain. The fact that a sizable portion of the component supply chain is imported currently presents a significant challenge for participants in the majority of segments and sub-segments.
- Limited emphasis on "Innovate in India": Since imports have dominated the market, there hasn't been a lot of opportunity for "local innovation." Except for a few, most have viewed India as one of several export markets concentrating primarily on broad sales and distribution.

9.7.3 Global Environment

- Doing business is still more difficult in India than it is in the majority of other nations. Currently, products need to receive multiple agency approvals due to complex regulatory requirements. Time delays result from this on several levels, both nationally and statewide.
- Restricted labor laws and a shortage of skilled workers: India has enough human capital, but its labor laws are unfavorable, and its trained workers are insufficient to install, operate, repair, and service equipment.
- Exorbitant financing costs: This sector requires a lot of capital. The cost of financing is currently between 14 and 18%, which puts more financial strain on companies looking to establish manufacturing facilities in India (Strategic information group, 2021).

9.8 Government Initiatives

An essential component of India's healthcare system is the medical device sector. The Indian medical device industry has become even more recognized with the mass production of medical devices and diagnostic kits during the COVID-19 pandemic. These kits include respirators, rapid antigen test kits, real-time reverse transcription polymerase chain reaction (RT-PCR) kits, infrared (IR) thermometers, personal protective equipment (PPE) kits, and N-95 masks.

The *Atma Nirbhar Bharat Abhiyaan* campaign aims to make India self-sufficient. It is the vision of a new India, as Hon. Prime Minister Shri Narendra Modi envisioned. In addition to seeing India become a manufacturing powerhouse, India's standing as the second-largest global manufacturer of medical devices will be strengthened by Atma Nirbhar Bharat and a dependable supplier of high-quality goods for international supply chains (Kale and Wield 2018). The programs the Indian government has introduced to encourage the production of medical devices remain:

Production Linked Incentive Schemes: The Indian government has designated the manufacturing ecosystem for medical devices as a priority area for its flagship "Make in India" program. The Production Linked Incentive Scheme (PLI) promoting Domestic manufacturing of Medical Devices and the introduction of pharmaceuticals is intended to support India's goal of becoming a center for the production of medical equipment worldwide.

Plan for Promotion of Medical Devices Parks: To provide common infrastructure, build a robust manufacturing ecosystem for medical devices, and lower production costs, four medical device parks have been approved for construction by the Indian government (Invest in India). To lessen India's reliance on imported medical equipment and technology, policymakers must devise a plan of action.

In addition to these programs, the Indian government has launched many other measures to support the medical device industry, mainly research and development (R&D) and 100% FDI for medical devices to expand the market.

- A budget of Rs. 86,200 crore (US$ 11.3 billion) was allotted for the pharmaceutical and healthcare sector in the Union Budget 2022–23
- To boost the acceptance of Indian medical products in the international market, the government declared plans to create a new medicines, cosmetics, and medical devices bill in October 2021
- The PLI plan for medical devices, in which the government announced in October 2021 that 13 businesses had been approved and is anticipated to increase domestic production in the nation (Marken et al 2020)
- The Quality Council of India (QCI) and the Association of Indian Manufacturers of Medical Devices (AiMeD) launched the Indian Certification of Medical Devices (ICMED) 13,485 Plus program in June 2021 to validate the safety, efficacy, and quality of medical devices
- On March 25, 2021, the Department of Pharmaceuticals released a revised notification concerning the Public Procurement Order (PPO) that included 19 medical devices in the updated PPO guidelines. Improved domestic medical device manufacturing (supporting "Make in India") and a 4000 crore (US$ 538.62 million) reduction in import costs are anticipated as a result (Kumar and Kumar 2021)

9.9 Conclusion

Medical device businesses should develop India as a manufacturing hub for domestic and foreign markets, encourage indigenous production alongside India-based innovation, collaborate on the Make in India and Innovate in India initiatives, and create low- to medium-tech products to meet the needs of the marginalized domestic markets. Policymakers must establish an action plan to reduce India's reliance on imported healthcare products and technology. Over the past decade, the Indian medical device business has undergone significant growth and change. Several variables, including the increased incidence of chronic diseases, rising affluence, and improvements in medical technology, have contributed to the sector's upward trajectory. Government initiatives have also boosted the industry to improve healthcare infrastructure and increase access to affordable medical care. The Indian medical market will likely expand further in the coming years; therefore, insightful companies must be ready.

References

Anon (2023) COVID-19 revealed true potential of the Indian medical devices industry—India business and trade. [online] Available at: https://www.tpci.in/indiabusinesstrade/blogs/covid-19-revealed-true-potential-of-the-indian-medical-devices-industry. Accessed 15 Nov 2023

Bureau EN (2021) Medtronic opens its first surgical robotics experience center (SREC) in Asia Pacific. [online] Express Healthcare. Available at: https://www.expresshealthcare.in/news/medtronic-opens-its-first-surgical-robotics-experience-center-srec-in-asia-pacific/431241/. Accessed 15 Nov 2023

Corpbiz (2023) Import and export of medical devices—an overview. [online] Available at: https://corpbiz.io/learning/import-and-export-of-medical-devices. Accessed 15 Nov 2023

Dang A, Sharma JK (2019) Economics of medical devices in India. Value in Health Regional Issues 18:14–17

Davey SM, Brennan M, Meenan BJ, McAdam R (2011) Innovation in the medical device sector: an open business model approach for high-tech small firms. Technol Anal Strat Manag 23(8):807–824

Deloitte India (n.d.) Medical devices | Deloitte India | Life sciences & healthcare. [online] Available at: https://www2.deloitte.com/in/en/pages/life-sciences-and-healthcare/articles/medical-devices-making-in-india.html. Accessed 15 Nov 2023

van Eck CF, Chen AF, Klatt BA, D'Antonio J, Fu F (2009) The classification of implants: class I, II, III. J Long-Term Eff Med Implant 19(3)

Hoda F, Verma R, Arshad M, Siddiqui AN, Khan MA, Akhtar M, Najmi AK (2020) Materiovigilance: concept, structure and emerging perspective for patient's safety in India. Drug Research 70(09):429–436

IIT-Delhi's centre to give push to medical device startups (2022) The Times of India. [online] 8 July Available at: https://timesofindia.indiatimes.com/city/delhi/iit-ds-centre-to-give-push-to-medical-device-startups/articleshow/92733167.cms. Accessed 15 Nov. 2023

India Brand Equity Foundation (2023) Medical devices industry in India—market share, reports, growth & scope | IBEF. [online] Available at: https://www.ibef.org/industry/medical-devices.aspx. Accessed 15 Nov 2023

India Brand Equity Foundation (2022) Import of medical devices rose a record 41% in FY22: Report | IBEF. [online] Available at: https://www.ibef.org/news/import-of-medical-devices-rose-a-record-41-in-fy22-report

India Brand Equity Foundation (n.d.) The evolving opportunities and challenges in medical devices sector | IBEF. [online] Available at: https://www.ibef.org/blogs/the-evolving-opportunities-and-challenges-in-medical-devices-sector. Accessed 15 Nov 2023

India—Healthcare and Medical Equipment. [online] www.trade.gov. Available at: https://www.trade.gov/country-commercial-guides/india-healthcare-and-medical-equipment

Investindia.gov.in. (2018) Medical devices. [online] Available at: https://www.investindia.gov.in/sector/medical-devices

Jarow JP, Baxley JH (2015) Medical devices: US medical device regulation. Urol Oncol: Semin Orig Investig 33(3):128–132

Kalaiselvan V, Thomson M, Kumar P, Mishra N, Abirami V (2020) Available at: https://www.ipc.gov.in/images/Guideline_-_11-03-2020.pdf. Accessed 12 June 2023

Kale D, Wield D (2018) In search of the missing hand of 'collaborative action': evidence from the Indian medical device industry. Innov Dev

Kaye AD, Okeagu CN, Pham AD, Silva RA, Hurley JJ, Arron BL, Sarfraz N, Lee HN, Ghali GE, Gamble JW, Liu H (2021) Economic impact of COVID-19 pandemic on healthcare facilities and systems: international perspectives. Best Pract Res Clin Anaesthesiol 35(3):293–306

Kumar V, Kumar A (2021) Self-reliance in advanced diagnostic imaging for accessible, affordable, and quality health care in India. J Nat Sci Biol Med 12(2):129–129

Kumar A (2022) India's dependence on imported medical devices at 'alarming' level, industry group says. [online] Moneycontrol. Available at: https://www.moneycontrol.com/news/business/indias-dependence-on-imported-medical-devices-at-alarming-level-industry-group-says-8890141.html. Accessed 15 Nov 2023

Markan S, Nath R, Sharma J, Chekuri DK (2020) Indian Medical Device sector-blue print & regulatory policy roadmap. Int J Drug Regul Aff 8(2):25–35

Martin JL, Norris BJ, Murphy E, Crowe JA (2008) Medical device development: the challenge for ergonomics. Appl Ergon 39(3):271–283

Marucheck A, Greis N, Mena C, Cai L (2011) Product safety and security in the global supply chain: issues, challenges and research opportunities. J Oper Manag 29(7–8):707–720

Medical device manufacturing in India—a sunrise. (n.d.). Available at: https://pharmaceuticals.gov.in/sites/default/files/medicaldevicemanufacturinginindia-asunrise-170221053503%20%281%29.pdf

Medinnova (2022) Indian medical device industry—key drivers of growth. [online] Medinnova. Available at: https://www.medinnovasystems.com/indian-medical-device-industry-key-drivers-of-growth. Accessed 15 Nov 2023

Pib.gov.in. (n.d.) Cabinet approves the policy for the medical devices sector. [online] Available at: https://pib.gov.in/Pressreleaseshare.aspx?PRID=1919984. Accessed 15 Nov 2023

Strategic Information Group (2021) Medical device industry faces many challenges—strategic. [online] Available at: https://www.strategic.com/blog/the-medical-device-industry-faces-many-challenges. Accessed 15 Nov 2023

The boom in India's healthcare firms isn't showing signs of slowing. The Economic Times. Retrieved from https://economictimes.indiatimes.com/industry/healthcare/biotech/healthcare/the-boom-in-indias-healthcare-firms-isnt-showing-signs-of-slowing/articleshow/88966321.cms

The Indian Express (2022) Ultrahuman announces wearable smart ring to track metabolism. [online] Available at: https://indianexpress.com/article/technology/gadgets/ultrahuman-announces-wearable-smart-ring-to-track-metabolism-8010840. Accessed 15 Nov 2023

Thomas (n.d.) Challenges and opportunities in medical device manufacturing. [online] blog.thomasnet.com. Available at: https://blog.thomasnet.com/challenges-and-opportunities-in-medical-device-manufacturing.

www.cipla.com. (n.d.) Cipla launches Spirofy®: strengthens its lung leadership position by facilitating early diagnosis for COPD | Cipla. [online] Available at: https://www.cipla.com/press-releases-statements/cipla-launches-spirofy-indias-first-pneumotach-based-portable-wireless-spirometer-for-diagnosis-of-copd-and-asthma. Accessed 15 Nov 2023

www.precedenceresearch.com (n.d.) Medical devices market size, report 2020 to 2027. [online] Available at: https://www.precedenceresearch.com/medical-devices-market

Chapter 10
Impact of COVID-19 Pandemic on Mental Health Status Among Middle-Aged and Elderly Individuals

Deepali Sinha, Manish Kumar, Charu Chhabra, and Vishwajeet Trivedi

10.1 Introduction

The global proliferation of the coronavirus has ushered in an unparalleled crisis for mental health on a global scale. The corona virus pandemic has emerged being a black swan event, prompting governments across the world to adopt a myriad of strategies in an attempt to mitigate its impact (Lee 2020). Consequently, measures such as lockdowns, which necessitated people staying at home and adopting remote work arrangements, have induced significant levels of fear, apprehension, and concern among the general population, with certain demographic groups, particularly the elderly and middle-aged individuals, experiencing heightened distress. The rapidly evolving situation has wrought profound changes in people's daily lives, emotions, and behaviours, all of which exert a direct and pronounced influence on their mental well-being (Singh et al. 2020).

The coronavirus infection is an extremely contagious illness triggered by a novel coronavirus. Severe Acute Respiratory Syndrome Coronavirus 2 (SARS-CoV-2) stands as the root cause behind this global health crisis (Wilken et al. 2017). While the pandemic has led to severe illnesses among geriatric population and those suffering

D. Sinha
SAAOL Ortho Care, New Delhi, India

M. Kumar (✉) · V. Trivedi
Department of Physiotherapy, School of Medical and Allied Sciences, G D Goenka University, Gurugram, Haryana, India
e-mail: mkumar.physio@gmail.com

C. Chhabra
School of Physiotherapy and Rehabilitation Sciences, K. R. Mangalam University, Gurugram, Haryana, India

N. A. Siddiqui et al. (eds.), *Advances in Disaster Management, Volume 2*, Springer Transactions in Civil and Environmental Engineering,
https://doi.org/10.1007/978-981-96-4047-8_11

from medical co-morbidities, the majority of individuals infected with the viral infection have been inflicted with minor to substantial symptoms of the respiratory distress (Sohrabi et al. 2020). Importantly, the pandemic has not solely affected physical health; it has also inflicted considerable harm upon mental well-being.

Individuals have found themselves ensnared in vicious cycles of deteriorating mental health, which in turn affects the body's immune system, thereby compromising physical health (Chakraborty and Maity 2020). The lockdown measures imposed in various countries, including India, have profoundly impacted mobility in essential areas such as grocery shopping, accessing pharmacies, engaging in recreational activities, participating in retail, commuting to work, visiting parks, and attending workplaces (Zhang et al. 2021). This period of uncertainty has left the youth grappling with uncertainties surrounding their future career prospects, while entrepreneurs have been plagued by anxieties related to market stability and the prospect of returning to normalcy.

The pandemic's pervasive uncertainty and fear have weighed heavily on the minds of middle-aged and elderly individuals, given their heightened vulnerability to infections and concerns about mortality. Mass unemployment and reduced wages have eroded the quality of life for many families, amplifying feelings of insecurity and giving rise to increased levels of panic, helplessness, and anxiety among these age groups (Khanna et al. 2020; Ghosh et al. 2020).

In the light of these challenges, the importance of good mental health and fostering a positive outlook cannot be overstated, as it can aid in better coping with the threat of COVID-19 and bolster the immune system (Kochhar et al. 2020). Moreover, individuals with poor health status, weakened immune function, diminished organ function, a higher probability of underlying diseases, and inadequate attention to personal health are at an elevated risk of coronavirus susceptibility, not only among the elderly but also within the middle-aged demographic. This heightened susceptibility includes the risk of severe COVID-19 outcomes, leading to mortality and subsequent "what happens after me" concerns, which profoundly affects the mental well-being and overall health of these individuals (Chen et al. 2021).

Despite the gravity of these issues, there is limited literature on the specific impact of the coronavirus outbreak on the mental health of middle-aged and elderly individuals. This research was undertaken with the prime aim of assessing the effects of the corona virus outbreak on the mental health status of middle-aged and elderly individuals and exploring potential correlations between Perceived stress levels and Generalised anxiety disorder among them.

10.2 Materials and Methods

A cross-sectional study was performed on 100 subjects recruited in the study using convenience and snowball sampling. Informed consent was obtained from all the subject, and the subjects was divided into two group on the basis of age: Group A (middle age with mean age 48.08 ± 3.03) and Group B (elderly with mean age 63.86

± 5.14). The outcome measures taken were PSS-10: Perceived stress scale used to assess the perception of stress and GAD-7: Generalised anxiety disorder scale for evaluating anxiety level. A web-based questionnaire through a survey link was established and was circulated using various online portals. A background demographic information sheet consisting of subject name, age, gender, marital status, and education qualification, and they were asked about their current occupation in the questionnaire. Thereafter history and a brief analysis of data was done following the filling of by subjects. Data analysis was done to obtain results of the study.

10.3 Result

Descriptive statistics were taken in which 100 samples were divided in two groups with mean age of 48.08 ± 3.03 for middle aged (Group A) and mean age of 63.86 ± 5.14 for elderly (Group B), which is given in Table 10.1.

Table 10.2 states the Mean ± SD score of Group A and Group B of Perceived stress scale and Generalised anxiety disorder scale.

It was reported that 88% of participants from Group A and 82% from elderly population exhibits moderate level of perceived exertion (Table 10.3), whereas 42% of middle-aged individuals showed moderate level of anxiety under Generalised Anxiety level scale and remaining fall under mild anxiety level (24%), minimum anxiety level (18%), and severe anxiety level (16%). Group B exhibits the higher side of population under moderate to mild level of anxiety which contributes around 46% and 42% of the total group population, respectively. Remaining 12% of population showed minimum level (6%) and severe level (6%) of anxiety levels on scale (Table 10.4).

Karl Pearson was used to determine the relationship between Perceived stress levels and Generalised anxiety disorder level. Hence, it is clear from the Figs.4 and 5 that there is strong positive linear correlation between Perceived stress level and

Table 10.1 Descriptive statistics of age among group A and group B

Group	% of responses	Mean AGE ± SD	No. of males	No. of females
Middle aged (group A)	50%	48.08 ± 3.03	22	28
Elderly (group B)	50%	63.86 ± 5.14	24	26

Table 10.2 Mean ± SD score of PS scale and GAD scale among group A and group B

Group	PSS-10 (perceived stress scale-10)	GAD-(generalised anxiety disorder scale)
	MEAN ± SD	MEAN ± SD
Group A (middle aged)	42.58 ± 4.12	10.08 ± 4.6
Group B (elderly)	21.82 ± 4.6	9.5 ± 3.3

Table 10.3 Data show the level of perceived exertion among both the groups of participants

Range of stress (PSS-10 Score)	Level of perceived stress	Group A (no. of middle-aged individuals)	Group B (no. of elderly individuals)
Score 0–13	Low stress	2	4
Score 14–26	Moderate stress	44	41
Score 27 ABOVE	High perceived stress	4	5

Table 10.4 Data show the level of anxiety disorder among both the groups A and B

Range of anxiety (GAD-7 score)	Level of anxiety disorder	Group A (no of middle-aged individuals)	Group B (no of elderly individuals)
Score 0–4	Minimum anxiety	9	3
Score 5–9	Mild anxiety	12	21
Score 10–14	moderate anxiety	21	23
Score 15–21	Severe anxiety	8	3

Table 10.5 Correlation between both Group A and Group B

Karl Pearson coefficient of correlation (r)	Group A (middle aged)	Group B (elderly)
r value	0.70	0.65
p-value	0.0001	0.0001

Generalised anxiety disorder level in Group A (middle-aged individuals) and Group B (elderly individuals). The result was found to be significant in Group A ($p = 0.0001$) and Group B ($p = 0.0001$) with $r = 0.70$ and 0.65, respectively, as shown in Table 10.5.

10.4 Discussion

The current study aimed to explore the profound impact of the coronavirus outbreak on the mental health status of both middle-aged and elderly individuals. This research employed the Perceived stress scale and the Generalised anxiety disorder scale to evaluate the mental health status of 100 participants categorised into two groups: Group A, comprising middle-aged individuals, and Group B, consisting of elderly participants. The results unveiled a robust positive correlation between Perceived stress levels and Generalised anxiety disorder levels within both demographic groups, highlighting the significant association between the two in the coronavirus pandemic.

The utilisation of the PSS-10 Scale and the GAD-7 in this study was grounded in their established reliability, validity, and sensitivity. The PSS-10 Scale was shown to be easily comprehensible to individuals regardless of their educational background (Baik et al. 2019). Additionally, it exhibited adequate reliability and sensitivity. Likewise, the GAD-7, as demonstrated by Bernd Löwe and Oliver Decker (2006), was found to be a dependable and valid self-report measure for assessing anxiety in the general population (Löwe et al. 2008). Its unidimensional structure accurately reflected anxiety levels across various subject groups. Therefore, these scales were employed with confidence as reliable tools for measuring the impact of the COVID-19 pandemic on the mental health status of middle-aged and elderly individuals.

This study's findings align with prior research, shedding light on the adverse effects of pandemic-induced changes on the mental health and overall well-being of college students. Son et al. (2020) reported a substantial negative impact of the COVID-19 pandemic on academic, health, and lifestyle-related outcomes among college students, with an increased prevalence of stress and anxiety. A similar observation was made regarding college students in the UK, where moderate levels of stress were documented by the PSS (Son et al. 2020).

Interestingly, our study also resonates with recent research by Bruine de Bruin et al. (2020), which found that although the elderly perceived themselves to be at a higher risk of mortality if they contracted COVID-19, they exhibited better mental health and psychological functioning compared to their younger counterparts (Bruine de Bruin 2021). This suggests that age may have a nuanced role in how individuals perceive and cope with pandemic-related stressors. In line with this, Wang and colleagues reported no significant difference in anxiety levels between age groups in their longitudinal study (Wang et al. 2020).

While this study provides valuable insights into the relationship between Perceived stress and Generalised anxiety disorder levels among middle-aged and elderly individuals during the COVID-19 pandemic, several limitations must be acknowledged. First, the study's cross-sectional design limits our ability to establish causality. Longitudinal studies could provide more nuanced insights into the dynamics of mental health during prolonged crises like pandemics. Future research could extend this investigation by considering other variables such as social support, coping strategies, and access to healthcare services, which can significantly influence mental health outcomes during crises. Additionally, exploring interventions and strategies to mitigate the impact of Perceived stress and anxiety on mental health among different age groups is an essential avenue for future research.

10.5 Conclusion

This study concluded that there is high impact of COVID-19 pandemic on mental health status among middle-aged and elderly individuals. The result also conclude that middle-aged individuals are facing more stress than elderly individuals. There is significant relationship between stress and anxiety level, which shows that along with increase in stress level there is also increase in anxiety level among these group at the time of COVID -19 pandemic.

Chapter 11
Impact of Physical–Cognitive Training on Diabetes-Associated Decrements of Cognition and Balance

Aashi Bhatnagar, Vinika Chaudhary, Sonia Pawaria, Vishakha, and Shikha

11.1 Introduction

Diabetes mellitus (DM) stands out as a pervasive global health challenge, marked by rising prevalence and affecting both developed and developing nations (Goyal and Jialal 2018). In 2021, the global prevalence of diabetes was 10.5%, projected to reach 12.2% by 2045, impacting approximately 783.2 million individuals (Pradeepa and Mohan 2021). India, hosting 87 million cases, has earned the designation of the world's diabetes capital (Sun et al. 2022). The surge in type 2 diabetes (T2DM) is attributed to various factors, including aging, sedentary lifestyles, and dietary changes, contributing to a growing global burden (Issaka et al. 2018). Diabetes not only leads to severe complications such as retinopathy, nephropathy, and cardiovascular diseases but is also associated with cognitive impairment, affecting individuals across different age groups, including middle-aged populations (Demir et al. 2021). Cognitive decline is accentuated in those diagnosed later in life, with a 19% increase in midlife diabetes-associated cognitive decline over the period of two decades (Chakraborty et al. 2021). Factors such as disease duration and elevated HbA1c levels further exacerbate cognitive deterioration (Kodl and Seaquist 2008). Cognitive impairment in individuals with T2DM extends beyond attention and memory deficits, encompassing areas like executive function, processing speed, and motor control

A. Bhatnagar · Vishakha (✉)
Department of Physiotherapy, School of Medical and Allied Sciences, G D Goenka University, Gurugram, Haryana, India
e-mail: dagarvishakha9339@gmail.com

V. Chaudhary · S. Pawaria
Faculty of Physiotherapy, Shree Guru Gobind Singh Tricentenary University, Gurugram, Haryana, India

Shikha
Shri Badrinath Multispecialty Hospital, Gurugram, Haryana, India

N. A. Siddiqui et al. (eds.), *Advances in Disaster Management, Volume 2*, Springer Transactions in Civil and Environmental Engineering,
https://doi.org/10.1007/978-981-96-4047-8_12

(Shaikh et al. 2019). This cognitive decline is linked to continuous glycemic changes and structural alterations in brain neuronal function (McCuskey and McCuskey 1984). The association between diabetes, cognitive impairment, and compromised mobility underscores the intricate relationship between cognitive and motor functions, impacting activities of daily living (Li et al. 2018). The heightened fall risk in T2DM individuals is attributed to sensory function decline, reduced metabolic muscle function, and impaired executive functions, necessitating a comprehensive understanding of effective intervention (Horak 2006). The Montreal Cognitive Assessment (MoCA) is crucial for assessing mild cognitive impairment in individuals with type 2 diabetes (T2DM), revealing varying rates of cognitive decline across age groups (Dong et al. 2016). In tandem, the Stand and Reach Test (SRT), Five Times Sit to Stand Test (5XSST), and Timed Up and Go Test (TUG) provide essential insights into the physical aspects of T2DM, highlighting flexibility, lower extremity strength, and functional mobility (Bahureksa et al. 2016). These combined assessments offer a comprehensive understanding of both cognitive and physical challenges in individuals with diabetes. Physical exercise, particularly balance training, has shown promise in improving cognitive function and reducing fall rates in older adults with T2DM (Dixon et al. 2017). The Square Step Exercise (SSE), a simultaneous training method combining cognitive and physical aspects, presents a progressive program designed to improve balance and cognitive function (Silva et al. 2018). While effective in older adults, its impact on individuals with diabetes remains an understudied area. Hence, ongoing research aims to evaluate SSE's effectiveness in improving balance and cognitive impairments in middle-aged patients with diabetes, providing insights into potential interventions for this population.

11.2 Methodology

This study employed an experimental design and determined the sample size based on a previous study's effect size which was 0.9. A total of 40 diagnosed community-dwelling type 2 diabetic patients were included using random sampling. They were divided equally into two groups: an experimental group and a control group. Inclusion criteria for the study were the following: pre-diagnosed patients of T2DM exceedingly more than 5 years but on regular medications within the age group 40-60 years fulfilling the criteria of score of MoCA $\leq$ 25 and TUG $\geq$ 10 seconds, both male and female. The exclusion criteria included any contraindication to physical exercise, any musculoskeletal or neurological disorder, any recent surgery, or any medications for neurological disorders.

The subjects of both groups were informed about the procedure of the study, and the written consent was signed. Pre-readings of the age, duration of diabetes, BMI, MoCA, TUG, 5XSST, and SRT were taken at baseline. The experimental group was given square-step training, while balance training including standing on one leg, tandem walking, changing direction on cue, etc., was given as a conventional method of treatment to the control group. The treatment session for both the groups

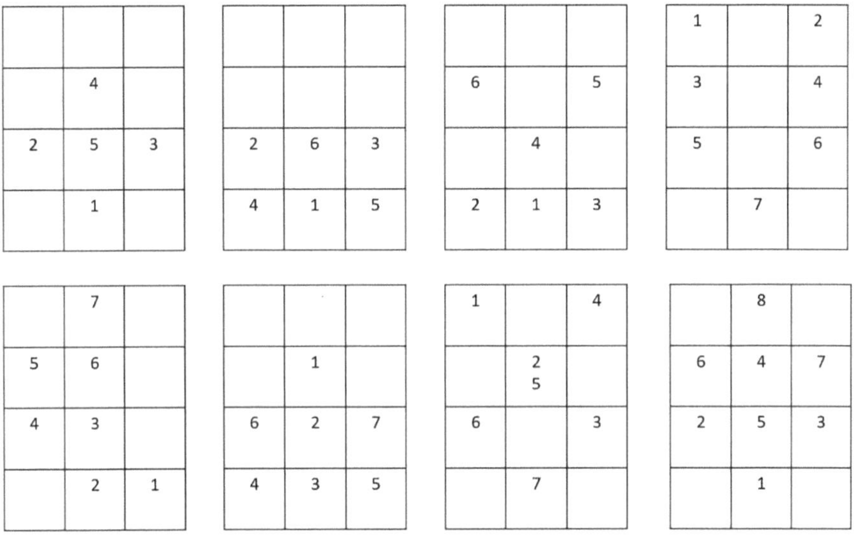

Fig. 11.1 Stepping patterns

was 40 min for 3 days/week for 4 weeks with 5 min of warm up and cool down for the experimental group and 10 minutes of warm up and cool down for the control group. All four variables, MoCA, TUG, 5XSST, and SRT were assessed after 4 weeks. The SSE training protocol for an experimental group involved reproduction of stepping patterns that were previously demonstrated, and participants were expected to observe and memorize them. They would then make an attempt to reproduce each stepping pattern. Stepping patterns were in different directions, and participants were asked to perform the same pattern at the same speed. Balancing tasks like tandem standing and clapping along with cognitive tasks like subtraction in series of seven or reverse counting were given while executing the stepping pattern. The control group was given dynamic balance and functional balance training (Fig. 11.1).

11.3 Statistical Analysis

Analysis of data was done by using IBM SPSS Statistics Software version 21. The Mean ± standard deviations of both groups were calculated and mapped in the form of a table to display the data's properties. Independent t-test was used to analyze the group differences in the Time Up and Go test (TUG), Five times Sit to Stand test (5 × SST), and Stand and Reach test (SRT) at baseline and at the end of 4th week, whereas paired t-test was used for within the group analysis of the same variables. Mann–Whitney U test was used to evaluate scores of Montreal Cognitive Assessment (MoCA) between the group and Wilcoxon signed rank test to analyze within the group differences.

11.4 Results

The study included 40 participants in both the groups. The characteristics of participants at baseline were found to be non-significant. The Mean ± standard deviation for age, duration of diabetes, BMI in experimental group were 52.50 ± 5.89, 8.52 ± 3.371, 29.04 ± 4.83, respectively, while in control group were 52.00 ± 5.554, 7.60 ± 2.604, 29.34 ± 2.979, respectively. Paired t-test was used to analyze the scores of TUG, 5XSST, and SRT, while MoCA was assessed using Wilcoxon test within the experimental group. There were significant changes seen in the scores of MoCA, TUG, 5XSST, and SRT with the significance value of 0.000*, 0.000*, 0.000*, 0.017*. Significant changes were seen in the scores of TUG, 5XSST, and SRT with the p value of 0.000*, 0.000*, and 0.013* within the control group, while no significant changes noted in the scores of MoCA. Independent t-test was used to analyze the scores of TUG, 5XSST, and SRT after 4 weeks of intervention and showed no significant differences in any of the variable. The score of MoCA was analyzed using Mann–Whitney U test which showed significant difference with the p value of 0.033* (Graphs 11.1, 11.2, 11.3 and 11.4).

Graph 11.1 Mean and standard deviation comparing pre- and post-scores of MoCA between experimental and control group

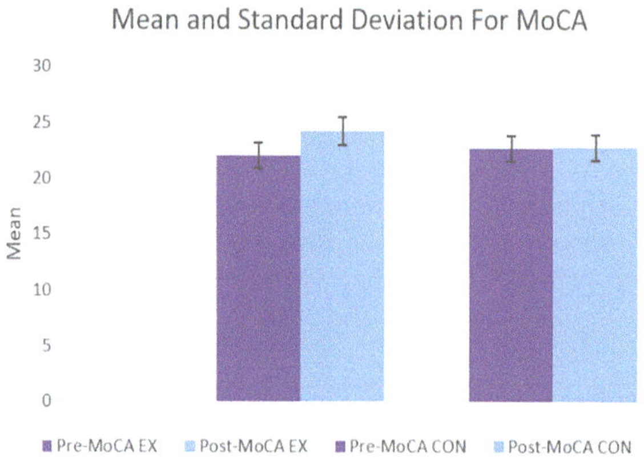

Graph 11.2 Mean and standard deviation comparing pre- and post-scores of TUG between experimental and control group

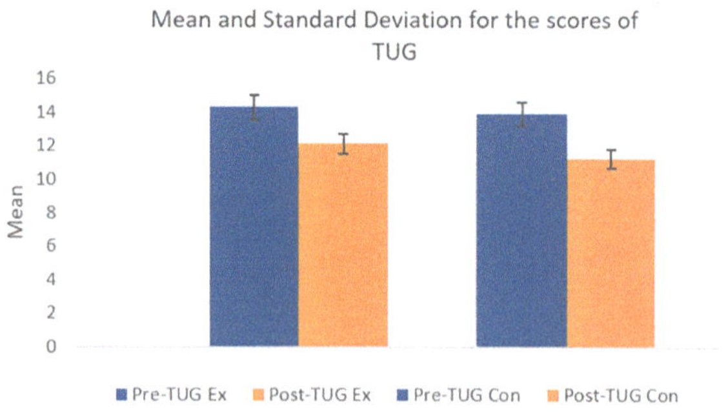

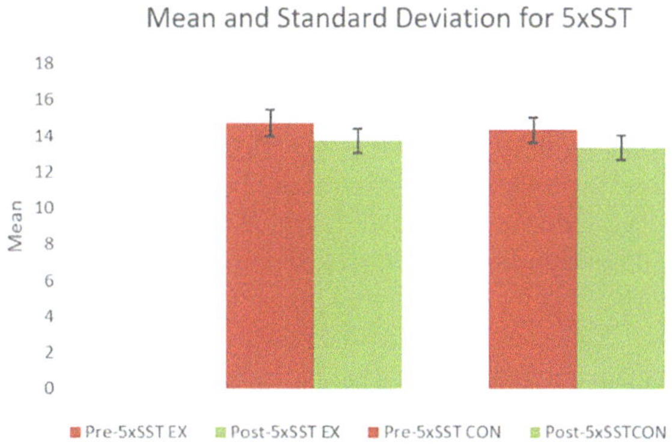

Graph 11.3 Mean and standard deviation comparing pre- and post-scores of 5XSST between experimental and control group

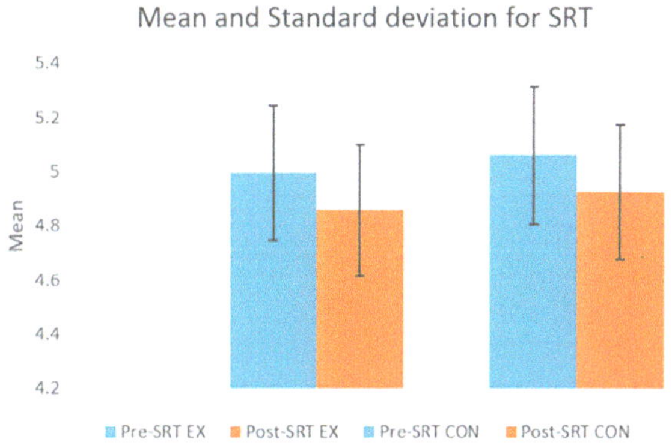

Graph 11.4 Mean and standard deviation comparing pre- and post-scores of SRT between experimental and control group

11.5 Discussion

The primary aim of this study was to assess the impact of physical–cognitive training on balance and cognitive function in middle-aged individuals with diabetes, focusing on a four-week intervention period. While prior research explored various exercises in elderly diabetic populations over longer durations, this study uniquely investigated the effects of physical–cognitive training in a middle-aged cohort. The results demonstrated that this training led to significant improvements in cognition, balance, functional lower limb strength, and flexibility within the experimental group, surpassing the outcomes of conventional balance training in the control group. The Montreal Cognitive Assessment (MoCA) scores revealed a marked difference after four weeks, highlighting the intervention's substantial positive impact on cognitive function.

These findings were consistent with other studies utilizing SSE in diabetic populations, such as Kraiwong et al. (2021) and Shellington et al. (Gregory et al. 2016). Notably, the study's middle-aged population experienced cognitive and balance improvements within a shorter timeframe, aligning with a review emphasizing the benefits of exercise in this demographic (Lin et al. 2020). The secondary analysis uncovered negative correlations between MoCA scores and age, diabetes duration, and BMI, emphasizing their impact on cognitive function. The study emphasized the intricate relationship between metabolic health and cognitive well-being, underscoring the need to consider both diabetes and obesity factors.

Furthermore, the study revealed that balance training alone in the control group did not significantly improve cognitive function, contrasting with previous research advocating for combined training approaches. The SSE intervention, targeting cognition, balance, and functional lower limb strength, demonstrated its effectiveness in enhancing these domains within a four-week period. The study also highlighted the bidirectional relationship between balance and muscle strength, emphasizing the importance of addressing both factors in interventions to enhance functional abilities and reduce fall risks. The protocol's safety and feasibility were evident, with no reported falls or injuries, although some participants required additional clarification for certain steps. Overall, the study suggested the potential benefits of integrating SSE interventions into diabetic care protocols for middle-aged individuals, offering a comprehensive approach to improving cognitive function, balance, and lower limb strength.

11.6 Conclusion

In conclusion, our study provides compelling evidence for the effectiveness of physical–cognitive training in improving cognitive function, balance, functional lower limb strength while to witness noticeable enhancements in flexibility, it is necessary to introduce more diverse activities that target range of motion in middle-aged individuals with diabetes. The significant improvements observed in cognition when compared to the control group suggest that physical–cognitive training has promising implications for enhancing cognitive function in this population. Our findings also demonstrate the potential of this training as a targeted intervention to address diabetes-associated cognitive and balance impairments.

11.7 Limitations of the Study

This training protocol is deemed inappropriate for individuals with pronounced cognitive or balance impairments. It fails to account for neuropathic changes influencing balance and lacks sophisticated measurement tools for assessing strength and flexibility, thereby restricting a comprehensive evaluation. Caution is strongly recommended when considering the application of this protocol to individuals with significant cognitive or balance challenges. Future refinements should contemplate the integration of tools to assess neuropathic changes and elevate precision in the measurement of strength and flexibility.

11.8 Future Scope of the Study

In order to optimize the outcomes of the SSE protocol, the integration of different domains of simultaneous training is highly recommended. By incorporating a diverse range of activities, the overall effectiveness of the protocol can be significantly enhanced. Moreover, the SSE protocol can be tailored to meet the specific needs of individuals with different degrees of cognitive impairment and balance issues. It is adaptable and can be modified to accommodate various levels of cognitive abilities and balance capabilities.

References

Bahureksa L, Najafi B, Saleh A, Sabbagh M, Coon D, Mohler MJ, Schwenk M (2016) The impact of mild cognitive impairment on gait and balance: a systematic review and meta-analysis of studies using instrumented assessment. Gerontology 63(1):67–83

Chakraborty A, Hegde S, Praharaj SK, Prabhu K, Patole C, Shetty AK, Mayya SS, Acharya RV, Hande HM, Prabhu MM, Upadhya D (2021) Age related prevalence of mild cognitive impairment in type 2 diabetes mellitus patients in the indian population and association of serum lipids with cognitive dysfunction. Front Endocrinol 12:798652

Demir S, Nawroth PP, Herzig S, Ekim Üstünel B (2021) Emerging targets in type 2 diabetes and diabetic complications. Adv Sci 8(18):2100275

Dixon CJ, Knight T, Binns E, Ihaka B, O'brien D (2017) Clinical measures of balance in people with type two diabetes: a systematic literature review. Gait posture 58:325–332

Dong Y, Kua ZJ, Khoo EYH, Koo EH, Merchant RA (2016) The utility of brief cognitive tests for patients with type 2 diabetes mellitus: a systematic review. J Am Med Dir Assoc 17(10):889–895

Goyal R, Jialal I (2018) Diabetes mellitus type 2

Gregory MA, Gill DP, Shellington EM, Liu-Ambrose T, Shigematsu R, Zou G, Shoemaker K, Owen AM, Hachinski V, Stuckey M, Petrella RJ (2016) Group-based exercise and cognitive-physical training in older adults with self-reported cognitive complaints: the multiple-modality, mind-motor (M4) study protocol. BMC Geriatr 16(1):1–14

Horak FB (2006) Postural orientation and equilibrium: what do we need to know about neural control of balance to prevent falls? Age Ageing 35(suppl_2):ii7-ii11

Issaka A, Paradies Y, Stevenson C (2018) Modifiable and emerging risk factors for type 2 diabetes in Africa: a systematic review and meta-analysis protocol. Syst Rev 7(1):1–10

Kodl CT, Seaquist ER (2008) Cognitive dysfunction and diabetes mellitus. Endocr Rev 29(4):494–511

Kraiwong R, Vongsirinavarat M, Rueankam M, Sumalrot T (2021) Effects of physical-cognitive training on physical and psychological functions among older adults with type 2 diabetes and balance impairment: a randomized controlled trial. J Exerc Rehabil 17(2):120

Li KZ, Bherer L, Mirelman A, Maidan I, Hausdorff JM (2018) Cognitive involvement in balance, gait and dual-tasking in aging: a focused review from a neuroscience of aging perspective. Front Neurol 9:913

Lin YH, Chen YC, Tseng YC, Tsai ST, Tseng YH (2020) Physical activity and successful aging among middle aged and older adults: a systematic review and meta-analysis of cohort studies. Aging (Albany NY) 12(9):7704

McCuskey PA, McCuskey RS (1984) In vivo and electron microscopic study of the development of cerebral diabetic microangiography. Microcirc Endothelium Lymphatics 1(2):221–244

Pradeepa R, Mohan V (2021) Epidemiology of type 2 diabetes in India. Indian J Ophthalmol 69(11):2932

Shaikh FA, Bhuvan KC, Htar TT, Gupta M, Kumari Y (2019) Cognitive dysfunction in diabetes mellitus. In: Type 2 Diabetes-from pathophysiology to modern management. IntechOpen

Silva NCBS, Gill DP, Gregory MA, Bocti J, Petrella RJ (2018) Multiple-modality exercise and mind-motor training to improve mobility in older adults: a randomized controlled trial. Exp Gerontol 103:17–26

Sun H, Saeedi P, Karuranga S, Pinkepank M, Ogurtsova K, Duncan BB, Stein C, Basit A, Chan JC, Mbanya JC, Pavkov ME (2022) IDF diabetes atlas: global, regional, and country-level diabetes prevalence estimates for 2021 and projections for 2045. Diabetes Res Clin Pract 183:109119

Chapter 12
Ocular Biometry as a Predictor of Anterior Chamber Angle Closure

Vidushi Gupta and Rakesh Nanjappa

12.1 Introduction

Glaucoma is one of the leading causes of irreversible blindness (Quigley 1996; Resnikoff et al. 2004). It is a major public health problem with a burden that is mostly underestimated because many people are unaware that they have glaucoma as most of them are asymptomatic (Tham et al. 2014). For clinical purposes, glaucoma is normally classified according to the presence or absence of obstruction by the iris to the outflow pathway of aqueous humour at the level of the proximal trabecular meshwork in the anterior chamber drainage angle (the angle between the peripheral cornea and the iris). Such cases are referred to as angle closure glaucoma and open-angle glaucoma, respectively. While the prevalence of primary open-angle glaucoma is greater than primary angle closure glaucoma (PACG) in the Western world (Kapetanakis et al. 2016; Weinreb et al. 2016), the primary angle closure accounts for 50% of global glaucoma blindness (Quigley and Broman, 2006).

V. Gupta · R. Nanjappa
Department of Optometry, School of Medical and Allied Sciences, G.D. Goenka University, Gurugram, India

V. Gupta
Department of Optometry, Dr. Shroff's Charity Eye Hospital, New Delhi, India

Present Address:
V. Gupta (✉)
Department of Optometry, School of Allied Health Sciences, Galgotias University, Greater Noida, India
e-mail: vidushi.gupta261@gmail.com

Present Address:
R. Nanjappa
Sudha Gopalakrishnan Brain Centre, Indian Institute of Technology, Madras, Chennai, India

N. A. Siddiqui et al. (eds.), *Advances in Disaster Management, Volume 2*, Springer Transactions in Civil and Environmental Engineering,
https://doi.org/10.1007/978-981-96-4047-8_13

Primary angle closure glaucoma (PACG) is a major cause of visual morbidity in East Asia (Foster et al. 1996, 2000; He et al. 2006). Recent studies in southern India found that the prevalence of PACG in Indians is also higher than that seen in European people (Dandona et al. 2000; Jacob et al. 1998; Ramakrishnan et al. 2003) with rates approaching those seen in Mongolia. The high prevalence of angle closure in China (Foster et al. 2000) and India means that PACG may be at least as common as primary open-angle glaucoma and therefore a major form of glaucoma worldwide. Eyes with primary angle closure tend to share certain biometric characteristics. These include shallow central anterior chamber depth, thick lens, anterior lens position, small corneal diameter and radius of curvature, and short axial length (Alsbirk, 1976; Lowe, 1970; Sihota et al. 2000).

In this study, we performed ocular biometry to record anterior chamber depth, lens thickness, and axial length of two groups: people with normal anterior chamber angle and those with a narrow angle, and through logistic regression and ROC analysis, we found that anterior chamber depth alone is a very good predictor of angle closure, and thus for the risk of developing primary angle closure glaucoma.

12.2 Methods

We conducted a prospective, comparative study on Indian subjects with angle closure attending a glaucoma clinic and normal control subjects from a population-based study. All participants underwent gonioscopy and A-scan biometry (IOL Master 700) to measure axial length, anterior chamber depth, and lens thickness. The ophthalmologist performing gonioscopy categorized patients with angle closure as primary angle closure suspect (PACS), primary angle closure (PAC), or primary angle closure glaucoma (PACG) based on gonioscopic findings, but for the purpose of grouping, all cases of angle closure were together grouped as ‘angle closure’.

Statistical analysis for comparison of ocular biometry parameters between the groups was done by students’ t-test. Predictive power of the biometry parameters was investigated and quantified by binary logistic regression and ROC analysis, with $\alpha = 0.05$. Data organization and statistical analysis were performed using MATLAB.

12.3 Results

We conducted a study to explore the potential of ocular biometry parameters in predicting angle closure. Initially, we compared the ocular biometry parameters (axial length, anterior chamber depth, and lens thickness) between two groups: normal subjects and those with angle closure. Our analysis revealed significant differences in all three ocular biometry parameters between the normal and angle closure groups (independent t tests, all $p < 0.001$; Fig. 12.1).

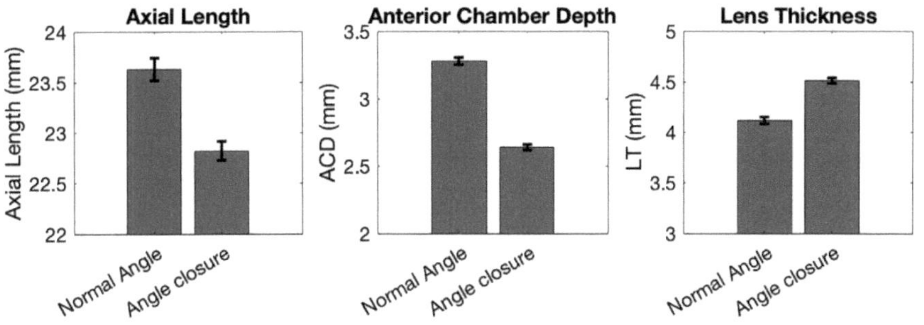

Fig. 12.1 Mean axial length (left), anterior chamber depth (centre) and lens thickness (right) for normal and angle closer groups. Error bars show s.e.m

The mean axial length was significantly shorter in the angle closure group (mean ± SD: 22.818.

± 1.331 mm) compared to the control group (mean ± SD: 23.628 ± 1.142 mm; $p < 0.001$). Similarly, the angle closure group exhibited shallower anterior chamber depth (mean ± SD: 2.639 ± 0.286 mm) compared to the control group (mean ± SD: 3.281 ± 0.267 mm, $p < 0.001$). Additionally, the lens was significantly thicker in the angle closure group (mean ± SD: 4.513 ± 0.4 mm) compared to the control group (mean ± SD: 4.114 ± 0.351 mm, $p < 0.001$).

To further investigate the predictive power of these parameters, we employed binary logistic regression, which allowed us to model the association between ocular biometry parameters and the likelihood of angle closure (Fig. 12.2). Anterior chamber depth showed a statistically significant negative association with angle closure (Estimate = – 9.3066, $p < 0.001$), suggesting that for each one-unit increase in anterior chamber depth, the log-odds of having angle closure decreased by 9.3066, holding other variables constant. However, axial length (Estimate = 0.16996, $p = 0.44216$) and lens thickness (Estimate = – 0.34454, $p = 0.61738$) did not exhibit statistically significant associations with angle closure. Overall, the model was statistically significant ($\chi^2 = 242$, $p < 0.001$) and explained a portion of the variability in angle closure (pseudo $R^2 = 0.5987$).

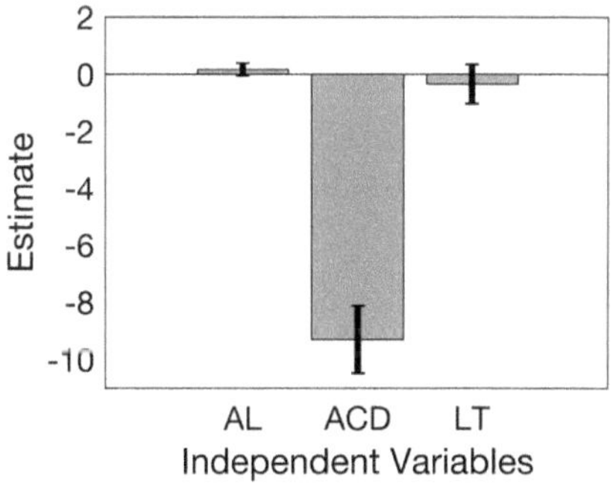

Fig. 12.2 Association of different ocular biometry parameters (AL: axial length, ACD: anterior chamber depth, LT: lens thickness) with likelihood of angle closure as modelled by binary logistic regression. Error bars show s.e.m

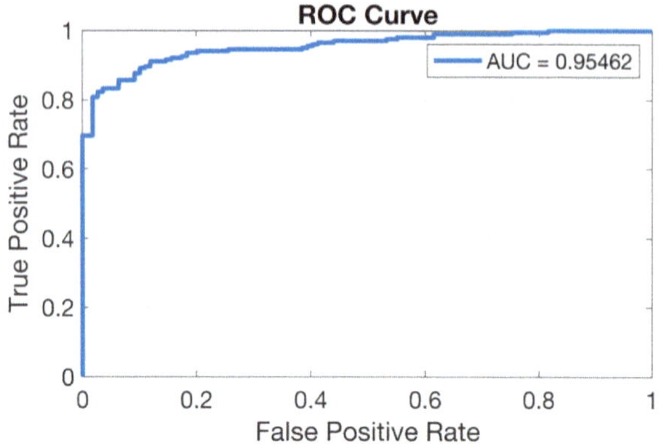

Fig. 12.3 ROC curve shows the relationship between true positive rate and true negative rate, to quantify the predictiveness of the logistic regression model as shown in Fig. 12.2. Area under the curve (AUC) was 0.95, reflecting a high discriminability power between normal and angle closure groups by the ocular biometry parameters

To quantitatively assess the predictive performance of our logistic regression model, we conducted receiver operating characteristic (ROC) analysis (Fig. 12.3). The ROC curve plotted the true positive rate (sensitivity) against the false positive rate (1-specificity) for different classification thresholds. The area under the curve (AUC) was calculated as 0.95462, indicating a strong discriminative ability of our logistic regression model in distinguishing between normal and angle closure cases. The AUC value close to 1 signified that the model had excellent discriminatory power.

To validate the performance of our logistic regression model and evaluate its generalization ability, we performed k-fold cross-validation with $k = 10$. The k-fold cross-validation involved partitioning the data into 10 subsets, training the model on 9 subsets, and testing on the remaining one in each iteration. The process was repeated 10 times to obtain robust performance estimates. The average AUC value across all folds was found to be 0.9556, further confirming the reliability and consistency of our model in predicting angle closure (Fig. 12.4).

Overall, our study demonstrated that ocular biometry parameters, particularly axial length, anterior chamber depth, and lens thickness, hold promising predictive potential for identifying individuals at risk of angle closure. The logistic regression model achieved a high AUC value, suggesting that these parameters can effectively aid in the early detection and management of angle closure, potentially mitigating the risk of irreversible vision loss associated with late diagnosis.

12.4 Discussion

Our study focussed on the predictive potential of ocular biometry parameters, specifically axial length, anterior chamber depth, and lens thickness, in the context of angle closure. The findings from our investigation hold important implications for early detection and intervention in individuals at risk of angle closure, addressing a critical gap in current glaucoma management.

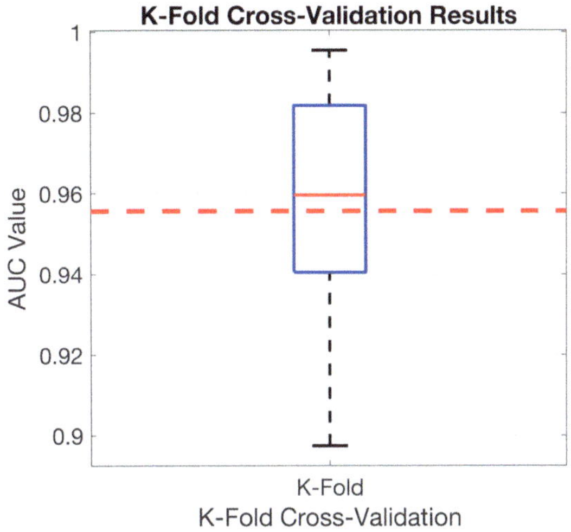

Fig. 12.4 *K*-fold cross-validation of logistic regression model, showing that average AUC remained closed to 0.96 on multiple iterations of training and testing the model. The top and bottom horizontal edges of the box show 25 percentile and 50 percentile marks, and the whiskers show full range of AUC values

The significant differences observed in ocular biometry parameters between normal and angle closure groups underscore the potential clinical relevance of these measures in identifying anatomical variations associated with angle closure. The increase in lens thickness emerged as a noteworthy predictor in our logistic regression model, aligning with existing literature linking thicker lenses to angle closure risk (Nongpiur et al. 2011). Additionally, the nuanced relationships uncovered for axial length and anterior chamber depth emphasize the multifaceted nature of angle closure pathogenesis, where alterations in multiple biometric parameters contribute to disease susceptibility.

The robust discriminative ability of our logistic regression model, as reflected by the high AUC of 0.95462, highlights the efficacy of ocular biometry parameters in predicting angle closure. The ability to differentiate between normal and angle closure cases with such precision suggests the clinical utility of these routine measurements in identifying individuals at heightened risk. The strong predictive performance is particularly encouraging given the asymptomatic nature of angle closure in its early stages, emphasizing the potential for these measures to serve as valuable screening tools.

The results of our *k*-fold cross-validation further bolster the credibility of our predictive model. The consistent average AUC of 0.9556 across multiple folds attests to the stability and generalization ability of the logistic regression model. This finding is crucial for ensuring that our model's predictive power extends beyond the confines of the initial dataset, reinforcing its reliability in real-world applications.

Despite the promising nature of our results, certain considerations should be taken into account, including the inclusion of diverse demographic groups and consideration of other potential confounding factors would enhance the external validity of our findings.

The application of our results in a clinical setting could potentially revolutionize glaucoma risk assessment. Integrating ocular biometry parameters into routine eye examinations may enable the early identification of individuals at risk of angle closure, facilitating timely interventions to prevent progression to glaucomatous optic neuropathy.

12.5 Conclusion

In conclusion, our study provides compelling evidence for the predictive power of ocular biometry parameters in identifying individuals at risk of angle closure. The integration of these routine measurements into glaucoma risk assessment protocols has the potential to significantly impact clinical practice, offering a proactive approach to glaucoma management and reducing the burden of irreversible vision loss associated with late-stage diagnosis. Future research should delve deeper into refining predictive models, considering additional risk factors, and validating findings in diverse populations to further enhance the precision and applicability of glaucoma risk prediction.

Acknowledgements Mr. Saptarshi Mukherjee provided guidance in identifying the research topic.

Conflict of Interest Statement There is no conflict of interest.

References

Alsbirk PH (1976) Primary angle-closure glaucoma. Oculometry, epidemiology, and genetics in a high risk population. Acta Ophthalmologica Supplementum 127:5–31

Dandona L, Dandona R, Mandal P, Srinivas M, John RK, McCarty CA, Rao GN (2000) Angle-closure glaucoma in an urban population in southern India: the Andhra Pradesh eye disease study. Ophthalmology 107(9):1710–1716. https://doi.org/10.1016/S0161-6420(00)00274-8

Foster PJ, Baasanhu J, Alsbirk PH, Munkhbayar D, Uranchimeg D, Johnson GJ (1996) Glaucoma in Mongolia: a population-based survey in Hövsgöl Province, Northern Mongolia. Arch Ophthalmol 114(10):1235–1241. https://doi.org/10.1001/archopht.1996.01100140435011

Foster PJ, Oen FTS, Machin D, Ng T-P, Devereux JG, Johnson GJ, Khaw PT, Seah SKL (2000) The prevalence of glaucoma in Chinese residents of Singapore: a cross-sectional population survey of the Tanjong Pagar district. Arch Ophthalmol 118(8):1105–1111. https://doi.org/10.1001/archopht.118.8.1105

He M, Foster PJ, Johnson GJ, Khaw PT (2006) Angle-closure glaucoma in East Asian and European people. Different diseases? Eye 20(1):3–12. https://doi.org/10.1038/sj.eye.6701797

Jacob A, Thomas R, Koshi SP, Braganza A, Muliyil J (1998) Prevalence of primary glaucoma in an urban south Indian population. Indian J Ophthalmol 46(2):81–86

Kapetanakis VV, Chan MPY, Foster PJ, Cook DG, Owen CG, Rudnicka AR (2016) Global variations and time trends in the prevalence of primary open angle glaucoma (POAG): a systematic review and meta-analysis. Br J Ophthalmol 100(1):86–93. https://doi.org/10.1136/bjophthalmol-2015-307223

Lowe RF (1970) Aetiology of the anatomical basis for primary angle-closure glaucoma. Biometrical comparisons between normal eyes and eyes with primary angle-closure glaucoma. Br J Ophthalmol 54(3):161–169

Nongpiur ME, He M, Amerasinghe N, Friedman DS, Tay W-T, Baskaran M, Smith SD, Wong TY, Aung T (2011) Lens vault, thickness, and position in chinese subjects with angle closure. Ophthalmology 118(3):474–479. https://doi.org/10.1016/j.ophtha.2010.07.025

Quigley HA (1996) Number of people with glaucoma worldwide. Br J Ophthalmol 80(5):389–393. https://doi.org/10.1136/bjo.80.5.389

Quigley HA, Broman AT (2006) The number of people with glaucoma worldwide in 2010 and 2020. Br J Ophthalmol 90(3):262–267. https://doi.org/10.1136/bjo.2005.081224

Ramakrishnan R, Nirmalan PK, Krishnadas R, Thulasiraj RD, Tielsch JM, Katz J, Friedman DS, Robin AL (2003) Glaucoma in a rural population of southern India: The Aravind comprehensive eye survey. Ophthalmology 110(8):1484–1490. https://doi.org/10.1016/S0161-6420(03)00564-5

Resnikoff S, Pascolini D, Etya'Ale D, Kocur I, Pararajasegaram R, Pokharel GP, Mariotti SP (2004) Global data on visual impairment in the year 2002. Bull World Health Organ 82(11):844–852

Sihota R, Lakshmaiah NC, Agarwal HC, Pandey RM, Titiyal JS (2000) Ocular parameters in the subgroups of angle closure glaucoma. Clin Experiment Ophthalmol 28(4):253–258. https://doi.org/10.1046/j.1442-9071.2000.00324.x

Tham Y-C, Li X, Wong TY, Quigley HA, Aung T, Cheng C-Y (2014) Global prevalence of glaucoma and projections of glaucoma burden through 2040: a systematic review and meta-analysis. Ophthalmology 121(11):2081–2090. https://doi.org/10.1016/j.ophtha.2014.05.013

Weinreb RN, Leung CKS, Crowston JG, Medeiros FA, Friedman DS, Wiggs JL, Martin KR (2016) Primary open-angle glaucoma. Nat Rev Dis Primers 2(1):16067. https://doi.org/10.1038/nrdp.2016.67

Chapter 13
Dermatoglyphics as an Indicator for Genetic Disorders: A Case–Control Study

Sheetal Malhan, Archana Khanna, and Farhan Khan

13.1 Introduction

The intricate patterned designs present on the epidermis of the fingers, palms, soles and toes have always been an area of interest since a long time whose origin cannot be dated back. There have been records indicating the acquaintance with these patterns, or dermatoglyphs, before the period of scientific analysis (Raizada et al. 2013).

The term dermatoglyphics was originated by Cummins and Midlo from the Greek words, 'derma' (skin) and 'glyphe' (carving). It is a collective term used for the study of the skin patterns present on the palms, fingers, soles and toes. The patterns formed by these epidermal ridges have been used to clinically diagnose the hereditary diseases (Raizada et al. 2013; Shirahatti and Dixit 2019).

According to the 'Samudra Shastra' of the ancient India, which describes the study of the ridges, they are said to be of three types, namely, Shankya, Chakra and Padma, which have been termed as Loops (L), Whorls (W) and Arches (A) by the current modern classification (Shirahatti and Dixit 2019). The distribution of ridge patterns on fingers and interdigital areas are counted. Whorls and Loops are classified as true patterns. In some areas like interdigital areas, inter-toe areas and thenar eminences,

S. Malhan (✉)
Department of Physiotherapy, School of Healthcare and Allied Sciences, GD Goenka University, Gurugram, India
e-mail: sheetal.malhan@gdgu.org

Department of Physiotherapy, Teerthanker Mahaveer University, Moradabad, India

A. Khanna
School of Allied Health Sciences, Sharda University, Greater Noida, India

F. Khan
Department of Physiotherapy, Integral University, Lucknow, India

N. A. Siddiqui et al. (eds.), *Advances in Disaster Management, Volume 2*, Springer Transactions in Civil and Environmental Engineering,
https://doi.org/10.1007/978-981-96-4047-8_14

frequencies of only true patterns are counted as these areas usually contain the loop patterns (Lĭ et al. 2006).

The *a-b* ridge count is the size of the second interdigital area (ID II) of the hand. The fetal pads in this area are the first to develop followed by the pads on the fingers. The ridge formation on the palm precedes that on the fingers, however, the progression of which is slower in comparison to the fingertips. The distal–radial to proximal-ulnar direction is followed in the process of differentiation of the ridges as well as the pad formation. Due to this fashion of development process the ridges in ID II area are among the first to form and develop over an extended period of time than the digital ones (Rose et al. 1987).

In humans, during the 6th week of gestation, the underlying mesenchyme is seen to witness down growths of epidermis as an effect of mesenchymal influence. Also, during this time, there is development of the volar pads which in turn determine the formation of the dermal ridges; which fully develop by 12–13th weeks. It can be deducted from this information that the genetic material embedded in the genome (faulty or perfect) is deciphered and exhibited by the patterns (Raizada et al. 2013; Shirahatti and Dixit 2019; Temaj et al. 2012).

Determination of a dermatoglyph is genetic which is influenced by various environmental factors in the early developmental phase of the fetus (Temaj et al. 2012). After birth, as these patterns are static i.e., not affected by the aging process, these may be used to study the genetic confluence. Hence, dermatoglyphs may serve as bioindicators of an altered fetal development (Raizada et al. 2013). Paired morphological structures exhibiting non-directional asymmetry, reflect decreased developmental buffering, which increases the susceptibility to developmental noise (Rose et al. 1987).

The significance of this science has the foundation based on the facts that these patterns are unique for every individual (even monozygotic twins), they remain throughout the life, and remain stable (age and environmentally) after 21 weeks of gestation (Durgawale et al. 1967).

Methods of recording Dermatoglyphs: Ink Method, Scotch-tape India-ink method, scanning prints, using scanning electron microscope, rubber and Plaster of Paris casts, Integrated Automated Fingerprint Identification System (IAFIS) and lipstick method (Shirahatti and Dixit 2019).

Cerebral palsy (CP) is 'a group of permanent disorders of the development and posture, causing activity limitation, that are attributed to non- progressive disturbances that occurred in the developing fetal or infant brain' (Cvjeticanin et al. 2018). According to the studies, only 10–15% of CP cases can be ascribed to intrapartal problems, whereas, prenatal risk factors account for approximately 75% of CP cases (Polovina-Prološčić et al. 2009).

The diagnosis of CP is predominantly based on the evaluation of the clinical presentation which may be evident after a delay of the motor milestones occurring at the age of about one and a half years of age. Such a delay in diagnosis hampers effects of the early intervention program as there is a development of abnormal motor patterns and tonicity. This delay can be minimized by adopting a diagnostic aid at the earliest possible time (at birth), which being capable of identifying a potential at- risk population, can lead to an effective medical treatment and rehabilitation, resulting in an efficient prognosis.

Dermatoglyphics being a non-invasive method, could be an appropriate indicator for screening and follow up of individuals having a predisposing genetic history (Shrivastava et al. 2016; Harikrishna et al. 2017).

13.2 Objectives of the Study

1. To find dermatoglyphic features in patients of Cerebral Palsy.
2. To compare dermatoglyphic features in normal and Cerebral Palsy patients.
3. To find the clinical utility of the findings of the study.

13.3 Methodology

A case- control study was conducted with forty five individuals (26 males and 19 girls), with diagnosed cerebral palsy, aged 4–18 years, treated at various outpatient facilities of different physiotherapy centers in North India. Forty five phenotypically healthy individuals (26 males and 19 females) from North India were categorized in the control group. The control group was matched for gender as the dermatoglyphic traits are dependent on this variable.

The inclusion criteria for the case group, comprised of individuals diagnosed with CP, of the age group between 4 and 18 years; for the control group it was healthy males and females, greater than 4 years of age. The individuals with current atopic dermatitis and any finger or palm abnormalities which interfered with the procedure were excluded from the study (Ezzati et al. 2014). The age has not been taken as a consideration of inclusion criteria as dermatoglyphs are not dependent on the ageing process, i.e., the fingerprint patterns remain same throughout the life, from birth till death.

An informed consent was taken from the caregivers or the participants after explaining the nature and purpose of the study. Ethical clearance was sought from the Institutional Ethical Board (PM/ETHICAL/PT/2023/001).

The participants were made to clean their hands thoroughly with soap & water and dry them with a towel. The digital and the palmar scans of both the left and right sides were taken by placing the palms (palm- down manner), on the flatbed of the electronic flatbed scanner (CanonLide300), without overpressure. The scans of the thumbs were taken separately to ensure the full digital print on the scans.

The 'atd' angle was measured using an online protractor tool (Angulus) from the palmar scans. The digital and palmar scans were used to assess and record the fingerprint patterns and the *a-b* ridge counts, based on the system given by Cummins and Midlo (Polovina-Prološčić et al. 2009).

For recording the observations, the following system was used: For left hand, 'L' is used and for right hand, 'R' is used. The following is the denotion of the digits:-1: Thumb, 2: Index Finger, 3: Middle finger, 4: Ring finger and 5: Little finger (Lĭ et al. 2006; Shrivastava et al. 2016; Ezzati et al. 2014).

13.4 Statistical Analysis

Statistical analysis was done using IBM SPSS Statistics for Windows, Version 20.0. Armonk, NY: IBM Corp. Chi-square test was used for comparing frequency of pattern types of right and left hand between control and study groups. Sample t test was used to find the difference in the quantitative variables, *a-b* ridge count and the 'atd' angle of both the left and the right hands between the two groups. *P*-values less than 0.05 were considered statistically significant.

13.5 Results

A total of 90 participants, 45 cases and 45 controls, were enrolled in the study. Table 13.1 shows the number and percentage of the fingerprint pattern types in the studied groups. The results of the statistical analysis present the predominance of the loop patterns in almost all of the digits followed by whorls and then arch in both the groups. Significant differences have been found in the pattern types in L1 ($p = 0.016$) and R4 ($p = 0.01$) digits, with the loop patterns being those of the highest frequency.

According to the Table 13.2, statistically significant differences can be seen in the left and right *a-b* ridge counts ($p = 0.000$) and the left and right 'atd' angles ($p = 0.000$).

13.6 Discussion

This study aimed at assessing the characteristic digitopalmar dermatoglyphic traits associated with the neurological disorder, Cerebral Palsy. Digitopalmar dermatoglyphic traits, both qualitative (fingerprint patterns) and quantitative ('atd' angle and *a-b* ridge count) were recorded and analyzed in the patients with Cerebral palsy, which were then compared with the healthy controls.

According to the results, statistically significant differences between the quantitative variables, namely, 'atd' angle and the *a-b* ridge count were found between the CP patients as compared to the controls.

In the present study, the 'atd' angles of both the hands were found to be lower in the CP group than the controls. Taussing et al. observed an increased 'atd' angle in

Table 13.1 Number and percentage of dermatoglyphic patterns in CP patients and control group

Control group							Experimental group								
	ARCH		LOOP		WHORL		ARCH		LOOP		WHORL		Pearson Chi-square	*p* value	Significance
	N	%	N	%	N	%	N	%	N	%	N	%	Value		
L1	4	8.9	27	60	14	31	1	2	39	86.7	5	11	8.245	0.016	Significant
L2	11	24.4	27	60	7	16	4	9	35	77.8	6	13	4.376	0.112	Not significant
L3	10	22.2	27	60	8	18	8	18	27	60	10	22	0.444	0.801	Not significant
L4	4	8.9	23	51	18	40	9	20	20	44.4	16	36	2.25	0.317	Not significant
L5	6	13.3	33	73	6	13	10	22	28	62.2	7	16	1.487	0.476	Not significant
R1	7	15.6	25	56	13	29	8	18	25	55.6	11	24	0.222	0.895	Not significant
R2	7	15.6	28	62	10	22	9	20	25	55.6	11	24	0.467	0.792	Not significant
R3	6	13.3	29	64	10	22	13	29	21	46.7	11	24	3.907	0.142	Not significant
R4	3	6.7	27	60	15	33	13	29	25	55.6	7	16	9.236	0.01	Significant
R5	7	15.6	28	62	10	22	13	29	26	57.8	6	13	2.874	0.238	Not significant

Table 13.2 Descriptive statistics of the quantitative variables between the studied groups

Variables	Control group	Experimental group	*t* value	*p*	Significance
LabRC	33.9 ± 5.6	19.3 ± 7.7	40.5	0.000	Significant
RabRC	32.9 ± 6.3	23.1 ± 8.4	18.5	0.000	Significant
Latd	50.1 ± 5.3	43.5 ± 10.9	63.2	0.000	Significant
Ratd	48.8 ± 6.1	47.6 ± 10.2	31.3	0.000	Significant

the patients with cystic fibrosis, whereas, a lower 'atd' angle was found by Kobylisky et al. in females with cystic fibrosis and a higher angle value in the males (Ezzati et al. 2014). A higher 'atd' angle was reported by Simsek et al. in female patients suffering from cerebral palsy. Similar findings were observed in the study done by Cvjetičanin et al., where the female CP patients had a higher value of 'atd' angle (Cvjeticanin et al. 2018).

On comparison of the *a-b* ridge counts, lower values were found to be associated with the CP group of patients. Similar findings were observed in the study done by Kobylisky et al. and Atefeh et al. (Ezzati et al. 2014) where mean *a-b* ridge count values were lower in the studied groups. However, the study done by Rezaeinezhad et al. showed higher values of *a-b* ridge count in the patients of type 1 diabetes mellitus patients as compared to the controls (Ezzati et al. 2014). Simsek et al. reported a decreased total ridge count in female cerebral palsy females as compared to the males. The data obtained by Cvjetičanin et al. in their study revealed a higher *a-b* ridge count in the female CP patients as compared to the males (Cvjeticanin et al. 2018).

When we have a look at the qualitative aspect in terms of the fingerprint patterns, the results show a higher frequency of loop patterns in all the fingers of both the hands. Statistically significant differences were found in the digits L1 and R4. The highest frequency was found to be of the loop patterns followed by whorls and the arches. Aesha et al. reported a substantial increase in the loop patterns (ulnar) in patients with Down's syndrome. However, a reduced number of radial loop patterns were found to be associated with patients of squamous cell carcinoma of head and neck (Shirahatti and Dixit 2019).

To the best of the authors' knowledge this is the only study which has attempted to study the qualitative aspect of the dermatoglyps in the patients suffering from cerebral palsy. Other studies done in this domain have focused largely on the quantitative aspects only. Such an attempt shall be useful in expanding the horizons of the application of the study of fingerprints in the diagnosis of the diseases having a genetic predisposition.

The earliest possible diagnosis of Cerebral Palsy based on the dermatoglyphic analysis proves to be a boon to the medical fraternity, providing a simple, cost efficient and bed-side determination of the at-risk population. The early diagnosis of the condition would help in the initiation of early intervention. When the at- risk child gets a therapeutic intervention at the early stages, the outcomes are favorable in decreasing the chances of developing abnormal motor patterns and thus facilitation of a near normal neuro- muscular development pattern.

The study is limited to a small sample size and also in the study of limited parameters of the qualitative dermatoglyphic traits. Future studies encompassing a sample across the geographic and socio-cultural domains with large data samples would help at reaching to more definitive conclusions.

13.7 Conclusion

The study highlights the importance of the use of finger print studies in screening the at- risk population as it has the advantage of being simple, cheap and available at the bed-side. The study establishes findings of distinct dermatoglyphic traits which are characteristic of Cerebral Palsy, these being, lower 'atd' angles, lower *a-b* ridge counts and an increased frequency of loop patterns specifically in the left thumb and the right fourth digit. However, further studies focusing on a larger sample are necessitated for strengthening the utility of dermatoglyphics for screening and diagnosing the genetic conditions.

References

Cvjeticanin M, Polovina-Proloscić T, Polovina S, Polovina A, Cvjeticanin T, Fučkar I (2018) Quantitative analysis of digitopalmar dermatoglyphics in thirty female cerebral palsy children. Int J Curr Res 11(01):507–602

Durgawale J, Karmalkar A, Sawant U (1967) Alterations in frequency of Ulnar loops and 'Atd' angle in congenital heart. Life 1000:3

Ezzati A, Batoei F, Jafari SA, Kiyani MA, Mahdavi-Shahri N, Ahanchian H, Tehranian S, Kianifar HR (2014) Dermatoglyphic patterns in cystic fibrosis children. Iran J Pediatr 24(5):609

Harikrishna R, Praveen K, Atifabdul B (2017) Dermatoglyphics and periodontal diseases—a possible relation for early prediction. Int J Adv Res 5:1332–1338

Lĭ H, Pan S, Donnelly M, Tran D, Qin Z, Zhang Y, Cheng X, Yin R, Lin W, Hoang V, Pham V (2006) Dermatoglyph Groups Kinh Vietnamese to Mon-Khmer. Int J Anthropol 21:295–306

Polovina-Prološčić T, Miličić J, Cvjetičanin M, Polovina A, Polovina S (2009) Comparison of digito-palmar dermatoglyphic traits in children with cerebral palsy and their close family members. Coll Antropol 33(3):925–931

Raizada A, Johri V, Ramnath T, Chowdhary DS, Garg RP (2013) A cross-sectional study on the palmar dermatoglyphics in relation to carcinoma breast patients. J Clin Diagn Res: JCDR 7(4):609

Rose RJ, Reed T, Bogle A (1987) Asymmetry of ab ridge count and behavioral discordance of monozygotic twins. Behav Genet 17:125–140

Shirahatti A, Dixit D (2019) A cross-sectional study of palmar dermatoglyphic pattern in patients with type 2 diabetes mellitus at a hospital in South India. Indian J Clin Anat Physiol 6(4):503–508

Shrivastava M, Mathur RK, Dhaneria V, Goyal S, Dhaneria V (2016) Dermatoglyphic study in bipolar disorder. Indian J Clin Anat Physiol 3(2):243–247

Temaj G, Skaric-Juric T, Narančić NS, Milicic J (2012) A detection of microevolutionary changes by the analysis of qualitative dermatoglyphic traits: an example of Albanians from Kosovo

Chapter 14
Exploring the Influence of Transcutaneous Supraorbital Stimulation on Migraine—A Scopic Systematic Review

Anuradha Dagar, Saurabh Kumar, and Vinika Chaudhary

14.1 Introduction

Migraine is a common headache indisposition that is categorized by recurring attacks of moderate or severe headache, often accompanied by other symptoms. The severe headache and other symptoms of migraine can make it difficult or impossible to carry out daily activities and can greatly reduce the quality of life of the people. World Health Organization (WHO) states that migraine is the second most common neurological disorder worldwide (Riederer et al. 2015). The International Headache Society describes migraine as a type of headache that can occur either with or without aura as well. Migraine with aura is identified by transitory neurological symptoms, such as visual disturbances, numbness, and weakness, which usually occurs before or sometimes even accompany the headache. Whereas, migraine without aura is marked by recurrent headache attacks with distinguished features and some other associated symptoms, such as nausea, vomiting, and photophobia and phonophobia (Russo et al. 2015). Another classification categorizes migraines into two types based on how often they occur: chronic and episodic. Chronic migraines are identified as those having 15 headache days or more per month and among those at least 8 days are migraine headache days. Less than 15 headache days per month are considered to be indicative with episodic migraines (Raghuveer et al. 2021). Serotonin and dopamine play a major part in the pathophysiology of migraines, which often centers on the trigeminal sensory pain pathways. However, single gene abnormalities have also been identified in certain uncommon migraine syndromes (Charles 2018). There are many anti-migraine medications available, and they have been shown to be effective. However,

A. Dagar (✉) · S. Kumar · V. Chaudhary
Faculty of Physiotherapy, SGT University, Gurugram, India
e-mail: anudagar6302@gmail.com

N. A. Siddiqui et al. (eds.), *Advances in Disaster Management, Volume 2*, Springer Transactions in Civil and Environmental Engineering,
https://doi.org/10.1007/978-981-96-4047-8_15

many people have trouble tolerating the side effects of these medications, which can make it difficult to take them regularly as a preventive measure. At least 50% of the migraineurs among the general population have been reported to be overusing the acute medications for migraine headaches (Deng et al. 2020). The effectiveness of prophylactic anti-migraine medication therapy is insufficient, and many of them have onerous side effects. According to studies, only 28.3% and 44.8% of participants with episodic and chronic migraines, respectively, had stopped taking anti-migraine medications due to an equal amount of side effects and lack of efficacy (Magis et al. 2013). The conventional approach to treating migraine headaches has typically been restricted to providing symptomatic relief; nevertheless, even with the most effective treatments, the success rates are still below 50%, and considerably lower for chronic forms of the disease (Hokenek et al. 2020). There is currently a lot of interest in emerging neurostimulation as a potential solution to the limits of medication (Jiang et al. 2018). The occipital and supraorbital region's electrical nerve activation is one of the most well-studied migraine treatments. These invasive neuromodulation methods do come with drawbacks, too, including the potential for lead migration, an inflammatory reaction, infection, and localized discomfort. Since non-invasive, transcutaneous stimulators have been developed during the past 10 years, there is now a more appealing and secure way to deliver the stimuli (Ordás et al. 2019). Transcutaneous supraorbital stimulation (TSNS), a non-invasive neurostimulation technique, has been approved by the Food and Drug Administration (FDA) for the treatment of adult migraines (Jiang et al. 2018). Neuromodulation is the technique of controlling the aberrant behavior of neuronal pathways that have developed as a result of the illness process by using specialized devices to stimulate the central nervous system or peripheral nervous system with electric, magnetic, or any other type of energy (Tiwari and Agrawal 2022). The treatment of the specific area of the body being targeted varies depending on the type of electrical nerve stimulation (ENS) being utilized, with an electrical signal supplied by voltage. Essentially, it is believed that these voltage-based impulses alter the actions of several neurotransmitters, such as dopamine, gamma-aminobutyric acid (GABA), serotonin, and others. The rhythm of neuronal firing is altered by the electrical regulation of neurotransmitters, which can also change the pain, frequency, and other conventional migraine-related factors (Patel et al. 2021).

Hence, in view of the need to explore alternatives for pharmacological treatment of migraine, this review aimed at studying the role and effectiveness of a non-invasive neuromodulation technique, i.e., transcutaneous supraorbital stimulation, in migraine as recorded in the available evidence on various databases.

14.2 Methodology

14.2.1 Search Strategy

A comprehensive search was carried out of the PubMed, MEDLINE, Research Gate, Scopus, and Google Scholar databases, and the studies were restricted to those written in the English language. All the relevant articles were independently screened and only those which satisfied the inclusion criteria were selected for further detailed analysis. The articles published between 2013 and 2023 were included in the search. The Medical Subject Headings (MeSH) phrases, keywords, and text terms relating to migraine, transcutaneous supraorbital stimulation and non-invasive neuromodulation were used to build the search strategy. To find any further research that might be relevant, the references of publications that were discovered through search of databases were examined. The screening for the articles is as shown in the PRISMA flowchart in Fig. 14.1.

14.2.2 Study Eligibility Criteria

1. Population—Participants of the age group 18–75 years who met the ICHD-3 BETA version criteria for migraine given by the International Headache Society were included also considering the intake of anti-migraine drugs.
2. Intervention—Intervention for migraine included non-invasive neurostimulation method, i.e., transcutaneous supraorbital stimulation on the forehead bilaterally to stimulate the supraorbital and supratrochlear branches of the trigeminal nerve.
3. Comparison—Comparison of tSNS with anti-migraine drugs, sham tSNS, percutaneous electrical nerve stimulation, transcutaneous vagus nerve stimulation, cervical mobilization and myofascial release.
4. Outcomes—Pain, Headache impact test-6 questionnaire (HIT-6), frequency and duration of migraine headache attacks.
5. Study design—the studies included were open label preliminary trials, RCT's and prospective exploratory clinical study.

14.2.3 Data Extraction and Synthesis

The data for the review was retrieved using the databases—PubMed, MEDLINE, Research Gate, Scopus and Google Scholar. The articles were chosen for examination after a thorough screening. The PEDro scale, which could be obtained directly from the PEDro database if it was available, or by evaluating each trial manually if it wasn't, was used to evaluate the quality of the data. The methodological quality of the clinical trials was also assessed using the Risk of Bias 2 (RoB 2) technique

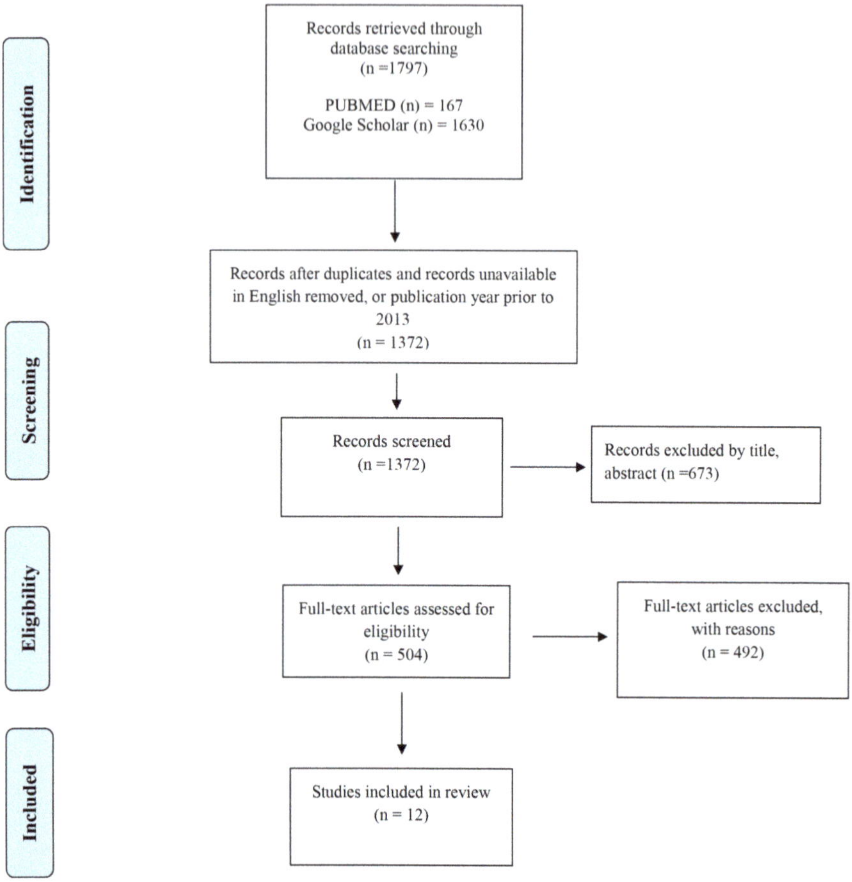

Fig. 14.1 PRISMA flowchart

developed by Cochrane. It facilitates the reader's ability to assess if the clinical trials have sufficient internal validity and statistical data to permit the interpretation of their results (Fig. 14.2; Table 14.1).

14.3 Results

After screening of all the relevant results from the search strategy, a total of 12 studies were retrieved and analyzed. To evaluate the methodological quality of the clinical trials included, the Physiotherapy Evidence Database (PEDro) scale and Cochrane's Risk of Bias 2 (RoB 2) tool were used. Out of the 12 studies included, 8 were randomized control trials with the mean PEDro score of 9.4 and the remaining 4 studies were open label preliminary clinical trials in which the PEDro scoring could

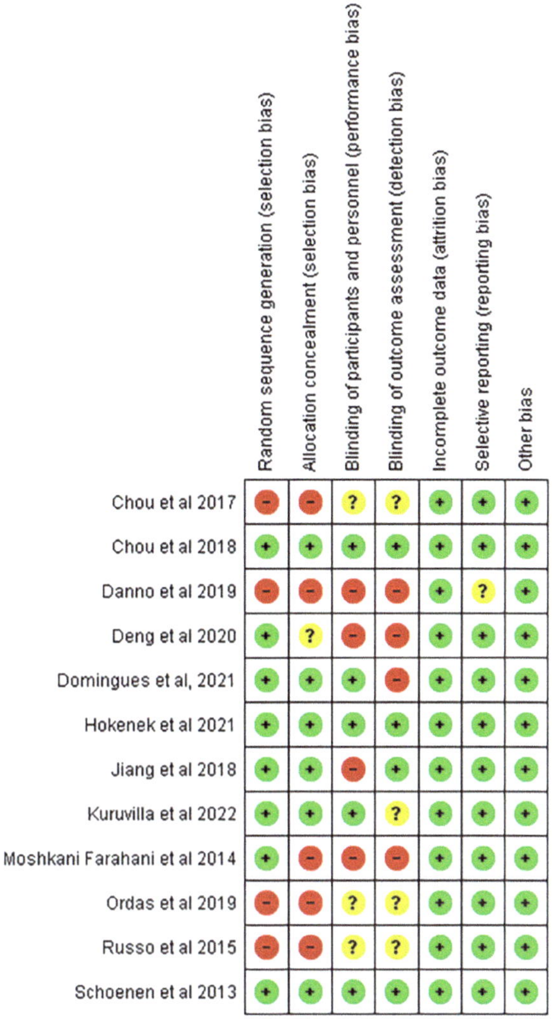

Fig. 14.2 Risk of bias summary: review authors' assessments of each risk of bias item (Risk of Bias scale) for every paper that is included

not be applied. RoB 2 tool was used for evaluating the risk of bias in the clinical trials which showed that the features with the worst methodological quality in the set of studies are the selection biases as per allocation concealment with a high risk of over 45% followed by the detection and performance bias because of non-blinding in assessment of outcomes with an approximate high risk of about 30%. The biases from the selection of reported results and incomplete outcome data had the best methodological quality with over 90% low risk which has been displayed in Fig. 14.3.

The studies analyzed for this review have shown a significant reduction in the pain severity, frequency and duration in both, migraine with and without aura after application of tSNS as a prevention and treatment method for migraine headaches.

Table 14.1 Summary of studies included in the review and their respective findings

S. No.	Author/Year	Population	Methods/Intervention	Outcome variables	Result	Conclusion
1.	Deng et al. (2020)	90 subjects Age = 18–65 years Patients with Episodic migraine with or without aura as per ICHD 3 beta version	PMES group ($n = 45$) received PMES B/L mastoid area (pulse width 90 µs, frequency 1.8 kHz, peak current 10 mA) for 45 min for 3 months STS group (n = 45) Transcutaneous supraorbital stimulation on B/L forehead (pulse width 250 µs, frequency 60 Hz, peak current 16 mA) for 20 min for 3 months	1. Migraine days 2. Severity of migraine 3. frequency of migraine attacks 4. HIT-6 questionnaire	In the third month, there was a ≥ 50% decrease in migraine days for 77.8% of patients in the PMES group and 62.2% of patients in the STS group. Between the two groups, there was no discernible difference in the change in the outcome variables. In the STS group, the Headache Impact Test-6 (HIT-6) change was more pronounced	Both PMES and STS therapy were useful in preventing migraines. PMES and STS both had similar levels of safety and effectiveness
2.	Hokenek et al. (2020)	151 Patients admitted to the emergency ward with migraine with or without aura Age = 18–50 years as per ICHD 3 beta criteria for migraine	83 patients randomized into 2 groups Group1 (n = 42) verum TENS Group 2 (n = 41) Sham TENS TENS treatment's (Headaterm TENS model YF-HT-W1) effect at the 20th minute and 120th minute after initiation of treatment.(frequency = 50 Hz, Pulse width = 125 µs) for 20 min	(1) Effect of TENS at the 20th minute and 120th minute after initiation of treatment (2) post-treatment Likert-type verbal scale scores of the patient's (3) Duration and usage of medicines since the start of their illness in months (4) Frequency of migraine episodes in days (whether they suffered from migraine with or without auras, and whether TENS therapy had an effect on their migraine symptoms was also evaluated)	It was discovered that giving the verum group therapy with the Headaterm TENS device reduced their discomfort by 69.66%. The VAS scores of the patients decreased statistically significantly from baseline to 20 min and then again from 20 to 120 min	Emergency department patients who suffer from acute migraines can benefit from the quick-acting and efficient treatment of TENS therapy

(continued)

Table 14.1 (continued)

S. No.	Author/Year	Population	Methods/Intervention	Outcome variables	Result	Conclusion
3.	Jiang et al. (2018)	154 patients of episodic migraine fulfilling ICHD 3 criteria for episodic migraine with or without aura Age = 18–65 years	Patients randomized into 3 groups: GROUP 1(Flunarizine alone)—a 5 mg dose daily as per Chinese migraine prevention guidelines GROUP 2 (tSNS alone) for 20 min daily (a pulse width of 250 μs, frequency of 60 Hz and a maximum intensity of 16 mA GROUP 3(Combined therapy)—tSNS + Flunarizine dose for 3 months	(1) Changes in monthly migraine days (2) Responder rate (3) Changes in migraine intensity (4) Intake of rescue medication (5) The satisfaction and adverse events	When compared to flunarizine monotherapy (46.15%) or tSNS monotherapy (39.22%) alone, the combination therapy's 50% responder rate (78.43%) was noticeably greater. In the combination group, there was a greater reduction in migraine intensity and rescue drug intake. Between the flunarizine group and the combination group, there was no difference in adverse events (P = 0.89)	tSNS is a useful treatment for migraines and can be a good choice for individuals who want medical intervention but are hesitant to take oral drugs or who have low migraine frequency and/or intensity for which prophylactic therapy is not recommended
4.	Danno et al. (2019)	100 Japanese subjects Age = 18–75 years Subjects fulfilling ICHD 3 criteria for migraine with or without aura and those on Acute/ prophylactic medication for migraine for the last 3 months were chosen	96 Subjects completed the 12 weeks treatment protocol The Cefaly device with a pulse width of 300 μs, frequency of 60 Hz, and maximum electric current of 16 mA was used for transcutaneous supraorbital stimulation for 20 min every 24 h for 12 weeks	(1) Changes in number of migraine days/month (2) Number of migraine attacks (3) Anti-migraine drug consumption	After 12 weeks, there was a substantial drop in the number of migraine days, with an average reduction of 1.32 days (p = 0.036), from 8.16 in the baseline period to 6.84 after 4 weeks After 12 weeks, there had been 3.94 migraine attacks, a substantial decrease from 5.33 during the baseline period (p = 0.0002). The average headache severity did not significantly alter, but the number of headache days over the previous four weeks and the use of acute medication did. Between the baseline period and the 12-week mark, 19.3% of 50% of respondents reported a 50% reduction in migraine days, while 28.9% of 30% of respondents saw the same drop	Cefaly device for neurostimulation is a secure and well-tolerated modality for Asian migraineurs and can also be used as an adjunct to the already existing anti-migraine interventions

(continued)

Table 14.1 (continued)

S. No.	Author/Year	Population	Methods/Intervention	Outcome variables	Result	Conclusion
5.	Ordas et al. (2019)	25 Chronic migraine patients Age = at least 18 years and meeting ICHD-3 criteria for CM (at least 6 months)	Patients given transcutaneous supraorbital stimulation for supraorbital and supratrochlear nerves (250 pulse width, 60 Hz frequency, and 16 mA current intensity) for a fixed period of 20 min for 3 months They were also instructed to fill headache diaries to record the intensity and duration of headache and acute medication intake	Mean of moderate or severe headache days from baseline to end of study (i) change in total headache days per month from baseline to the end of the study (ii) change in monthly symptomatic medication intake per month consumed for headache (iii) 30% responder rate (defined as a percentage of subjects having a reduction in monthly headache days with moderate or severe intensity of at least 30%) from baseline to the end of the study (iv) Change in the HIT-6 scale from baseline to the end of the study v) side effects	According to the per-protocol analysis, there was a nonsignificant decrease of 2.95 in the mean monthly total headache days and a significant four-day decrease in the mean monthly days with moderate or severe headache baseline from the beginning of the research to the conclusion (t test, P ¼ 0.0163). The intention-to-treat analysis revealed a significant 2.75 reduction in the mean monthly days with any headache (t test, P ¼ 0.016) and a nonsignificant 3.37 reduction in the mean monthly days with moderate or severe headache during the same time	Although the impact of tSNS on the treatment of CM may be negligible or debatable, it may have preventative qualities Double blind, sham-controlled investigations are necessary to confirm these findings

(continued)

Table 14.1 (continued)

S. No.	Author/Year	Population	Methods/Intervention	Outcome variables	Result	Conclusion
6.	Moshkani Farahani et al. (2014)	45 patients selected according to the criteria of the ICHD-III, with primary headaches developed after entering the health profession for at least 3 years and no associated underlying disease or trauma	45 patients randomly divided into 3 groups of NFB, TENS and controls NFB group- in each session migraine treatment protocol including suppressing Theta (4–8), strengthening the sensorimotor rhythm (SMR) (12–15 Hz), and suppressing high beta (21–30 Hz) was performed at T3 and T4 sites.(15 sessions) TENS-performed using a CEFALY-brand machine (STX-Med Sprl factory, Belgium) The current intensity had to be set according to patients' tolerance threshold (20 min daily sessions) for 20 sessions Headache diary was also maintained	Headache frequency, severity, duration	Treatment with NFB and TENS had caused significant decrease in the frequency, severity, and duration of headache in experimental groups There were significant differences between pain frequencies in experimental groups	The use of NFB and TENS have a marked potential and efficacy in treating migraine headaches with NFB being slightly better at reducing headache frequency and duration

(continued)

Table 14.1 (continued)

S. No.	Author/Year	Population	Methods/Intervention	Outcome variables	Result	Conclusion
7.	Schoenen et al. (2013)	84 patients Age = 18–65 years based on ICHD-2 criteria of migraine with or without aura with at least 2 attacks per month	67 patients were randomized into 2 groups: Experimental group (n = 34 verum group) received 45 days of neurostimulation (rectangular impulses, pulse width 250 μs, frequency 60 Hz, intensity 16 mA; CG: pulse width 30 μs, frequency 1 Hz, intensity 1 mA) for 20 min. daily for 3 months along with a headache diary Control group (n = 33 sham group) received sham transcutaneous supraorbital nerve stimulation for the same period under same parameters as the verum group	(1) Change in monthly migraine days between the run-in month and the third month of treatment (2) percentage of "responders (3) change in monthly migraine days between run-in and the average 3 months of treatment (4) change in monthly migraine attack frequency (5) change in monthly frequency of any headache (6) change in mean headache severity per migraine day (7) change in monthly acute anti-migraine drug use and in associated symptoms per migraine headache between run-in and third month of treatment (8) Percentage of patients stating at the end of the trial that they are very satisfied, moderately satisfied, or not satisfied with the treatment	The mean number of migraine days decreased significantly in the verum (6.94 vs. 4.88; p 5 0.023) along with monthly migraine attacks (p 5 0.044), monthly headache days (p 5 0.041), and monthly acute anti-migraine drug intake (p 5 0.007), but not in the sham group (6.54 vs. 6.22; p 5 0.608). The 50% responder rate was significantly greater (p 5 0.023) in the verum (38.1%) than in the sham group (12.1%). No adverse effects were reported in either group	As reported in the study, tSNS is effective as well as a safe for migraine treatment but still demands well-controlled studies to untwine the exact mode of action

(continued)

Table 14.1 (continued)

S. No.	Author/Year	Population	Methods/Intervention	Outcome variables	Result	Conclusion
8.	Russo et al. (2015)	24 consecutive Italian patients satisfying the ICHD-3 beta criteria for MwoA with attacks ≤ 5/month	tSNS was delivered with a 30 mm × 94 mm self-adhesive electrode (placed on the supratrochlear and supraorbital areas of forehead) Biphasic rectangular impulses; frequency: 60 Hz, pulse width: 250 μs and intensity: 16 mA. The duration of tSNS sessions was 20 min/day The patients were also instructed to maintain a daily headache diary	(1) change in migraine attacks per month (2) change in migraine days per month (3) Responders rate [(a) reduced average of pain intensity during migraine attacks; (b) significant reduced HIT-6 questionnaire rate] (4) Reduced intake of rescue medication during migraine attacks per month (5) percentage of: (a) Patients that are satisfied or not satisfied with the treatment (b) Patients showing compliance to the treatment	There has been a noteworthy decline in the monthly occurrence of migraine days ($p < 0.001$) and migraine attacks ($p < 0.001$). According to the study, 81 and 75% of patients experienced a 50% or greater decrease in monthly migraine attacks and migraine days, respectively. It has been demonstrated that there is a statistically significant decrease in the average pain intensity during migraine attacks ($p = 0.002$), in the HIT-6 rating ($p < 0.001$), and in the consumption of rescue medication ($p < 0.001$). Every patient exhibited excellent compliance and no noteworthy adverse effects	If a patient cannot or will not take daily medication, or if their migraine frequency and/or intensity are low enough that pharmaceutical preventative therapy are not necessary, tSNS may be a viable choice for treating migraine attacks in advance

(continued)

Table 14.1 (continued)

S. No.	Author/Year	Population	Methods/Intervention	Outcome variables	Result	Conclusion
9.	Chou et al. (2017)	30 patients were chosen according to the diagnostic criteria oh ICHD-3 beta version for episodic or chronic migraine with or without aura Age = 18–65 years	e-TNS was applied via the Cefaly VR neurostimulator device (CEFALY Technology, Seraing, Belgium) for a 60-min treatment session Pulse frequency of 100 Hz Pulse width of 250 μs at 16 mA	Primary outcome—the mean change in pain intensity after the one-hour treatment, compared to baseline. Secondary outcomes—mean change in pain intensity at two hours after the start of treatment compared to baseline, as well as the percentage of patients not requiring rescue medication at two and 24 h following the treatment	Mean pain intensity was significantly reduced by 57.1% after the one-hour e-TNS treatment (23.22 6 2.40; $p < 0.001$) and by 52.8% at two hours (22.98 6 2.31; $p < 0.001$)	A double-blind, randomized, sham-controlled experiment should be conducted to further investigate transcutaneous supraorbital neurostimulation as a potential safe and effective acute therapy for migraine attacks
10.	Kuruvilla et al. (2022)	538 patients diagnosed with episodic migraine with or without aura that fulfilled the ICHD-3 beta version Age 18–65 year with the diagnosis before 50 years of age	Verum group (n = 259) Sham group (n = 279) e-TNS device (CEFALY® Technology, Seraing, Belgium) given continuously for 2-h. Sham device—low frequency pulses of 3 Hz; the verum device-high frequency pulses of 100 Hz. During stimulation, the intensity increases linearly to reach a maximum of 16 mA after 14 min Total dose of current delivered during a 2-h session is 2.728 C A headache diary was also maintained	1. Pain freedom at 2 h 2. Absence of MBS at 2 h 3. Pain relief at 2 h 4. Absence of all MAS at 2 h 5. Rescue meds at 2–24 h 6. Pain freedom at 24 h 7. Pain relief at 24 h	Verum patients had a 7.2% greater percentage of pain-free patients at two hours (25.5%) than sham patients (18.3%; $p = 0.043$). Verum had a 14.1% greater rate of migraine-associated symptom resolution (56.4%) than sham (42.3%) ($p = 0.001$). At 2 h, verum (69.5%) had pain relief 14.3% higher than sham (55.2%; $p = 0.001$); at 2 h, verum (42.5%) had absence of all migraine-associated symptoms 8.4% higher than sham (34.1%; $p = 0.044$); and at 24 h, verum (22.8 and 45.9%) had sustained pain freedom and pain relief 7.0% and 11.5% higher than sham (15.8 and 34.4%; $p = 0.039$ and 0.006, respectively)	A 2-h e-TNS treatment is a safe and effective method. It is a non-invasive and non-pharmacological alternative to the acute treatment of migraine attacks in an at-home setting

(continued)

Table 14.1 (continued)

S. No.	Author/Year	Population	Methods/Intervention	Outcome variables	Result	Conclusion
11.	Domingues et al. (202[illegible])	159 patients Age group = 18–65 years with migraine with or without aura as per the ICHD 3 criteria	77 patients were randomized into 2 groups: Intervention group (n = 44) received TENS (pulse width 80 μs, frequency 55 Hz, continuous and intermittent (automatic) modes, ramp 2 s, pulse train 1 s (continuous), and 3 s (intermittent), with an application time of up to 20 min.) for 3 months Sham group (n = 33)—an active-placebo device was used which provided an electrical current for 30 s and no current for the next 15 s thereafter	(1) Pain intensity and its impact on functional disability by VAS and Wong and Baker face pain rating scale and Associated scale for pain and functional disability assessment (2) HIT-6 questionnaire (3) rescue mediation (4) health service usage	In comparison to the sham group, the intervention group's pain scores decreased statistically significantly. Additionally, there was a greater median difference ($p = 0.02$) in the functionality scores of the intervention group pre- and post-intervention. The quantity of analgesic rescues and the frequency of device use did not differ across the groups. The groups' HIT-6 scores did not differ from one another	The use of a self-applied, TENS device is safe and effective in relieving pain associated with migraine attacks. But there is a need to record the outcomes following use of this treatment option for use at home during migraine attacks
12.	Chou et al. (2018)	109 patients Age group 18–65, fulfilling the ICHD-3 beta version, with or without aura, headache persisting for at least 3 h and pain severity steady for at least 1 h prior to enrollment; no medication intake within 3 h previous to registration	106 patients were randomized into 2 groups: Verum: 52, sham: 54 Neurostimulation was applied via the e-TNS Cefaly device (CEFALY Technology, Seraing, Belgium) for a 1-h session Pulse frequency -100 Hz Pulse width- 250 ms Maximum dose of current delivered during 1-h session - 1.284 C. The intensity was increased linearly upto16 mA after 14 min and then kept constant for 46 min (placed on the forehead to excite the supratrochlear and supraorbital nerves bilaterally	(1) Mean change in pain score at the 1-h time point compared to baseline (2) Mean change in pain score at the 2-h and 24-h time points compared to baseline (3) The proportion of subjects who didn't use migraine rescue medication at 2 h and within 24 h after initiation of the e-TNS session	The verum group exhibited a considerably greater reduction in the major outcome measure compared to the sham group: -3.46 ± 2.32 versus -1.78 ± 1.89 ($p < 0.0001$), or 59% versus 30% ($p < 0.0001$) Migraine without aura attacks—There was a significant difference in pain reduction between verum and sham: mean visual analogue scale reduction at 1 h was -3.3 ± 2.4 for the verum group versus -1.7 ± 1.9 for the sham group (p ¼ 0.0006) Migraine with aura attacks—pain reduction was numerically greater for verum but didn't reach significance: the mean visual analogue scale reduction at 1 h was 4.3 ± 1.8 for the verum group versus 2.6 ± 1.9 for the sham group (p ¼ 0.060)	External trigeminal nerve stimulation produced a considerable reduction in headache pain after a one-hour therapy, and it was well tolerated. This suggests that it could be a safe and efficient acute treatment for migraine attacks

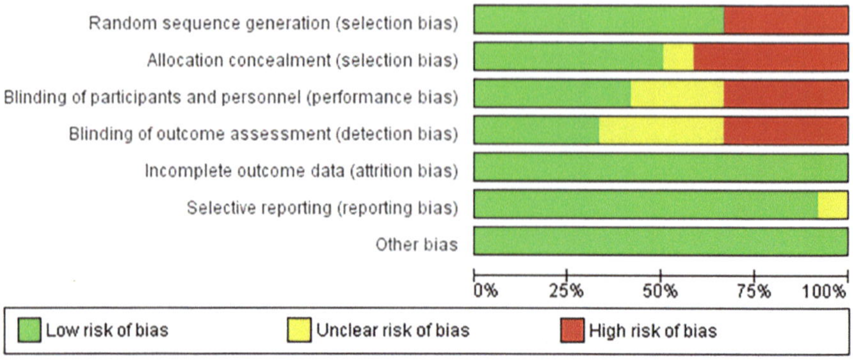

Fig. 14.3 Risk of bias graph: review authors' assessments of each risk of bias item presented as percentages in the included studies (Risk of Bias scale)

Research has shown that verum groups treated with tSNS had a response rate of over 50%. It was noted that in the experimental groups, the mean number of migraine days and the monthly intake of acute anti-migraine medications had been significantly decreased. The patient satisfaction was also recorded at the end of some studies which was undoubtedly higher among those in the verum group. It was shown that the use of tSNS did not have any adverse effects on the participants of the studies and their quality of life had also improved.

14.4 Discussion

This review encapsulates the feasibility and effectiveness of application of tSNS or transcutaneous supraorbital neurostimulation in migraine population. The various articles included in this review have thrivingly aided in investigating the role and influence of non-invasive neuromodulation, i.e., tSNS on reducing the number of migraine headache attack days, its intensity as well as duration along with consumption of anti-migraine drugs. It has bloomed as not only a safe but a cost-effective intervention for treating migraine patients.

In 2021, Hokenek et al. conducted a randomized control experiment which revealed that tSNS is a fast-acting treatment for acute migraine in patients who are hospitalized to the emergency room. Migraine is a condition that has detrimental effects on society, the individual, and the economy. The study demonstrated that there was significant decrease in patient's VAS scores, and that this device is effective within the first 20 min of treatment ($p < 0.001$). Significant statistical and clinical differences were observed between the verum and sham groups in the trial. The only presentable drawback was that in the emergency department, the patients were usually agitated and preferred instant injectable treatment methods and often avoided the electrical treatment. Therefore the study also depicted the need of better public education (Hokenek et al. 2020).

The study conducted by Jiang et al. was the first trial to combine neurostimulation and medicine for migraine prevention to improve clinical efficacy. Although combine therapy was shown to produce greater effects than tSNS or medication alone but the study managed to highlight the positive effects of neurostimulation. The results demonstrated a significant decrease in migraine severity following tSNS therapy. The study also demonstrated that tSNS can be used alone to improve parameters of migraine in addition to being utilized as a substitute for anti-migraine medications (Jiang et al. 2018).

Various open-label clinical trials conducted by Russo et al., Ordas et al., Danno et al., and Chou et al. with Cefaly device for supraorbital transcutaneous stimulation in the migraine population presented encouraging results after the application of tSNS in chronic migraine and episodic migraine with and without aura. The participants were asked to maintain a headache diary which recorded the changes in severity, frequency and duration of each headache episode. These trials made it noteworthy that tSNS holds a good efficacy which can add to the available pharmacological prophylactics for migraine population and also highlighted the need to consider the placebo/sham TENS effects in treating migraine (Russo et al. 2015; Ordás et al. 2019; Danno et al. 2019; Chou et al. 2017).

There are studies such as randomized control trials by Schoenen et al., Kuruvilla et al., Domingues et al. and Chou et al. which focused on evaluating the safety, portability, self-application and efficacy of tSNS in migraine respectively. The mean monthly migraine days decreased, and participants in the intervention group demonstrated a statistically significant improvement in their functional impairment and pain levels. When taken continuously for two hours during a migraine episode or once a day for 20 min over the course of three months, it was discovered to have good tolerability and to have no negative effects on patients (Schoenen et al. 2013; Kuruvilla et al. 2022; Domingues et al. 2021; Chou et al. 2018).

Some studies such as those conducted by Deng et al. and Moshkani Farahani et al. also compared the effect of tSNS with percutaneous mastoid electrical stimulation (PMES) in migraine and with Neurofeedback Behavioral therapy (NFB) in primary headache disorders respectively. The results showed that the alterations brought about by tSNS were comparable to those of PMES, albeit marginally less successful than NFB, although the tSNS group still experienced a greater reduction in the mean score of headache duration. Compared to the PMES group, those who got tSNS experienced a $\geq 50\%$ decrease in migraine days during the third month of treatment (Deng et al. 2020; Moshkani Farahani et al. 2014).

14.5 Conclusion

In this scoping review, it was found that transcutaneous supraorbital stimulation which is a non-invasive intervention, has a budding clinical efficacy in treating migraine be it acute or chronic or episodic. It can either be used as an adjunct to pharmacological management or individually as well. The evidence available till now with regards to its effectiveness is rather controversial or very mild and in need of further well controlled studies.

14.6 Limitations and Future Scope

It has now been made evident that although tSNS can indeed produce a decrease in the mean monthly days with moderate or severe headaches but its effect may either be mild or controversial. The current evidence available on the databases falls short in depicting the effect of tSNS on migraine in the long term follow ups. Hence, keeping in line the interest of better options for migraine treatment, more sham controlled, double-blind studies are essential to confirm the available findings especially in Indian population for which there is scarcity of evidence and also to explore whether tSNS holds the potential to bring better results in chronic migraine with further modifications in tSNS parameters.

References

Charles A (2018) The pathophysiology of migraine: implications for clinical management. Lancet Neurol 17(2):174–182. https://doi.org/10.1016/s1474-4422(17)30435-0

Chou DE, Shnayderman Yugrakh M, Winegarner D, Rowe V, Kuruvilla D, Schoenen J (2018) Acute migraine therapy with external trigeminal neurostimulation (ACME): a randomized controlled trial. Cephalalgia 39(1):3–14. https://doi.org/10.1177/0333102418811573

Chou DE, Gross GJ, Casadei CH, Yugrakh MS (2017) External trigeminal nerve stimulation for the acute treatment of migraine: open-label trial on safety and efficacy. Neuromodulation Technol Neural Interface 20(7):678–683. https://doi.org/10.1111/ner.12623

Danno D, Iigaya M, Imai N, Igarashi H, Takeshima T (2019) The safety and preventive effects of a supraorbital transcutaneous stimulator in Japanese migraine patients. Sci Rep 9(1). https://doi.org/10.1038/s41598-019-46044-8

Deng Y, Zheng M, He L, Yang J, Yu G, Wang J (2020) A head-to-head comparison of percutaneous mastoid electrical stimulator and supraorbital transcutaneous stimulator in the prevention of migraine: a prospective, randomized controlled study. Neuromodulation Technol Neural Interface 23(6):770–777. https://doi.org/10.1111/ner.13127.

Domingues FS, Gayoso MV, Sikandar S, Silva LM, Fonseca RG, Barros GAM (2021) Analgesic efficacy of a portable, disposable and self-applied transcutaneous electrical nerve stimulation device during migraine attacks: a real-life randomized controlled trial. Pain Practice. https://doi.org/10.1111/papr.13042

Hokenek NM, Erdogan MO, Hokenek UD, Algin A, Tekyol D, Seyhan AU (2020) Treatment of migraine attacks by transcutaneous electrical nerve stimulation in emergency department: a randomize controlled trial. Am J Emergency Med. https://doi.org/10.1016/j.ajem.2020.01.024
Jiang L, Yuan DL, Li M, Liu C, Liu Q, Zhang Y, Tan G (2018) Combination of flunarizine and transcutaneous supraorbital neurostimulation improves migraine prophylaxis. Acta Neurologica Scandinavica. https://doi.org/10.1111/ane.13050
Kuruvilla DE, Mann JI, Tepper SJ, Starling AJ, Panza G, Johnson MAL (2022) Phase 3 randomized, double-blind, sham-controlled trial of e-TNS for the Acute treatment of Migraine (TEAM). Sci Rep 12(1):5110. https://doi.org/10.1038/s41598-022-09071-6
Magis D, Sava S, d'Elia TS, Baschi R, Schoenen J (2013) Safety and patients' satisfaction of transcutaneous Supraorbital NeuroStimulation (tSNS) with the Cefaly® device in headache treatment: a survey of 2,313 headache sufferers in the general population. J Headache Pain 14(1). https://doi.org/10.1186/1129-2377-14-95
Moshkani Farahani D, Tavallaie SA, Ahmadi K, Fathi Ashtiani A, Sheikh M, Yahaghi E (2014) Comparison of neurofeedback and transcutaneous electrical nerve stimulation efficacy on treatment of primary headaches: a randomized controlled clinical trial. Iranian Red Crescent Med J 16(7). https://doi.org/10.5812/ircmj.17799
Ordás C, Cuadrado M, Pareja JA, de-las-Casas-Cámara G, Gómez-Vicente L, Torres-Gaona G, Begoña Venegas-Pérez, Álvarez-Mariño B, Ana Diez Barrio, Pardo-Moreno J (2019) Transcutaneous supraorbital stimulation as a preventive treatment for chronic migraine: a prospective, open-label study. In: Pain medicine. https://doi.org/10.1093/pm/pnz119
Patel K, Batchu S, Wang R, Bunachita S, Joshi A, Soni R, Pandya A, Patel U (2021) The use of electrical nerve stimulation to treat migraines: a systematic review. Cureus. https://doi.org/10.7759/cureus.17554
Raghuveer R, Marbate S, Ruchi S (2021) Effect of cervical mobilization and transcutaneous supraorbital nerve stimulation in migraine without aura. Int J Res Pharmaceutical Sci 12(4):2316–2324. https://doi.org/10.26452/ijrps.v12i4.4862
Riederer F, Penning S, Schoenen J (2015) Transcutaneous Supraorbital Nerve Stimulation (t-SNS) with the Cefaly® device for migraine prevention: a review of the available data. Pain Ther 4(2):135–147. https://doi.org/10.1007/s40122-015-0039-5
Russo A, Conte F, Marcuccio L, Giordano A, Tedeschi G, Tessitore A (2015) P019. Transcutaneous supraorbital neurostimulation in 'de novo' patients with migraine without aura: the first Italian experience. J Headache Pain 16(S1). https://doi.org/10.1186/1129-2377-16-s1-a136
Schoenen J, Vandersmissen B, Jeangette S (2013) Migraine prevention with a supraorbital transcutaneous stimulator: a randomized controlled trial. Neurology 80(8):697–704. https://doi.org/10.1212/WNL.0b013e3182825055
Tiwari V, Agrawal S (2022) Migraine and neuromodulation: a literature review. Cureus. https://doi.org/10.7759/cureus.31223

Chapter 15
A Deliberation on a Resilient Public Healthcare System During a Crisis Situation in the State of Haryana

Ankur Gupta, Alok Pandey, Kishore Kumar Morya, and Anureet Kaur

15.1 Introduction

"Public Health refers to all the organized public or private measures to prevent disease, promote health, and prolong life among the population as a whole. It provides conditions in which entire population can be healthy" (WHO) (Roopesh 2020).

American Public Health Association defines Public Health as the practice of preventing disease and promoting good health within the group of people, from small communities to entire countries. Public health caters to population at large rather than focussing on individual health activities (Jarvis et al. 2020).

The World Organization Health System Strengthening Glossary defines health system as a set of all activities which aims to promote, restore, and/or maintain health (WHR, 2000) and the people, institutions, and resources arranged together in accordance with established policies (Roopesh 2020). The World Health Report 2003 differentiates Healthcare System from Health System. It says that "Healthcare System refers to the institutions, people and resources involved in delivering the healthcare to individuals" (Roopesh 2020).

Public healthcare system (PHS) is a backbone of well-being and good living standard of people of any state. In a crisis situation such as pandemic, a sound public healthcare system is lifeline for the people of that state. A resilient and robust public

A. Gupta · K. K. Morya · A. Kaur (✉)
Scholar at School of Management, GDGU, Sohna, India
e-mail: anureet.kaur@gdgu.org

K. K. Morya
e-mail: kishore.morya@gdgu.org

A. Pandey
School of Banking and Finance, OPJGU, Sonipat, India
e-mail: alok.pandey@jgu.edu.in

N. A. Siddiqui et al. (eds.), *Advances in Disaster Management, Volume 2*, Springer Transactions in Civil and Environmental Engineering,
https://doi.org/10.1007/978-981-96-4047-8_16

healthcare system is a manifestation of effective and efficient management of a healthcare ecosystem provided by a government. This paper aims to deliberate on public healthcare system of the state of Haryana. The paper has delved upon the elements of a resilient public healthcare system and followed a systematic investigation of the system. First part of the paper introduces the concepts of public health, health system, and healthcare system. Second part finds the historical facts about Indian health and healthcare system though literature review. Objectives of the research have been formulated. Third part presents research methodology used followed by data collection and data analysis. Last part of the paper entails findings, conclusions and recommendations, and limitations of the study.

15.2 Background of Public Health and Public Healthcare System in India

The Commonwealth Fund has described Indian Health System with respect to Public and Private, two different sectors. Majority of the population of India have access to government hospitals, dispensaries, Primary Health Centres (PHCs), and Community Health Centres (CHCs). The healthcare system in India has been divided into three levels, primary, secondary, and tertiary level (Fig. 15.1). PHCs and sub-centres constitute primary level, CHCs and smaller sub-district hospitals make up secondary level, and General Hospitals and Medical Colleges at city level complete the tertiary level. Table 15.1 depicts the status of public health infrastructure since the year of 1951 to 2018. As per the latest figures as of March 2022, India has 157, 935 functional sub-centres, 31,053 functional PHCs, 6064 functional CHCs. There are total of 25,778 public hospitals, 43,486 private hospitals, and 654 medical colleges in India.

The public health system is governed by governments of 28 states and 8 union territories independently. Due to independent systems, there is significant variation in delivery models, insurance coverage, availability, and access of public health to the large proportion of the population. The Indian healthcare system, thus, is a gigantic network of public and private sector that is catering to approximately 1.4 billion people. Government of India has taken multiple initiatives to enhance the healthcare system through National Health Mission, National Rural Health Mission, National Urban Health Mission, Ayushman Bharat Scheme, and independent schemes by the state governments (Kumar, 2023). A complete structure of Indian Public Health System is depicted in Fig. 15.1. The figure entails smallest unit, which is Community Health Worker (CHW) of public health system and other entities such as Auxiliary Nurse Midwifery (ANM), Rural Healthcare Practitioners, etc. It presents availability of District Hospitals, Medical Colleges at tertiary level Human Resources for Health (HRH), sub-district hospitals and CHCs at secondary level, and PHCs & SHCs at primary level (Chokshi et al. 2016).

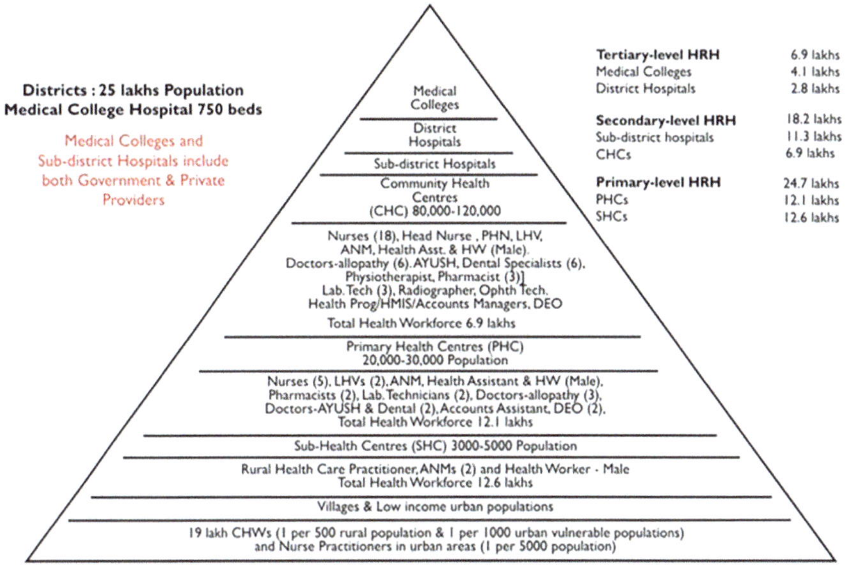

Fig. 15.1 Indian public health system. *Source* National Rural Health Mission, Ministry of Health and Family Welfare, GOI (Chokshi et al. 2016)

Table 15.1 Public health infrastructure in India (1951–2018)

Item	1951	1981	2000	2018
Hospitals (Gov.)	2694	6805	15,888	25,778
Bed (Govt.)	117,000	504,538	719,861	713,986
PHCs	725	9115	22,842	25,743
Sub-centres	–	84,736	137,311	158,417
CHCs	–	761	3043	5624
Dispensaries	6600	16,745	23,065	27,951

Source www.cbhidhs.nic.in

Indian Healthcare System is an arrangement of multiple agencies, which are working at various level of health requirements of the country. Broadly it has been divided into two categories (i) Public Healthcare and (ii) Private Healthcare. The above flow chart depicted in Fig. 15.2 demonstrates the entirety of Indian Public Healthcare system.

The State of Haryana: Haryana is a state located in northern part of India. It covers 44,212 km^2 which is about 1.4% of the total area of the country. Chandigarh is the capital city of Haryana. Haryana has relatively good public healthcare system. It has a network of 71 public hospitals, 120 CHCs, 407 PHCs, 2727 sub-centres that serve total population of approximately 25.35 million (Census 2011). Health department in Haryana is headed by Director General, Health Service and supported by Additional

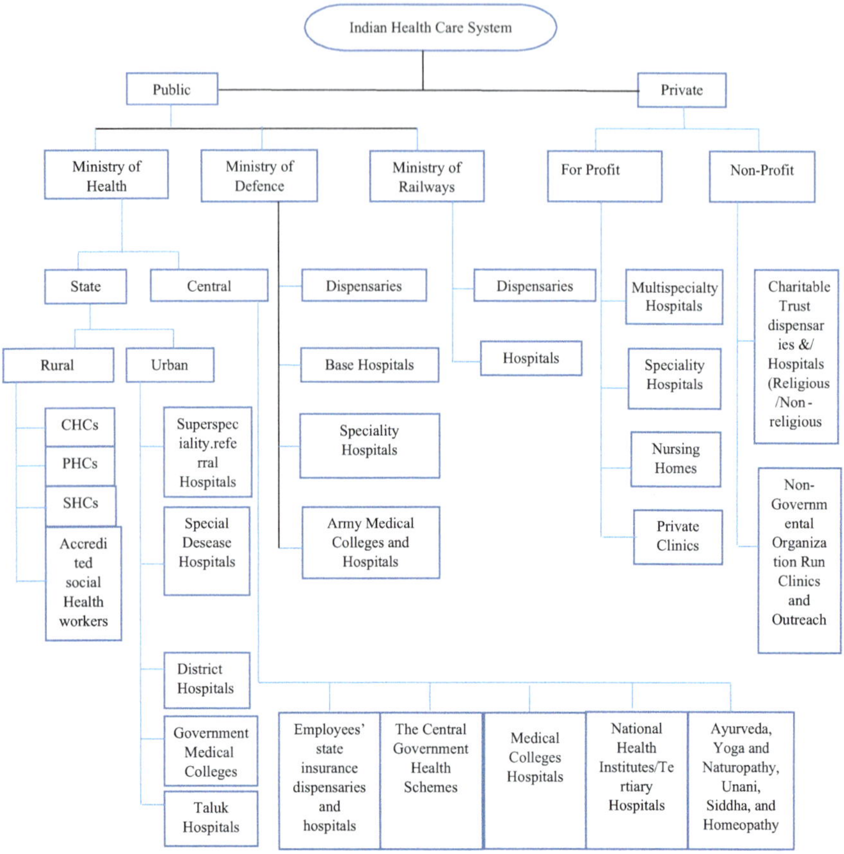

Fig. 15.2 Indian healthcare system. *Source* Drishtiias.com (2023)

Director General, Seven Directors, One Additional director (Administration), one Joint Director (Administration), and Deputy Director at the Head Office (Health Department Haryana 2023) (Gupta et al. 2023). Health Services at district level is headed by Civil Surgeon and Public Hospitals are headed by Principal Medical Officers, Medical Superintendents, and Senior Medical Officers. CHCs and PHCs are headed by Senior Medical Officers and Medical Officers in-charge, respectively.

15.3 Resilient Healthcare System

A system which can sustain disruptions due to unpredictable shocks, events, and policies by adapting quickly to new environment is a resilient healthcare system. It learns from the experience to become more prepared and better performing system (OECD 2023). Health system resilience minimizes the negative consequence of disruptions

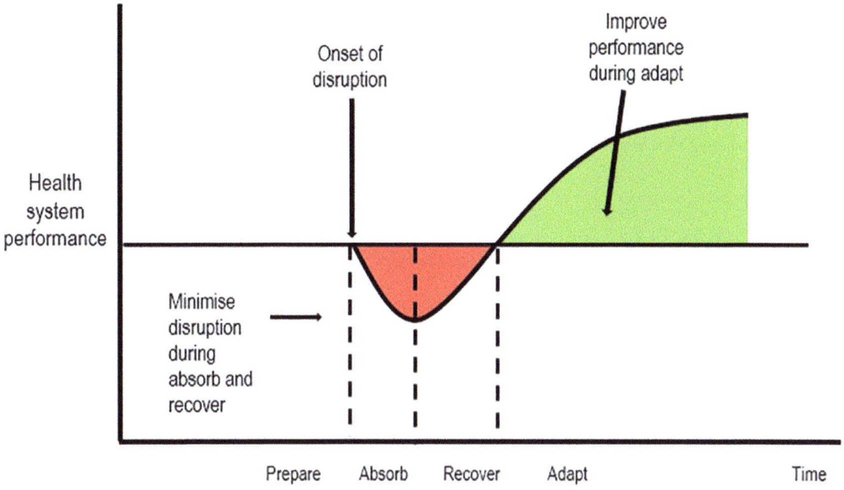

Fig. 15.3 The disruption cycle: the four stages of a response over time. *Sources* Adapted from OECD (2019), "Resilience-based Strategies and Policies to Address Systemic Risks," and National Research Council (2012), Disaster resilience: A national imperative

and recover with agility. To truly establish a resilient healthcare system requires a multidisciplinary approach. Following figure aptly demonstrates various dimensions of a resilient healthcare system (Fig. 15.3).

15.4 Factors of Framework of Public Health System

World Health Organization (WHO) uses an analytical framework to describe health systems. The framework has following six core components or factors which are considered as building blocks of a sound health system (WHO 2010) (Roopesh 2020):

i. Leadership and Governance
ii. Service Delivery
iii. Health Workforce
iv. Health Information System
v. Medical Products, Vaccines and Technologies
vi. Health Care Financing.

Each of the above factors has been defined by more than one underlined variables. These variables have been identified after comprehensive and incisive discussions were held with government executives who had spent years in the management of medical healthcare systems at various levels of state of Haryana. Leadership and Governance has 12 underlining variables (items). Financing and Service Delivery have 3 and 4 variables, respectively. This is followed by Medical Product and Information which have 4 and 3 variables. Factors of Health Workforce and Safety contain 6 and 5 variables, respectively. These factors and variables (items) have been depicted in Table 15.2: Framework of Public Health System that has been presented in the Chap. 8 of Data Analysis and Findings.

15.5 Research Problem Statement

Determining the relative importance of factors of framework of Health System defined by WHO in the state of Haryana, India, with respect to effective and efficient healthcare system in the crisis situation.

15.6 Research Objectives

RO1: To explore and understand meaning of health system and healthcare system.
RO2: To get awareness about Public Health and Public Healthcare System (PHS) in India.
RO3: To explore the factors of framework of PHS defined by WHO.
RO4: To deliberate on resiliency of factors of framework of PHS in the state of Haryana.
RO5: To find out the underline variables of a resilient PHS in the state of Haryana.

15.7 Research Methodology: Data Collection

This study has used the exploratory and descriptive research design. Information about public health system and public healthcare system has been explored through published articles and research papers, and blogs. Problem statement was formulated and objectives were defined in sync with the purpose of the research. The research process followed a survey method in which a structured questionnaire having responses based on the Likert Scale was used. The questionnaire administered to government employees who were managing the healthcare system of the state. Reliability and validity of the questionnaire were tested using Cronbach's Alpha and Covariance matric found to be satisfactory. Data collection was completed through judgemental sampling in which respondents were considered well informed and well experienced for the topic of research. Factors of resilient public healthcare system were contemplated after comprehensive deliberations with experts in the field of management of public healthcare. Dimensions of public healthcare system were chosen which were classified as the factors contributing in enhancing the resilience of the system. These factors are Leadership and Governance, Financing, Service

Table 15.2 Framework of resilient public healthcare system

Factor	Items under each factor	Average ratings	Mean std. err.
Leadership and Governance	Having a surveillance system that senses the crisis, monitors the system continuously, and detects errors and problems quickly	4.56	0.157
	Establishing strategic policies and plans to face the outbreak (i.e. adopting new rules and plans to cope with the crisis and make the system more integrated, establishing specific rules to prioritize patient management)	4.31	0.151
	Implementing a multidisciplinary approach within the health system (i.e. establishing a close collaboration between different departments to avoid centralized decision-making)	4.56	0.157
	Adjusting performance indicators based on the new situation (i.e. morbidity, mortality rate, etc.)	4.00	0.158
	Increasing or decreasing the number of staff based on the needs and priorities to manage the critical situation (i.e. recruiting or reassigning staff)	4.38	0.155
	Multi-sector collaboration for decision-making	4.13	0.155
	Adjusting service coverage indicators (i.e. number of beds, visited patients per hour, etc.)	4.31	0.176
	Maintaining equity (i.e. ensuring that all patients have access to services and receive treatment)	4.50	0.158
	Ability to adapt to new changes (i.e. rules, protocols, and technologies	4.38	0.155
	Strengthen trust (among management, staff, and patients)	4.50	0.158

(continued)

Table 15.2 (continued)

Factor	Items under each factor	Average ratings	Mean std. err.
	Expanding the tasks and responsibility of health professions (i.e. in response to workload and changes in the duties according to the new situation)	4.44	0.157
	Managing community involvement (i.e. using people in the community with suitable skills to reduce the pressure on the system)	4.56	0.157
Financing	Developing strategies to optimize the use of available resources and protecting the resources	4.43	0.157
	Using diverse financial resources (i.e. receiving extra funds or equipment from outside of the health system such as charities, etc.)	4.37	0.154
	Minimizing the risk of being underfunded	4.31	0.176
Service delivery	Establishing strategies and rules to maintain the quality of services at the same level as before	4.25	0.144
	Avoiding interruption of clinical activities (both elective and non-elective)	3.69	0.151
	Using a well-established care pathway to manage patients and treatment	4.25	0.144
	Applying strategies to classify and prioritize patients based on their conditions	4.37	0.154
Medical products	Sustaining the inventory system (always provide and maintain the specific amount of medical equipment and medicines.)	4.38	0.221
	Forecasting and detecting shortage of medicine and medical products	4.56	0.157

(continued)

Table 15.2 (continued)

Factor	Items under each factor	Average ratings	Mean std. err.
	Changing the inventory system strategies during the crisis (i.e. inventory reserves)	4.63	0.155
	Considering alternative suppliers to provide demands	4.50	0.158
Information	Availability of data to anticipate possible problems that may arise within the system during the crises (anticipation based on internal and external data)	4.44	0.157
	Continuous collection of information before and during the outbreak about the quality and comprehensiveness of services (regular monitoring of the performance by collecting data)	4.38	0.155
	Establishing a system to analyse the information and release the results periodically (analysing internal and external data to facilitate the procedures, improving the system's performance, and detecting errors)	4.44	0.157
Health workforce	Having multi-skilled healthcare professionals	4.50	0.158
	Providing extra emotional and mental support for the workforce	4.50	0.158
	Providing proper facilities to staff to feel more comfortable during long shifts, under challenging conditions and high workload.	4.56	0.157
	Training health professionals periodically to keep them up to date (i.e. up-to-date staff with new information).	4.69	0.151
	Allocating a sufficient number of staff in each department according to burden of work	4.63	0.155

(continued)

Table 15.2 (continued)

Factor	Items under each factor	Average ratings	Mean std. err.
	Providing compensation for damages incurred by the staff during the crisis (i.e. financial, physical, etc.)	4.50	0.158
Safety	Equipping the system for safety reasons	4.50	0.158
	Establishing risk management strategies within the system	4.69	0.151
	Providing guidelines and protocols to increase safety in the workplace (i.e. to face infected patients)	4.75	0.144
	Training staff to increase their awareness about safe behaviours during the crisis	4.68	0.150
	Immunizing of the system against infectious diseases (i.e. vaccination)	4.69	0.151

Source Adopted from WHO (2010)

Delivery, Medical Products, Information, Health Workforce, and Safety. Multiple items under each of the factors have been rated on the Likert Scale. Descriptive research design has been used to find out which factors have been rated higher in defining the resilient public healthcare system. Descriptive statistics techniques such as central tendencies, cross-tabulation, and charts have been used to analyse the data. Ratings of the factors of framework of public health system have been used to find out the relative importance of each of these factors.

15.8 Data Analysis and Finding

Data analysis was envisaged in such a way that objectives of the research are achieved. Exploration of extant literature helped in understanding health and healthcare system. Objective number two was aimed to get awareness about the public healthcare system in India. This has been achieved by exploring the data available on websites of government agencies and ministries. Objective three has been achieved by exploring the framework of public health system defined by World Health Organization. To deliberate on framework of public health system, descriptive statistics have been used. Factors that defines the framework have been rated by the respondents. Average rating and standard error of each item under each variable have been demonstrated in Table 15.2: Framework of public healthcare system has been analysed. Items with highest average rating have been identified. Standard error of each of the highest rated item has also been discussed. Findings from the data analysis demonstrated that for the factor Leadership and Governance, two items; (i) Having a surveillance system that sense the crisis, monitors the system, and detects errors and problems quickly; (ii) implementing a multidisciplinary approach within the healthcare system scored the highest rating. For Financing developing strategies to optimize the use of available resources, applying strategies to classify and prioritize patients based on their conditions came out at the top for service delivery. Medical product changing the inventory system strategies during the crisis scores the highest rating. Availability of data to anticipate possible problems that may arise within the system during the crisis and establishing a system to analyse the information and release the results periodically were given highest rating under the factor information. Training health professionals periodically to keep them up to date scored top rating under the factor of health workforce. Providing guidelines and protocols to increase safety at the workplace was given top priority under the factor of safety. Among the factors Leadership and Governance got the highest rating followed by Health Workforce. A resilient public healthcare system is an outcome of the meticulous planning and adapting to uncertainty caused by shocks. This requires visionary leadership and governance that can guide system to identify, understand the problems, and then provide solutions. It needs robust financing that can develop strategies to optimize available resources so that shocks can be absorbed. Disparate variables under other factors are also resulting in a resilient public healthcare system. These are exhibited in the Table 15.2.

15.9 Conclusion

A resilient public healthcare system is an outcome of composite and integrated activities and instruments represented by defined factors and variables defined in the framework of public health system by WHO. A successful leadership and governance have been manifested by having a surveillance system that senses the crisis, monitors the system continuously, and detects errors and problems quickly along with implementing a multidisciplinary approach within the health system. Financing a resilient system requires developing strategies to optimize the use of available resources and protecting the resources. For a resilient system, service delivery must apply strategies to classify and prioritize patients based on their conditions. There are other factors that constitute the important variables for a resilient healthcare system. First, medical product that has the inventory system strategies during the crisis. Second availability of data to anticipate possible problems that may arise within the system during the crisis and establishing a system to analyse the information and release the results periodically. Third, training health professionals periodically to keep them up to date under the factor of health workforce. Fourth, providing guidelines and protocols to increase safety at the workplace under the factor of safety. Among the factors Leadership and Governance got the highest rating followed by Health Workforce.

15.10 Limitations and Future Scope

The study conducted is primarily descriptive in nature. There were limited number of respondents who understood the meaning and relevance of resilience healthcare system. Instrument to measure resilience of public healthcare system has not been developed before; therefore, it requires broader acceptability. Conclusive research using causal research design is the way ahead for the researchers to design a framework of a resilient public healthcare system.

References

Chokshi M, Patil B, Khanna R, Neogi SB, Sharma J, Paul VK, Zodpey S (2016) Health systems in India. J Perinatol (online) 36(S3):S9–S12. https://doi.org/10.1038/jp.2016.184

Drishtiias.com (2023) Available at: https://www.drishtiias.com/images/uploads/1590578965_image1.png. Accessed 6 Nov 2023

Gupta A, Pandey A, Bansal P, Kaur A, Prakash P (2023) Health care systems during pandemic: role of government and regulatory authorities. J Indian Med Assoc 121(8):88–90

Jarvis T, Scott F, El-Jardali F, Alvarez E (2020) Defining and classifying public health systems: a critical interpretive synthesis. Health Res Policy Syst 18(1). https://doi.org/10.1186/s12961-020-00583-z

Kumar A (2023) The transformation of the Indian healthcare system. Cureus [online] 15(5). https://doi.org/10.7759/cureus.39079

Roopesh (2020) Public health system: definition and types [online] communitymedicine4all. Available at: https://communitymedicine4all.com/2020/11/04/public-health-system-definition-and-types/

World Health Organization (2010) Health system building blocks [online] Who.int. Available at: https://extranet.who.int/nhptool/BuildingBlock.aspx

Chapter 16
Formulation and Evaluation of Colon-Targeted pH-Sensitive Microbeads for the Treatment of Inflammatory Bowel Disease

Sunny Jakhar, Dinesh Chandra Bhatt, Ankita Sharma, and Shaveta Ahalwat

16.1 Introduction

Inflammatory bowel disease (IBD) is a serious disorder as it involves the chronic inflammation and ulcers of the gastro-intestinal tract and small intestine. IBD majorly includes Ulcerative colitis and Crohn's disease (Teruel et al. 2020). Ulcerative colitis involves inflammation of the lining of large intestine (colon) and rectum, whereas, Crohn's disease is characterized by inflammation of the lining of your digestive tract, and small intestine which often involves the deeper layers of the digestive tract (Lee et al. 2020).

Most of the conventional drug delivery systems for treating these diseases are failing as the drugs do not reach the site of action in appropriate concentrations (Prasanth and Mathew 2012). Accordingly, an effective and safe therapy for these colonic disorders is a challenging task to the pharmaceutical technologists. Colon specific drug delivery systems not only can deliver the drug in its intact form as close as possible to the target site but can also significantly reduce the incidence of adverse side effects associated with conventional drug delivery system (Seyedian et al. 2019).

S. Jakhar · D. C. Bhatt · A. Sharma · S. Ahalwat (✉)
Department of Pharmaceutical Sciences, Guru Jambheshwar University of Science and Technology, Hisar, India
e-mail: shaveta.ahalwat@gmail.com

A. Sharma
Gurugram Global College of Pharmacy, Farrukh Nagar, Gurugram, India

S. Ahalwat
Department of Pharmacy, School of Medical and Allied Sciences, G. D. Goenka University, Gurugram, India

N. A. Siddiqui et al. (eds.), *Advances in Disaster Management, Volume 2*, Springer Transactions in Civil and Environmental Engineering,
https://doi.org/10.1007/978-981-96-4047-8_17

Some potent anti-inflammatory agents and oral systemic corticosteroids (e.g., prednisone, prednisolone) have been used to induce remission in IBD patients for more than 60 years (Bruscoli et al. 2021). Other mediators, including as amino acid metabolites and interleukins, are also inhibited by prednisolone and also being studied for its immune-suppressing, anti-neoplastic, and vasoconstrictive properties. Prednisolone either used alone or in combination with other agents (Placha and Jampilek 2021). Due to its high solubility in gastric fluid, it is rapidly absorbed from the stomach and hence could not reach in the desirable concentration to the colon, therefore, a higher dose is required to treat the colonic disorders (Amidon et al. 2015). Moreover, conventional tablets of prednisolone are associated with numerous adverse effects like gastric irritation, nausea, vomiting, etc. Therefore, a colon-targeted formulation of prednisolone could be developed for specific drug delivery to colon (Rashid et al. 2016).

Biodegradable microbeads are one of the promising approaches to deliver materials in an effective, prolonged and safe manner for controlled release of the drug. Biodegradable polymers like pectin, chitosan, sodium alginate offers a sustained release drug delivery systems that have potential for colonic drug delivery (Idrees et al. 2020).

Pectin is used in the formulation of microbeads as it is less susceptible to degradation in the gastrointestinal tract than alginate. The degradation of pectin occurs mainly in the colon by pectinolytic enzymes secreted by microorganisms. As a result, pectin has increasingly gained acceptance as the carrier polymer for sustained release and site-specific delivery dosage forms, such as beads, pellets, tablets, and films (Rubinstein et al. 1993; Ashford et al. 1994).

In the present study, prednisolone loaded sodium alginate microbeads (P-SA-MB) were prepared at fixed amount of prednisolone and calcium chloride with varying concentrations of chitosan, sodium alginate, pectin, and HPMC by ionotropic method and coated with Eudragit RS100 for providing pH sensitivity to the formulation to prevent the degradation in the stomach and achieves maximum drug release to the colon. All the formulations were characterized for encapsulation efficiency, flow properties and in vitro drug release to find out the best optimized formulation. The final optimized formulation was evaluated for FTIR, DSC, XRD for their crystalline/amorphous behavior. The surface morphology of the microbeads was seen under the scanning electron microscope (SEM).

Table 16.1 Formulations batches for preparation of sodium alginate microbeads

Formulation code	Sodium alginate (w/v)	Calcium chloride (w/v)	Drug (g)
F1	5	3	0.02
F2	6	3	0.02
F3	8	3	0.02

16.2 Materials and Methods

16.2.1 Materials

Eudragit RS100, Pectin, Sodium Alginate and Chitosan were procured from Hi-media Laboratories Ltd., Mumbai. Ethanol and HPMC were procured from Loba Chemie Ltd., Mumbai, India. The drug Prednisolone was purchased from Yarrow pharmaceuticals Ltd., Mumbai.

16.2.2 Methods

16.2.2.1 Preparation of Microbeads

Prednisolone loaded microbeads in varying concentrations of sodium alginate (5% w/v, 6% w/v, 8% w/v) were prepared using ionotropic gelation method (Table 16.1) (Khan et al. 2010). In brief, accurately weighed prednisolone (20 mg) was added to different concentrations of sodium alginate solution and was added drop wise into100 ml of 3% w/v aqueous calcium chloride solution through a 26 G syringe needle. The solution was stirred uniformly throughout the process by keeping it at 100 rpm speed on magnetic stirrer. Distance between surface of solution and tip of needle was kept constant. The formed beads were kept under stirring for 30 min and then separated by filtration. The separated beads were washed with distilled water and dried at 50° for 4 h and then at room temperature for 2 days for complete drying. The blank formulation was prepared using the same method without adding the prednisolone drug (Khan et al. 2010).

16.2.2.2 Preparation of Polymer-Reinforced Beads

Polymer reinforced beads were prepared to enhance the stability of integrity of the microbeads using varying concentration of HPMC, chitosan and pectin (Table 16.2). HPMC reinforced alginate beads were formulated by adding prednisolone in sodium alginate (6% w/v) solution and varying concentrations of HPMC (0.5% w/v and 1% w/v) was mixed respectively and stirred for 30 min and then drop wise added to

Table 16.2 Formulation batches for preparation of polymer-reinforced microbeads

Formulation code	Sodium alginate (w/v)	Calcium chloride (w/v)	HPMC (W/V)	Chitosan (w/v)	Pectin (w/v)	Drug (g)
F4	6	3	0.5			0.02
F5	6	3	1			0.02
F6	6	3		0.5		0.02
F7	6	3		1		0.02
F8	6	3			0.5	0.02
F9	6	3			1	0.02

agitated 3% w/v calcium chloride solution (100 ml). Chitosan reinforced alginate beads were formulated by adding prednisolone in sodium alginate solution (6% w/v) mixed with varying concentrations of chitosan (0.5% w/v and 1% w/v) respectively. The above solution was dropped through a needle into 3% calcium chloride solution. Pectin reinforced alginate beads were formulated by adding prednisolone in sodium alginate (6% w/v) solution and varying concentrations of pectin (0.5% w/v and 1% w/v) respectively and then added drop wise to 3% w/v calcium chloride solution. Beads were left for 30 min, washed twice with distilled water and dried at 50° for 4 h. Prepared microbeads were placed at room temperature for 2 days for complete drying (Pandey et al. 2012).

16.2.2.3 Coating of Drug-Loaded Microbeads

The enteric coating of all the drug-loaded micro beads was performed by spray coating in a conventional coating pan. The enteric coating solution was prepared by mixing Eudragit RS100 (2.98 g) in a mixture of isopropyl alcohol and acetone (20:14.5v/v) for 1 h using a magnetic stirrer. After 1 h, 0.5 ml of triethylcitrate (TEC) was added and stirring was continued for 30 min. Coating was done with a pan rotating at 20 rpm, with an inlet air temperature 60 to 70 °C and outlet air temperature of 35–40 °C. The coating solution was sprayed through an atomizer at a pressure of 2 bars and a flow rate was controlled at 2 rpm. The process was continued until the desired weight gain was achieved. After successful coating, all prepared microbeads were kept in hot air oven for 30 min at 60 °C (Patil et al. 2012).

16.2.2.4 Percent Encapsulation Efficiency

Prednisolone entrapped microbeads were investigated for percent encapsulation efficiency. The % EE was determined by subjecting the drug-loaded microbeads to sonication with solvent for disruption of beads followed by centrifugation. The beads were settled at the bottom and hence discarded. The supernatant was collected and analyzed spectrophotometrically at 244 nm for un-entrapped drug. The %EE of the drug was determined using the following equation (Setia and Ahlawat 2016; Ahalwat et al. 2023).

$$\%\mathrm{EE} = \frac{(\text{Total drug} - \text{Free drug})}{\text{Total drug}} * 100$$

16.2.2.5 In Vitro Drug Release Study

In vitro drug release studies were performed using USP dissolution test apparatus type-2. The dissolution studies were performed in 900 ml of dissolution medium at a temperature of 37±0.1°C under continuous stirring at 100 rpm using a pH progression method. Different dissolution media were used to mimic the behavior of the formulation in gastro intestinal tract. The simulated gastric fluid pH 1.2 was used for first 2 h, phosphate buffer saline pH 6.8 for next 3 h and pH 7.4 for rest of the time. Aliquots of 5 ml were withdrawn periodically and replaced with fresh medium to maintain the sink conditions. Prednisolone content in the aliquots was assayed spectrophotometrically (UV-1601, Shimadzu, Japan) at 244 nm. The percent of drug released was plotted versus time (Ahalwat et al. 2023).

16.2.2.6 Evaluation of Rheological Properties

Angle of Repose

Angle of repose was calculated using the following formula employing fixed funnel method (Ahalwat et al. 2022).

$$\text{Angle of repose } \theta = \tan -1(H/R)$$

where, H = Height of the heap; R = Radius of the heap

Bulk and Tapped Density

In a 10 ml measuring cylinder, precisely weighed amount of microbeads was added. The volume of the sample was then noted. The cylinder was then tapped 100 times and then volume was further noted down after tapping.

$$\text{Bulk Density} = \frac{\text{Weight of microcapsules}}{\text{Volume of microcapsules before tapping}}$$

$$\text{Tapped Density} = \frac{\text{Weight of microcapsules}}{\text{Volume of microcapsules after tapping}}$$

Carr's Index

It is also known as Compressibility index and it shows the compressibility of a powder. It is calculated by using the following formula (Ahalwat and Bhatt 2022).

$$\text{Carr's Index} = \frac{(\text{Tapped density} - \text{Bulk density})}{\text{Tapped density}} * 100$$

Hausner's Ratio

The Hausner ratio is a number that is correlated to the flowability of a powder or granular material.

$$\text{Hausner's ratio} = \frac{\text{Volume before tapping}}{\text{Volume after tapping}}$$

16.2.2.7 Optimization of Sodium Alginate Microbeads

The final optimized P-SA-MB formulation was finalized based on the maximum percent encapsulation efficiency, in vitro drug release and good flow property.

16.2.2.8 Fourier Transform Infrared Spectroscopy (FTIR) Analysis

FTIR analysis of drug, physical mixture and optimized batch was carried out using potassium bromide (KBr) pellet method. Weighed amount of 1 mg drug was triturated with approximately 10 mg of dry potassium bromide and subjected to the formation of pellet using hydraulic press and analyzed using FTIR spectrophotometer (Perkin Elmer Spectrum, BX II) in range between 4000 and 400 cm^{-1} (Ahalwat and Bhatt 2022).

16.2.2.9 Differential Scanning Calorimetry (DSC) Analysis

DSC Thermogram of drug and optimized batch was acquired by using DSC analyzer (Q-10, TA Instruments Waters). DSC was utilized to find out the melting point and thermal stability of both the drug and formulation. In this analysis, 5 mg of drug sample was kept in DSC pan and was heated under nitrogen condition at a constant rate of 10 °C per minute over a temperature range of 20–300 °C (Rohilla et al. 2019).

16.2.2.10 Powder X-Ray Diffraction (XRD) Analysis

Powder X-ray diffraction (XRD) patterns were studied using X-Ray diffractometer (Miniflex-II), Rigaku, Japan using Cu radiation source ata voltage of 30 kV, and current of 15 mA. The powder samples were packed in the X-ray sample holder from the top before analysis. These samples were continuously spun and scanned at a rate of 1°/min over a range of 5–70 °C (Rohilla et al. 2020).

16.2.2.11 In Vitro Drug Release Kinetics

The in vitro drug release data was fitted into the various mathematical (kinetic) modelsi.e. zero order kinetics, first order kinetics, Higuchi square root model, Korsmeyer–Peppas model to investigate the mechanism of drug release from the formulation (Joon et al. 2020).

16.2.2.12 Scanning Electron Microscopy (SEM)

The enteric coated microbeads were characterized for surface morphological characteristics using High Resolution Field Emission Scanning Electron Microscope with EDS (FE-SEM) (Ahalwat et al. 2023).

16.3 Result and Discussion

16.3.1 Optimization and Evaluation of Drug-Loaded Sodium Alginate Microbeads

Microbead formulation of prednisolone was prepared by Ionotropic gelation technique. The results of the encapsulation efficiency and in vitro drug release of the formulations are shown in Table 16.3. The encapsulation efficiency of the 9 batches was found to be in the range of 43.03–63.06% and in vitro drug release was found in the range of 68.43–91.24%. The maximum % encapsulation efficiency and % drug release found for batch F2 and F6 respectively. These two formulations were further subjected to investigation of flow properties.

Table 16.3 Formulation and evaluation of P-SA-MB (Batches F1 to F9)

Formulation code	Sodium alginate (w/v)	Calcium chloride (w/v)	HPMC (W/V)	Chitosan (w/v)	Pectin (w/v)	Encapsulation efficiency (%)	Drug release (%)
F1	5	3				53.30	79.73
F2	6	3				63.06	75.12
F3	8	3				43.03	68.43
F4	6	3	0.5			59.80	76.86
F5	6	3	1			53.60	81.46
F6	6	3		0.5		61.60	91.24
F7	6	3		1		62.50	87.00
F8	6	3			0.5	56.80	82.62
F9	6	3			1	57.10	84.71

Table 16.4 Characteristics flow properties batch F2 and F6

Characteristics formulation ↓	Angle of repose, θ (degree)	Bulk density (g/ml)	Tapped density (g/ml)	Carr's index (%)	Hausner's ratio
F2	21.95 ± 0.5	0.35 ± 0.01	0.43 ± 0.05	13.19 ± 0.49	1.22 ± 0.07
F6	19.82 ± 0.92	0.40 ± 0.02	0.44 ± 0.01	9.09 ± 0.01	1.11 ± 0.02

16.3.2 Evaluation of Rheological Properties

16.3.2.1 Angle of Repose

Values of angle of repose for the batches F2 and F6 comes out to be 21.95° and 19.82° respectively. Both formulations showed good flow property as there values were below 25 as shown in Table 16.4.

16.3.2.2 Bulk and Tapped Densities

Bulk density of both the formulations F2 and F6 comes out to be 0.35 and 0.40 g/ml and tapped density were found to be 0.43 and 0.44 respectively. Powder with greater density has good flow property. So, F6 batch had good flowability.

16.3.2.3 Carr's Index

Carr's index value less than or equal to 10 states excellent flow property whereas 12–16 are of good flow property.The Carr's index values of the F2 and F6 microbead formulation was found to be 13.19 and 9.09 respectively representing the good flow property of F6 batch compared to F2.

16.3.2.4 Hausner's Ratio

The value of Hausner's ratio of the F2 and F6 batch comes out to be 1.22 and 1.11 respectively (Table 16.4). Hausner's ratio having values between 1.00 and 1.11 depicts excellent flow property whereas; value 1.22 shows fair flow property.

16.3.3 Optimization of Sodium Alginate Microbeads

Based on the maximum encapsulation efficiency, drug loading and good flow property, formulation batch F6 was selected an optimized batch and subjected to further evaluation.

16.3.4 In Vitro Drug Release Study

The optimized microbead formulation was evaluated for drug release profiles in simulated gastric fluid pH 1.2 for 2 h, phosphate buffer saline pH 6.8 for next 3 h and pH 7.4 for the remaining time up to 24 h to find out the in vitro drug release pattern. The cumulative drug release was found to be 91.24%. Figure 16.1 illustrate the release of drug at different time points in the different dissolution medium. Eudragit RS100 coating restricted the release of drug in simulated gastric fluid on average with maximum release of 6%. In simulated intestinal fluid, an average of 15% drug release was there for 3 h. More than 75% of the drug was release in Simulated Colonic Fluid.

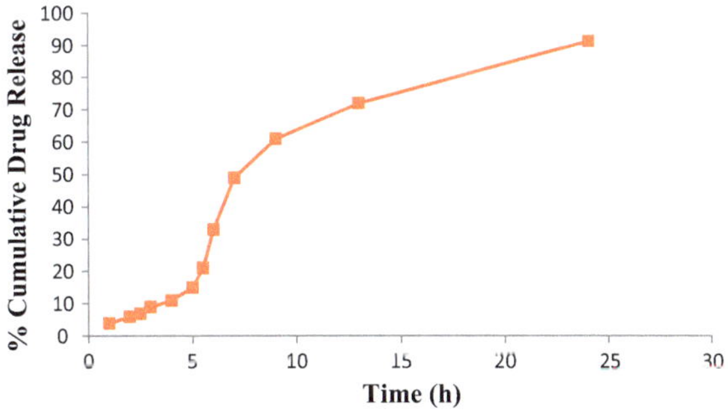

Fig. 16.1 In vitro drug release profiles of optimized batch F6 batch of P-SA-MB

16.3.5 In Vitro Drug Release Kinetic Study

To study the release kinetics and mechanism of drug release, the obtained data were fitted into different kinetic models like zero order, first order, Higuchi square root and Korsmeyer-Peppas plot (Table 16.5). The regression coefficient (R^2) was calculated for the optimized batch (F6) (Fig. 16.2). It was observed that the drug release from the formulation follows first order kinetics as it had the highest R^2 value. Further, the release exponent (n) was calculated from Korsmeyer–Peppas plot which indicates the release of prednisolone occurs through non-Fickian diffusion-controlled mechanism.

Table 16.5 Kinetics of the drug release

Models	Zero order R^2	First order R^2	Higuchi kinetics R^2	Korsmeyer–Peppas model	
				R^2 'n'	N
F6	0.9496	0.9832	0.9272	0.9532	0.9655

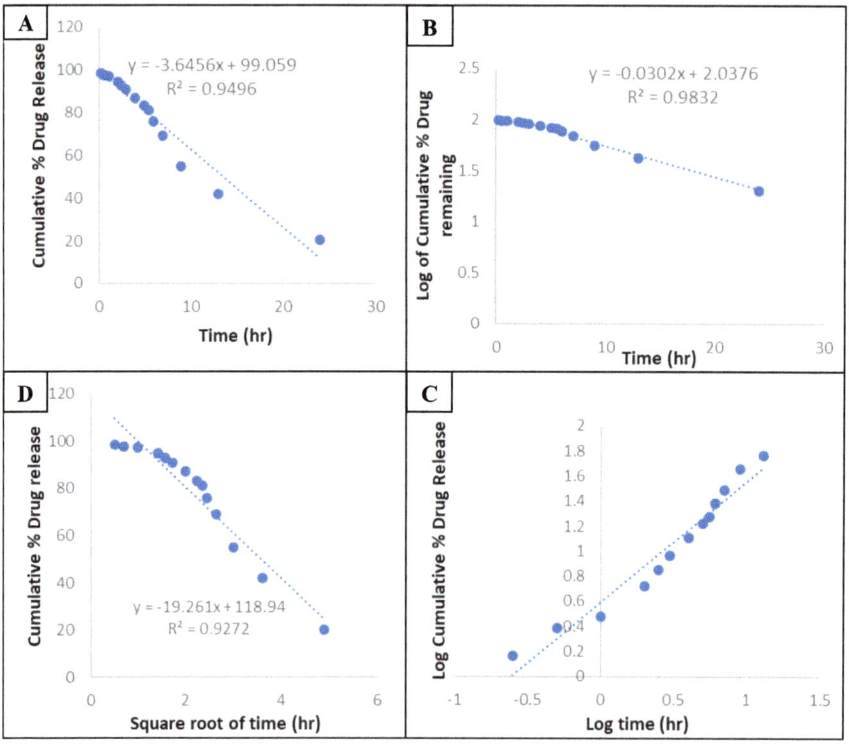

Fig. 16.2 Drug release kinetic plots **a** zero order plot **b** first order plot **c** Higuchi plot **d** Korsmeyer–Peppas plot

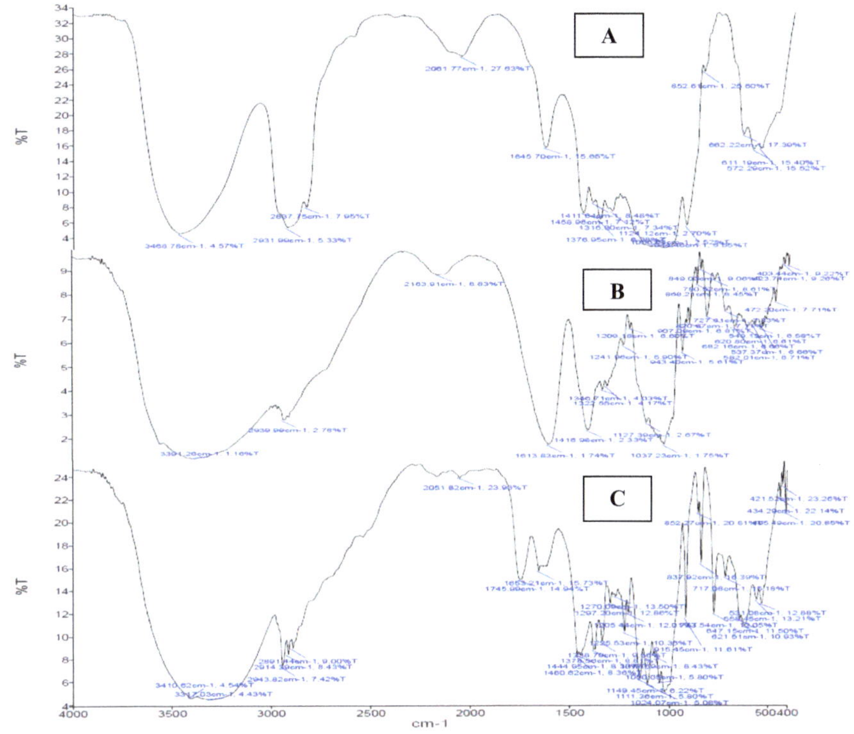

Fig. 16.3 FTIR Spectra of **a** prednisolone **b** physical mixture **c** optimized batch

16.3.6 Fourier Transform Infrared Spectroscopy

The FTIR spectrum of pure prednisolone demonstrated the O–H peak at 3357cm^{-1}, peak of carbonyl group at 1654 cm^{-1}, C–H stretch at 2887.75 and C–C stretch at1645.70 cm^{-1}as shown in Fig. 16.3. This represented the identification of the drug. All the functional group peaks of the drugs were found at their place representing the compatibility of the drug with the polymers. Some minor shifting was also observed in some peaks in the spectrum of the physical mixture as well as in the formulation represented the solubility of drug within the polymers as well as in the formulation.

16.3.7 Differential Scanning Calorimetry (DSC)

DSC thermogram of the drug and optimized formulation was evaluated and shown in Fig. 16.4. The melting point of the drug was found to be 185.47 °C, which was similar to the standard reference peak leads to the identification of the drug. A sharp endothermic peak with high enthalpy of 416.58 J/g represents the high crystalline

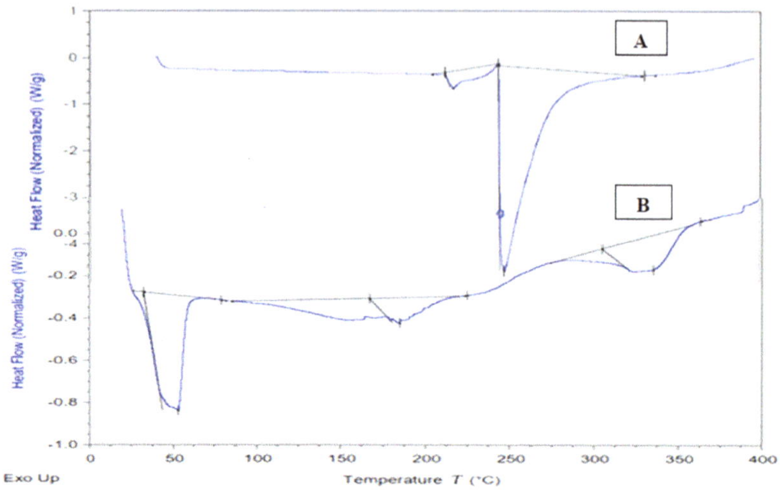

Fig. 16.4 DSC Thermogram of **a** pure drug **b** formulation

nature of the drug. This peak was diminished/becoming narrow in DSC thermogram of the formulation representing the amorphous nature of the formulation.

16.3.8 X-Ray Diffraction (XRD) Analysis

XRD analysis was performed out the crystalline/amorphous nature of the drug and optimized microbead formulation. XRD spectra of drug and optimized batch are shown in Fig. 16.5. The XRD spectrum of pure prednisolone showed several sharp and intense peaks indicating the crystalline nature of pure drug. These peaks were completely diminished in XRD spectrum of the optimized drug representing the amorphous nature of the formulation.

16.3.9 Scanning Electron Microscopy (SEM)

Surface morphology of the P-SA-MB was determined with the help of Scanning Electron Microscopy ((JSM- 7610FPlus, JEOL) at 50× and 200× magnification as shown in Fig. 16.6a, b respectively. The average particle size of the optimized batch was found to be 831 ± 29 µm. Spherical shaped microcapsules with rough surface were observed as a result of drying state which involves loss of water and shrinkage of microcapsules.

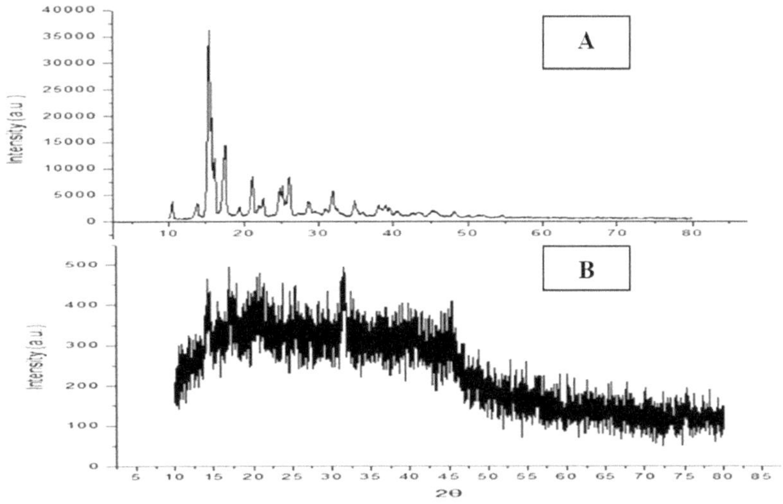

Fig. 16.5 X-Ray diffraction pattern of (A) prednisolone (B) microbeads

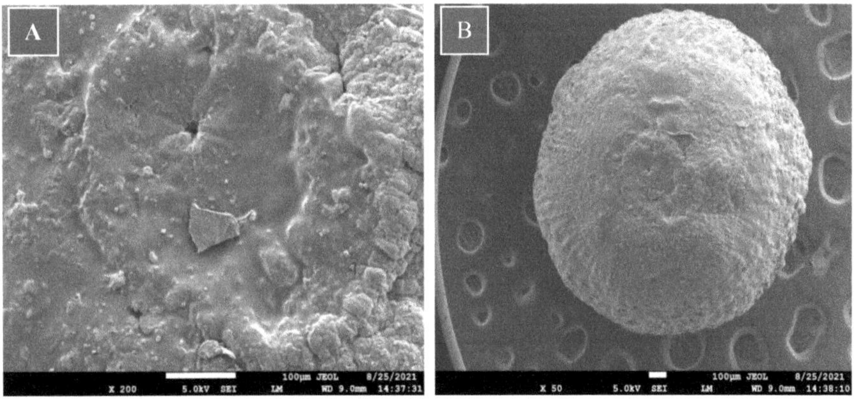

Fig. 16.6 Scanning electron microscopy of optimized batch **a** 50× **b** 200×

16.4 Conclusion

A pH sensitive sodium alginate microbeads were prepared by inotropic gelation method and successfully coated with Eudragit RS100 to protect them from gastric fluid and specifically released in the colonic location. Nine formulations were prepared using three different polymers and evaluated for maximum encapsulation, good flow properties and in vitro drug release. The encapsulation efficiency of all the 9 batches was found to be in the range of 43–63%. The cumulative drug release was found to be in the range of 68 to 91%. Maximum drug release of 91.24% was shown by F6 formulation at the colonic pH which indicates colon specific drug release. The drug release data of the optimized batch was fitted into different kinetic models and was found to follow first order kinetics as it shows the maximum R^2 value of 0.9669 indicating the concentration dependent release of drug from the porous matrix. The average particle size was found to be 831 ± 29 μm. The microbeads were found to be spherical in shape with good flowability and compressibility. The porous structure of the microbeads resulted in rapid drug release at the target site on the basis of pH variability. So, it concluded that pH sensitive sodium alginate microbeads prepared herein, was able to achieve site-specific drug delivery to the colon following oral administration and could be further investigated in future for improved application in elimination of the adverse effects on non-targeted organs.

References

Ahalwat S, Bhatt DC (2022) Development of novel lipid matrix for improved sustained release effect of a hydrophilic drug via response surface methodology. J Drug Deliv Sci Technol 1(67):102993

Ahalwat S, Bhatt DC, Rohilla S (2022) Fabrication of lipid combination for bioavailability enhancement of isoniazid. Res J Pharm Technol 15(12):5589–5596

Ahalwat S, Bhatt DC, Rohilla S (2023) Quality by Design (QbD) based formulation optimization of isoniazid loaded novel nanostructured lipid carriers for controlled release effect. J Pharm Innov 21:1–6

Ahalwat S, Bhatt DC, Rohilla S, Jogpal V, Sharma K, Virmani T et al (2023) Mannose-functionalized isoniazid-loaded nanostructured lipid carriers for pulmonary delivery: in vitro prospects and in vivo therapeutic efficacy assessment. Pharmaceuticals 16(8):1108

Amidon S, Brown JE, Dave VS (2015) Colon-targeted oral drug delivery systems: design trends and approaches. AAPS PharmSciTech 16:731–741

Ashford M, Fell J, Attwood D, Sharma H, Woodhead P (1994) Studies on pectin formulations for colonic drug delivery. J Control Release 30(3):225–232

Bruscoli S, Febo M, Riccardi C, Migliorati G (2021) Glucocorticoid therapy in inflammatory bowel disease: mechanisms and clinical practice. Front Immunol 3(12):691480

Idrees H, Zaidi SZ, Sabir A, Khan RU, Zhang X, Hassan SU (2020) A review of biodegradable natural polymer-based nanoparticles for drug delivery applications. Nanomaterials 10(10):1970

Joon P, Sharma K, Nagpal L, Ahalwat S, Sanduja M, Jogpal V (2020) Synthesis of substituted aromatic carboxylic acid derivatives and their biological evaluation. Eur Chem Bull 12(5):6438–6446

Khan MS, Sridhar BK, Srinatha A (2010) Development and evaluation of pH-dependent micro beads for colon targeting. Indian J Pharm Sci 72(1):18

Lee SH, Bajracharya R, Min JY, Han JW, Park BJ, Han HK (2020) Strategic approaches for colon targeted drug delivery: an overview of recent advancements. Pharmaceutics 12(1):68

Pandey AK, Choudhary N, Rai VK, Dwivedi H, Kymonil KM, Saraf SA (2012) Fabrication and evaluation of tinidazole microbeads for colon targeting. Asian Pac J Trop Dis 1(2):S197–S201

Patil P, Chavanke D, Wagh M (2012) A review on ionotropic gelation method: novel approach for controlled gastroretentive gelispheres. Int J Pharm Pharm Sci 4(4):27–32

Placha D, Jampilek J (2021) Chronic inflammatory diseases, anti-inflammatory agents and their delivery nanosystems. Pharmaceutics 13(1):64

Prasanth VV, Mathew ST (2012) Colon specific drug delivery systems: a review on various pharmaceutical approaches. J Appl Pharm Sci 30:163–169

Rashid M, Kaur V, Hallan SS, Sharma S, Mishra N (2016) Microparticles as controlled drug delivery carrier for the treatment of ulcerative colitis: a brief review. Saudi Pharm J 24(4):458–472

Rohilla S, Bhatt DC, Ahalwat S (2019) Effect processing variables on the characteristics of itraconazole hollow microspheres. Int J App Pharm 11(6):108–115
Rohilla S, Bhatt DC, Ahalwat S (2020) Fabrication of potential gastroretentive microspheres of itraconazole for stomach-specific delivery: Statistical optimization and in vitro evaluation. J Appl Pharm Sci 10(3):119–127
Rubinstein A, Radai R, Ezra M, Pathak S, Rokem JS (1993) In vitro evaluation of calcium pectinate: a potential colon-specific drug delivery carrier. Pharm Res 10:258–263
Setia A, Ahlawat S (2016) Quetiapine fumarate loaded mucoadhesive nanoemulsion: formulation, optimization & brain delivery via intranasal administration. Drug Deliv Lett 6(2):64–76
Seyedian SS, Nokhostin F, Malamir MD (2019) A review of the diagnosis, prevention, and treatment methods of inflammatory bowel disease. J Med Life 12(2):113
Teruel AH, Gonzalez-Alvarez I, Bermejo M et al (2020) New insights of oral colonic drug delivery systems for inflammatory bowel disease therapy. Int J Mol Sci 21(18):6502

Chapter 17
A Review: Contemporaneous Trend of Combinatorial Nano Therapy in Cancer Treatment

Laxmi Rani, Garima Jain, Pooja Mathur, Shaveta Ahalwat, and Shailendra Bhatt

17.1 Introduction

Cancer is major threat to global human well-being representing the second main cause of mortality after heart disease among lower and middle-income nations killing large number of population annually in past several periods (Kashyap et al. 2021; Oliveira Júnior et al. 2018; Khatoon et al. 2020). Approximately 19.3 million new patients and 10.3 million deaths were reported in 2020 (Talib et al. 2022). Combating this disease among this income group is complicated due to the lack of resources. According to the World Health Organization report, 27 million reported patients and a 17.5 million patient mortality is estimated annually by 2050 (More et al. 2021; Briguglio et al. 2020). Breast, cervical, and colorectal cancer screening programs are intended to minimize delayed diagnosis and cancer-specific fatalities which play a vital role in reducing overall mortality of cancer. The advantages of early detection of malignant disorders through screening have numerous advantages: enhanced quality of life, low risk of progression of the disease to advanced or metastatic stages, less harsh possible treatments, high probability of cure, and decreased cancer-specific mortality (Poppel et al. 2022).

For different types of cancer, chemotherapeutic agents are the standard treatment but are now associated with significant systemic toxicity and less therapeutic efficacy (Alemi et al. 2018). The long-term use of anticancer drugs elevates the level of some factors, i.e., nuclear factor kappa B (NFKB) a transcriptional factor together with changes in local immunity, stroma, and tumor microenvironment responsible for enhancing drug resistance in cancer cells. This is attributed to changes in drug

L. Rani (✉) · G. Jain · P. Mathur · S. Ahalwat · S. Bhatt
School of Healthcare and Allied Sciences, G. D. Goenka University, Gurugram, India
e-mail: Laxmi.rani@gdgu.org

N. A. Siddiqui et al. (eds.), *Advances in Disaster Management, Volume 2*, Springer Transactions in Civil and Environmental Engineering,
https://doi.org/10.1007/978-981-96-4047-8_18

metabolism, elevated rates of drug efflux, intercellular and intracellular drug sequestration in organelles, mutations of drug targets, cell death inhibition, epigenetic factor, and impaired drug uptake in the tumor cells (Praveena et al. 2022; Cree and Charlton 2017; Su et al. 2021). Radiotherapy and immunotherapy are other common cancer treatments that are linked to a variety of negative drug reactions associated with high doses of chemotherapeutic agents and non-selective therapeutic choices (ADRs). In spite of the numerous attempts made to promote early diagnosis, using novel therapeutic approaches, and other advancements in the treatment, emergence of chemoresistance, high healthcare costs, and secondary toxicity is still a problem in cancer treatment. Additionally, other side effects of the chemotherapeutic agents, such as nausea, vomiting, headache, anorexia, musculoskeletal pain, alopecia, and neuropathy further raise the cost of the treatment. (Khan et al. 2021; Vladu et al. 2022).

Combined chemotherapy of chemical cytotoxic compounds with natural plant-based bioactive compounds has been discovered to be more successful than individual chemotherapeutic treatment in terms of improved cytotoxicity, reduced adverse effects, and prevented the emergence of multidrug resistance due to their synergistic effect (Feng et al. 2020). Multi-drug nano-systems have been utilized in numerous therapeutic models to decrease the typical adverse events of chemotherapy (Nguyen et al. 2022). According to recent studies, combination therapy by using plant-based bioactive compounds (Fig. 17.1) can enhance the therapeutic effect of the drug at lower doses, reduce the cytotoxic effect, and prove to be the most effective method to combat drug resistance. They also have the ability to re-sensitize chemotherapy-resistant cells through a variety of molecular processes, which represent a developing field of cancer research (Castañeda et al. 2022).

Phytochemicals have gained great interest as a possible anticancer drug since they simultaneously target numerous carcinogenic signaling pathways in both in vitro and in vivo systems. Several phytochemicals like phenolics, organosulfur compounds, terpenoids, carotenoids, and, alkaloids, as anti-tumor agents have gained more interest in preclinical and clinical research (Mohapatra et al. 2022). Phytochemicals were used in the management of different types of cancers (Fig. 17.2). More importantly, phytochemicals can alter cellular processes like apoptosis, cell proliferation, motility, and invasion and have great anti-carcinogenic effects. Adversely, they have limited formulation potential due to factors such as low stability, poor aqueous solubility, inadequate absorption, and rapid metabolism. This problem can be overcome by delivering these compounds along with chemotherapeutic drugs, or other phytonutrients loaded in nanotechnology-based delivery systems. The development of nanotechnology in this instance has led to innovative formulations that maximize the potential use of phytonutrients in anticancer therapy by improving their solubility and stability increasing their blood half-life and enabling site-targeted dispersion (Fig. 17.3) (Briguglio et al. 2020; Wei et al. 2019).

Combination Therapies for Different Types of Cancers

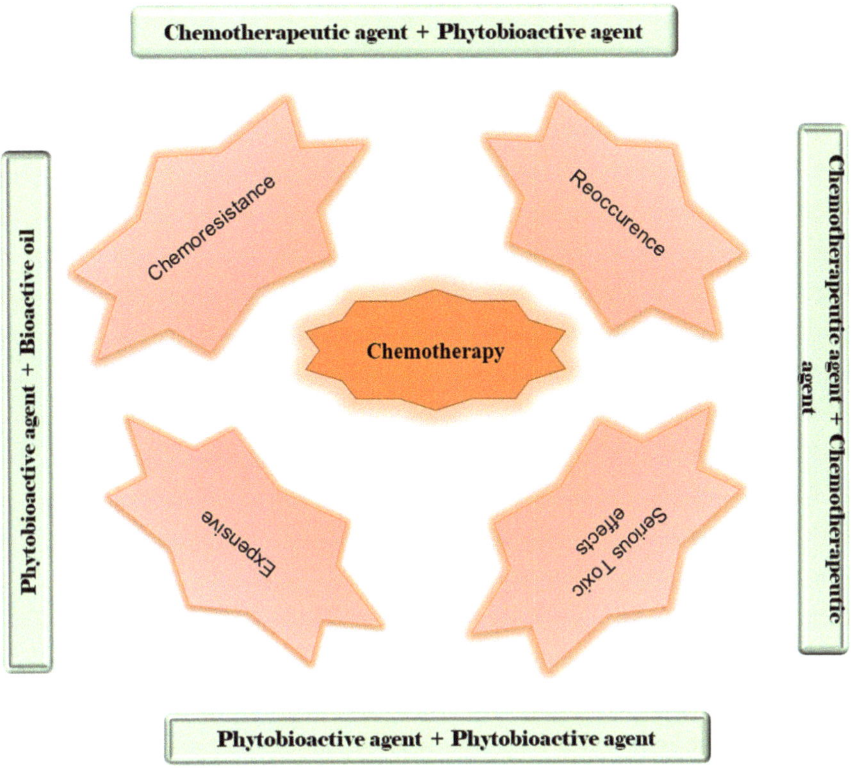

Fig. 17.1 Drawback and alternative of chemotherapy

17.1.1 Breast Cancer

Breast cancer is the most common kind of cancer in women, caused by uncontrolled cell division or tissue growth in the mammary gland and ducts, and is the second most frequent cause of cancer-related mortality rate worldwide (Noel et al. 2020; Ahmed et al. 2022; Alyami 2021; Patel et al. 2022). It is increasingly the primary cause of health problems for women globally, accounting for 22–27 percent of all cancer cases. Early diagnosis, monitoring, and management, however, help to reduce the death rate in cancer patients (Guo et al. 2022). The current approach for treating breast cancer consisting of chemotherapy, radiation, and surgical removal, is unreliable, intrusive, and expensive. Normal gene functions are also disrupted by radiation therapy and chemotherapy (Mohammadinejad et al. 2022). These treatments have a greater number of adverse effects, such as psychological disorders, structural abnormalities, toxicities, long-term adverse effects, drug resistance by tumor cells, and disruption to developing healthy tissues (Kashyap et al. 2021). Additionally, there are numerous problems encountered with conventional drug delivery during breast cancer treatment, (i) ineffective contact with metastatic sites, (ii) inadequate

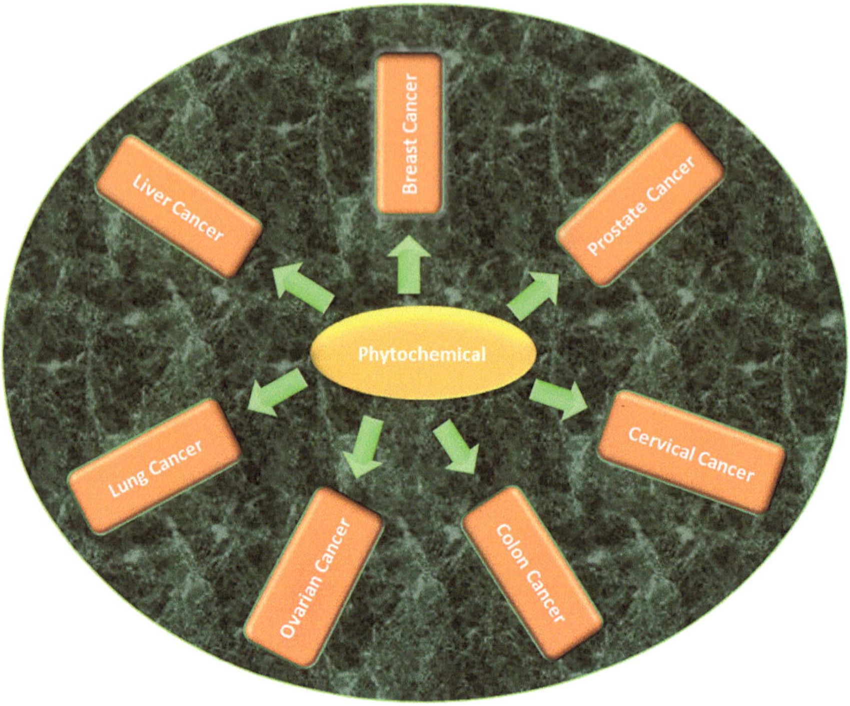

Fig. 17.2 Role of phytochemicals in various types of cancer

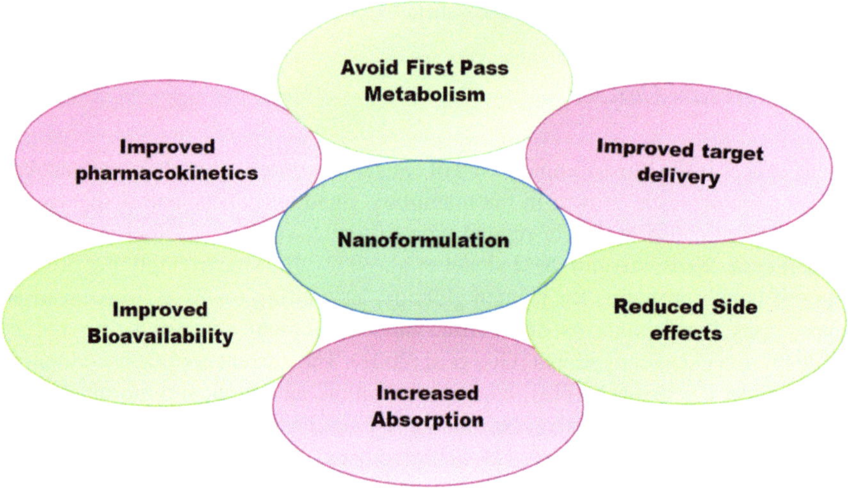

Fig. 17.3 Some major advantages associated with the nano-formulations

targeting for breast cancer, (iii) drug resistance at the level of the tumor microenvironment and cellular level, (iv) difficulty in elimination of cancer stem cells, (v) and inappropriate physical and chemical characteristics of the drug (Afzal et al. 2021). The use of natural chemotherapy sensitizers in synergistic combination treatment is emerging as an encouraging method for combating multiple drug resistance and minimizing the negative effects of chemotherapeutic medicines. Consequently, combining chemotherapeutic medications with natural bioactive substances proves to be a beneficial therapeutic approach. Co-delivering multiple treatments through the same carrier improves cancer therapy by coordinating their contact with the medications and generating a synergistic effect in tumor cells (Li et al. 2022). For instance, the previous study has demonstrated the combination of Dextran-coated magnetic nanoparticles loaded with Curcumin-Naringenin were prepared using chemical co-precipitation of the magnetite phase in an inert atmosphere were tested against the MCF-7 breast cancer cells in a rat model, and the result has shown the suppressed cell growth and induced apoptosis via the production of reactive oxygen species (ROS) (Askar et al. 2021). Similarly, the co-encapsulated PEGylated PLGA nanoparticles of artemisinin and chrysin in combination or alone were investigated against T-47D cells at different concentrations which shows that the combination of artimisinin and chrysin-loaded nanoparticles demonstrated dose-dependent cytotoxicity against T-47D cells and cancer cell development was significantly suppressed and had an anti-proliferative activity (Khoshravan Azar et al. 2022). In another study, the individual and synergistic action of polyphenolic combination of Quercetin, Curcumin, and Berberine was investigated using MDA-MB-231 cell line and MDA-MB-468 cell line as TNBC cell line models, which demonstrated the combined treatment (QUE + CUR + BBR) had a significant and synergistic impact on EMT proteins and cell motility. Additionally, clonogenic potential, innate cisplatin, the CD44+/CD24-population resistance in TNBC cells were all considerably affected by the COMQUE + CUR + BBR therapy (Kashyap et al. 2022). Certain Combination of Phytoconstituents and Chemotherapeutic agents for the treatment of Breast Cancer and their target cell lines were summarized in Table 17.1.

17.1.2 Lung Cancer

The second-most frequent common cancer in both men and women is reported at 1.59 million deaths rate annually globally. Among many symptoms, cough is the most dominant symptom and rest of the symptoms include hemoptysis, shortness of breath, systemic symptoms, such as anorexia or weight loss, etc. (Kim et al. 2022; Huang et al. 2022). The two types of lung cancer are: small-cell lung cancer (SCLC) and non-small-cell lung cancer (NSCLC) represent approximately 85–88 percent while others account for almost 13 percent to 15 percent of all lung cancer cases (Ma et al. 2019; Ishola and La'ah 2020). Although platinum-based therapies are most commonly used for general treatment but long-term survival of the patients is still poor due to the development of resistance as a result of prolonged or improper use of

Table 17.1 Different combination of phytoconstituents and chemotherapeutic agents for the treatment of breast cancer and their target cell lines

Phytochemical	Chemotherapeutic agent	Nano formulation	Target Cell/ Cell line	References
Curcumin	Paclitaxel	PTX-Sln@CUD) Nanoparticles	MCF-7	Wei et al. (2022)
		Heparin and Poloxamer P403 Hybrid Nanocarrier	MCF-7	Nguyen et al. (2022)
		Hyaluronic acid-hybrid Nanopaticles	MCF-7	Yang et al. (2017)
		Solid lipid nanoparticles	MCF-7	Baek and Cho (2017)
Curcumin	Doxorubicin	mPEG-PLGA-PGlu Nanoparticles	MCF-7	Yuan et al. (2018)
		–	MCF-7, MDA-MB-231 Cell line	Wen et al. (2019)
Resveratrol	Docetaxel	Polymeric Micelles	MCF-7 breast cancer cells	Guo et al. (2019)
Resveratrol	Exemestane	Shell-Crosslinked Zein Nanocapsules	MCF-7, 4T1 Cells	Elzoghby et al. (2017)
		Nanoparticles	MDA-MB-231	Karami et al. (2022)
Curcumin	Cisplatin	–	MDA-MB-231	Yang et al. (2021)
Resveratrol	Cisplatinum	Nanoemulsion	MCF-7	Meng et al. (2016)
Baicalein	Paclitaxel	Hybrid nanosuspension	MCF-7	Wang et al. (2019)
Butilinic acid	Paclitaxel	–	MCF-7, MDA-MB-231	Cai et al. (2018)
		–	AU-565, BT-474 Cell line	Zhang et al. (2016)
Berberine	Lapatinib	–	MCF-7	Zhao et al. (2016)
Berberine	Cisplatin	–	MCF-7	Pan et al. (2017)
Berberine	Doxorubicin	–	MCF-7	Mawalizadeh et al. (2021)

(continued)

Table 17.1 (continued)

Phytochemical	Chemotherapeutic agent	Nano formulation	Target Cell/ Cell line	References
Quercetin	5-Fluorouracil	Jenus nanometer	MCF-7/Adr	Zhou et al. (2023)
Quercetin	Doxorubicin	Phytosome	MCF-7	Minaei et al. (2016)
		–	MCF-7, MDA-MB-231	Abd-Alhaseeb et al. (2022)
Evening primrose oil	Tamoxifen	–	MCF7, T47D	Khamis et al. (2018)
Hesperidin + piperine + bee venom	Tamoxifen	Polymeric nanoparticles	–	Amirsaadat et al. (2021)
Silbinin	Metformin	PLGA-PEG Nanoparticles	T-47D	Hassani et al. (2022)
Artemisinin	Metformin	Nanoemulsion	T-47D	Sandhu et al. (2017)
Naringenin	Tamoxifen	–	MCF-7	Kaur et al. (2022)
Erucin	Paclitaxel	–	T-47D, L929	Sato et al. (2015)
Luteolin	Doxorubicin	–	MCF-7	Hassani et al. (2022)
Curcumin	Emodin	–	MDA-MB-231, MDA-MB-435 cells	Guo et al. (2022)
Curcumin	Thymoquinone	–	MDA-MB-231, MCF-7 Cells	Saddiq et al. (2022)
Curcumin	Quercetin	–	MCF-7	Srivastava and Srivastava (2019)

(continued)

Table 17.1 (continued)

Phytochemical	Chemotherapeutic agent	Nano formulation	Target Cell/ Cell line	References
Curcumin	Quercetin and berberine	–	MDA-MB-231, MDA-MB-468	Kashyap et al. (2022)
Thymoquinone	Resveratrol	–	MCF-7, T47D	Alobaedi et al. (2017)
Thymoquinone	Ferulic acid	–	MDA-MB-231	Al-Mutairi et al. (2021)
Piperine	Thymoquinone	–	EMT6/P	Talib (2017)
Berberine	Luteolin	–	MCF-7, MDA-MB-468, MDA-MB-231, T47D	Ponnusamy et al. (2020)
Hesperidin	Luteolin	–	MCF-7	Magura et al. (2022)
Curcumin	Chrysin	Nanocapsule	MDA-MB-231	Javan et al. (2019)
		Nanofibers	T44D	Rasouli et al. (2020)
Epigallocatechin	Piperine	Nanosuspension	MCF-7	Dahiya et al. (2018)
Naringenin	Quercetin	–	MCF-7	Rhman et al. (2022)
Naringenin	Curcumin	Magnetic nanoparticle	MCF-7	Askar et al. (2021)
Artemisinin	Chrysin	PEGylated PLGA nanoparticles	T-47D	Khoshravan Azar et al. (2022)

chemotherapeutic agents, which reduces the effectiveness of cancer treatment. Thus, cancer chemotherapy will be complicated due to the emergence and recurrence of the cancer as well as the cessation of the first-choice medication therapy (Ma et al. 2019; Ishola and La'ah 2020; Li et al. 2022; Chen et al. 2022; Okamoto et al. 2020). The concept of combination of natural compounds and anticancer medications as described in Table 17.2 can be proved to improve therapeutic outcomes due to the regulation of several signal pathways, natural compounds can boost immunity, have greater efficacy towards cancer cells, and decrease toxicity (Yang et al. 2021). In this regard, synergistic effect of natural substances with strong antitumor activity and minimal toxicity with conventional anticancer medications provides combination treatment of DNA-damaging chemotherapeutics with signaling inhibitors (Hashem et al. 2022). For instance, the combination of Cisplatin and Oridonin loaded layer-by-layer nanoparticles were developed and investigated on A549/DDP cell xenografts of mice and compared with single drug nanoparticles and free drug, and the results indicated that these nanoparticles could significantly increase antitumor activity and cell toxicity against A549/DDP cells in a lung cancer (Fan et al. 2021). In another example, Guo et al. examined the combined effect of Paclitaxel (PTX), Cetuximab (CET), and 5-Demethylnobiletin (DMN) loaded nanostructured lipid carrier and examined the formulation for synergistic effects, cellular uptake, and cell viability on different cells A549 cells and paclitaxel-resistant A549 cells, MRC-5 cells in mouse model and concluded that the cells' uptake of 5-demethylnobiletin loaded lipid carrier formulation was found to be 35.5% and the combined formulation has remarkable higher cellular uptake of 65.8% in A549 cells. Hence, the combination treatment exhibited the most significant in vivo tumor inhibition efficacy when compared with the single treatment (Guo et al. 2019).

17.1.3 Colon and Colorectal Cancer (CRC)

Colorectal cancer is a multifactorial illness induced by intricate mechanisms including altered signal transduction pathways, genetic mutations, oxidative stress, inflammatory responses, dysbiosis, food and lifestyle defects, and many more (Wang et al. 2020). Around 70 percent of primary CRC cases develop in the colon, making it the most deadly type of cancer worldwide in both genders in 2020 (Du et al. 2022). It could be divided into three clinical entities, i.e., right-sided colon cancer, left-sided colon cancer, and rectal cancer (Hodges et al. 2022). The patients with right-sided colon cancer had worse prognosis than the patients with left-sided colon cancer at similar stages and had a worse chance of survival (Kishiki et al. 2019). Distal or left-sided colorectal cancer includes the cancer in descending, sigmoid colon, and rectum, while right-sided colorectal cancer in ascending, caecum, and transverse colon up to the splenic flexure (Lee et al. 2015; Neamtu et al. 2022; Bracci et al. 2021). Although combinatorial treatment of chemotherapeutics with diagnosis has significantly improved the cancer survival, but its clinical applicability and therapeutic effect are still limited by the severe side effects associated

Table 17.2 Different combination of phytoconstituents and chemotherapeutic agents for lung cancer and their target cell lines

Phytochemical	Chemotherapeutic agent	Nano formulation	Target Cell/ Cell line	References
Curcumin	Paclitaxel	Liposome	A549/A549-T Cells	Feng et al. (2020)
Curcumin	Doxorubicin	Nanoparticles	A549/ADR cells, HUVEC cells	Hong et al. (2019)
Curcumin	Docetaxel	Nanostructured lipid carrier	NCI-H460 cells	Rawal et al. (2020)
Resveratrol	Docetaxel	Lipid-polymer Hybrid Nanoparticles	NSLC cells	Song et al. (2018)
Baicalein	Taxol	Self-Assembled nanoparticles	A549/PTX Cells	Wang et al. (2015)
Baicalein	Cisplatin	–	A549/CDDP Cells	Yu et al. 2017)
		–	A549 and H460 cells	Lu et al. (2018)
Galangin	Cisplatin	–	A549/DDP Cells	Yu et al. (2018)
Epigallocatechin gallate	Cisplatin	–	A549/DDP cell lines	Zhang et al. (2015)
Epigallocatechin gallate	Gefitinib	–	HCC827-Gef cells	Liu et al. (2020)
Allicin	Cisplatin	–	A549 Cells	Pandey et al. (2020)
Allicin	5-fluorouracil	–	DLD-1	Ţigu et al. 2020)
Emodin	Cisplatin	–	A549/DDP cell	Teng et al. (2021)
		–	A549 and H460 cells	Peng et al. (2021)
Silbinin	Metformin	Nanoparticles	A549 Cells	Mogheri et al. (2021)
Curcumin	Quercetin	–	A549 Cells	Srivastava and Srivastava (2019)
Thymoquinone	Quercetin	–	A549 Cells	Alam et al. (2022)

with chemotherapy, radiation therapy and their resistance (Gogoi et al. 2022). To overcome these difficulties, combinations of two or more polyphenols with other phytonutrients as summarized in Table 17.3 are being used for treatment, impacting multiple targets and improving treatment efficacy and reducing drug therapy and drug resistance (Zhao et al. 2022; Zhang et al. 2022). Among the various polyphenols, curcumin, epigallocatechin gallate, resveratrol, quercetin, apigenin, luteolin, gallic acid, anthocyanins, and capsaicin have comparatively better therapeutic effects and utilized as prophylactic treatments in CRC (Gogoi et al. 2022). For instance, the study of Jang et al. has examined the antitumor effect of luteolin with oxaliplatin combination utilizing xenograft mouse of colorectal cancer. The subcutaneous implantation of human colorectal carcinoma HCT-116 cells by intraperitoneal administration of oxaliplatin or in a combination given to tumor-bearing mice three times for three weeks. It was seen that comparing luteolin or oxaliplatin alone, or the combination of luteolin and oxaliplatin resulted in a synergistic inhibition of the progression of HCT-116 tumor cells (Jang et al. 2022). In another study, diverse mixtures of curcumin, sulforaphane, and dihydrocaffeic acid were assessed for their effects on the feasibility of colon cancer cells. A synergistic effect at a 90% cytotoxicity level was produced by a combination of sulforaphane, and dihydrocaffeic acid (1:1), and a mixture of curcumin and sulforaphane (1:4) produced synergistic action at all cytotoxic levels. Also, curcumin and sulforaphane mixture in (9:2) produced synergistic effect against Caco-2 cells. The most effective mixture was found to be sulforaphane and dihydrocaffeic acid (1:1), which was more synergistic and cytotoxic to cancer cells than normal cells (Santana-Gálvez et al. 2020). A combination of Erlotinib and Curcumin-loaded nanoparticles was found to be successful in reducing the expression of $\alpha_v\beta_3$ integrin and boost gene expression of PDK4 when evaluated in an erlotinib-resistant SW480 cancer cell line (Javadi et al. 2018).

17.1.4 Prostate Cancer

Among developed nations, new cases of prostate cancer, approximately 20% of all newly diagnosed cancers in males have been diagnosed in 2019 and will be increased up to 2022 (Qu et al. 2021). According to statistics, 9.8% of all cancer-related mortality rates in men were caused by prostate cancer (Cassell et al. 2019; Shore et al. 2022). Early diagnosis has some advantages, such as a high likelihood of cures, gentle treatment options, a slower tumor progression to metastatic and advanced stages, and an enhanced quality of life (Eickelschulte et al. 2022). Race, inherited disorders, and family history are recognized as potential risks for prostate cancer. Modifiable risk variables may affect both the probability of developing prostate cancer and the chances of dying from the condition, but there is little evidence suggesting prevention apart from early diagnosis can effectively lower the mortality caused by prostate cancer (Gandaglia et al. 2021). Aside from genetic mutations, numerous epigenetic modifications such as histone modifications (acetylation, phosphorylation, and ubiquitylation), DNA methylation, nucleosomal remodeling,

Table 17.3 Nano formulation for combination of phytoconstituents and chemotherapeutic agents for colon and colorectal cancer and their target cell lines

Phytochemical	Chemotherapeutic agent	Nano formulation	Target Cell/Cell line	References
Curcumin	Paclitaxel	Polymeric micelles	–	Gao et al. (2015)
Curcumin	Irinotecan	–	LoVo, HT-29 CRC cell lines	Huang et al. (2017)
Curcumin	5-fluorouracil	–	HCT-116 cells	Lu et al. (2020)
Resveratrol	5-fluorouracil	–	HCT-116, HCT-116 R	Buhrmann et al. (2018)
		–	HCT-116 and DLD1	Chung et al. (2018)
		–	SW-620, HT-29, HCT-116 and SW480 Cells	Moutabian et al. (2022)
Epigallocatechin	5-fluorouracil	–	HCT-116, DLD1 Cell line	La et al. (2019)
		Polymeric nanoparticle	HT-29 & CT-26	Wang et al. (2019)
Quercetin	5-fluorouracil	–	HCT-116 and Caco-2 cell lines	Terana et al. (2022)
Quercetin + luteolin	5-fluorouracil	–	HT-29 Cells	Erdoğan et al. (2022)
Piperine	Celecoxib	–		Srivastava et al. (2021)
Tangeretin	5-fluorouracil	–	HCT-116 cells	Dey et al. (2020)
Allicin	5-fluorouracil	–	SK-MES-1	Ţigu et al. (2020)
Luteolin	Oxiplatin	–	HCT-116 cells	Jang et al. (2022)
Curcumin	Quercetin	–	HCT-116 cells	Srivastava and Srivastava (2019)
Thymoquinone	Allicin	–	$CaCO_2$ cell lines	Mahdy et al. (2020)
Baicalein	Luteolin	–	LoVo/Dx cells	Palko-Labuz et al. (2017)
Curcumin	Chrysin	Polymeric nanoparticles	$CaCO_2$ cell lines	Lotfi-Attari et al. (2017)
Epigallocatechin	Piperine	Nanosuspension	HL60, SCC40 and colo205	Dahiya et al. (2018)

(continued)

Table 17.3 (continued)

Phytochemical	Chemotherapeutic agent	Nano formulation	Target Cell/Cell line	References
Curcumin	Piperine	Emulsomes	HCT-116 Cells	Bolat et al. (2020)
Curcumin	Luteolin	–	CL-188 (LS174T) and DLD-1 cells	Aromokeye and Si (2022)

and chromosomal looping have a substantial effect on the development, prognosis, diagnosis, and treatment of prostate cancer (Pandareesh et al. 2021). The use of artificial intelligence or nanotechnology, bispecific antibodies, cytokine inhibitors, immune checkpoint inhibitors, photodynamic therapy, focal cryotherapy, irreversible electroporation, and various sources of energy-based focal therapy are just a few of the detection and intervention techniques being used for clinical application (Man et al. 2022). While, Docetaxel, Abiraterone acetate, Cabazitaxel, and Enzalutamide with androgen deprivation therapy served as the mainstays of systemic management of Prostate Cancer for years, and docetaxel and cabazitaxel may appear to be more affordable than anti-hormonal drugs, but have poor tolerability and are more toxic for metastatic prostate cancer (Tsaur et al. 2022; Shayegan et al. 2022). Docetaxel resistance is considered to be strongly affected by abnormal AR and PI3K/Akt signaling activity, but quercetin showed the reversing effects of Docetaxel resistance in Docetaxel-resistant prostate cancer cells both in vitro and in vivo by affecting the development of the upregulation of P-glycoprotein, PI3K/Akt signaling pathways activation and androgen receptor. Moreover, the combination of Quercetin and Docetaxel reduced cancer development and strongly suppressed proliferation (Liskova et al. 2021). According to Tian et al., Genistein and Celecoxib loaded nano-liposomal formulation selectively induce apoptosis, demonstrating that a reduction in concentration of cellular GSH and increased reactive oxygen species (ROS) formation, inhibition of Glut-1 receptors as well as COX-2 synthesis are the main processes responsible for the reduction of prostate cancer cells as summarized in Table 17.4 (Tian et al. 2019).

17.1.5 Liver Cancer

It is one of the deadliest cancer and third-largest cancer-related mortality and has steadily increased over the past few decades and become one of the major causes of death worldwide (Cao et al. 2022; Xiao et al. 2019). It is called hepatocellular carcinoma, and resulted from chronic hepatitis B virus infection (Feng et al. 2021; Rahman et al. 2023; Hou et al. 2022). It caused new cases and 625,000 deaths globally in 2018 and became the fourth main cause of cancer-related deaths each year. According to estimates, liver cancer accounts for the predominant histological linked to human papillomavirus (HPV) infection (Nadile et al. 2022). Cervical cancer develops in the uterine cervix typically penetrates surrounding tissues and spreads to distant places via lymphatic or blood vessels (Cheng and Huang 2021). There are two primary types of cervical cancer, among which squamous cell carcinoma and adenocarcinoma, and account for the common cases seen in clinics (Zhang et al. 2019). Screening may lead to a timely diagnosis of cancer as a secondary objective, which would allow for earlier treatment and decrease the risk of mortality from cervical cancer (Bouvard et al. 2022). Exosomes are a subset of short extracellular vesicles, lipid bilayered, spherical particles with sizes ranging from 30 to 140 nm. They serve as long-range signaling effectors and mediators of tumor-cell interaction, coordinating a number of hallmark implications such as tumor development, advancement, angiogenesis, metastasis, treatment resistance, and tumor immunity in cancer (Bhat et al. 2018). The first-line therapeutic chemotherapeutic agent Cisplatin seems to develop chemo resistance, limiting the therapeutic efficacy, even though chemotherapy has been thought of as the standard treatment with advanced or recurring cervical cancer (Chen et al. 2022; Bhattacharjee et al. 2022). Despite the fact that patients initially responded favorably to cisplatin, it is common to observe growing resistance over the course of treatment, which depresses the efficiency of successive secondary platinum-based chemotherapeutics. According to many research studies, treatment of cancer with cisplatin alone may not be as effective as combined with additional treatments (Burmeister et al. 2022). Drug-based chemotherapy is now severely constrained by chemoresistance to cisplatin, doxorubicin, and 5-fluorouracil. A promising option to overcome this resistance has been found and involves the combination of cisplatin with some other phytochemicals. Due to their anti-proliferative properties, antioxidant, and anti-inflammatory, phytochemicals have the potential to be effective chemopreventive and chemotherapeutic treatments for a variety of malignancies, including cervical cancer as summarized in Table 17.6. (Shoaib et al. 2022). In the previous study, Sesamol, a naturally occurring phenolic chemical, was shown to have chemosensitizing effects in HeLa cell lines that received a 10 μM dose of Sesamol and were subjected to 7.5 nM paclitaxel in vitro. Additionally, Sesamol administration before paclitaxel elevated intracellular reactive oxygen species levels and improved apoptosis via mitochondrial membrane potential modifications. In addition, compared to paclitaxel-only treatment, cervical cancer cells treated with sesamol plus paclitaxel had a higher percentage of tail DNA (Xiong et al. 2022). Similarly, Resveratrol significantly increased Cisplatinum-based cell apoptosis and cell cycle arrest while inhibiting SiHa cell line proliferation. The improved anti-cancer benefits of the cis-platinum and resveratrol combination were initiated by the SIRT3 relative activation of anti-oxidative pathway. The anti-cancer efficacy of the combined Cisplatinum and Resveratrol was decreased by SIRT3 si-RNA interference (Jiang et al. 2022).

Table 17.6 Nano formulations for combination of phytoconstituents and chemotherapeutic agents for cervical cancer treatment and their target cell lines

Phytochemical	Chemotherapeutic agent	Nano formulation	Cell/cell lines	References
Curcumin	Cisplatin	Lipid-polymer Hybrid Nanoparticles	HeLa Cells	Li et al. (2017)
Resveratrol	Cisplatinum	–	SiHa Cell line	Jiang et al. (2022)
Epigallocatechin gallate	Cisplatin	–	HeLa cell lines	Kilic et al. (2015)
Caffeic acid	Cispatin	–	HeLa, and CaSki, SiHa and C33A (HPV-negative)	Koraneekit et al. (2018)
Eugenol	Cisplatin	–	HeLa cell lines	Fathy et al. (2019)
Emodin	Vinblastine	–	HeLa cell lines	Trybus et al. (2022)
Luteolin	Quercetin	–	(A431-P) cells	Lin et al. (2017)
Epigallocathechin gallate	Piperine	Nanosuspension	HeLa cell lines	Dahiya et al. (2018)
Epigallocatechin gallate	Rutin	Nanorods	HeLa cell lines	Saha et al. (2022)
Sulforaphane	Eugenol	–	HeLa cell lines	Hussain et al. (2012)

17.1.7 Ovarian Cancer

Ovarian cancer is the fifth most prevalent cause of death in women from malignancies of the female genital tract in the United States. According to American Cancer Society, approximately 12,810 women die from ovarian cancer and 19,880 new cases were found in 2022 (Psilopatis et al. 2023; Opp et al. 2023). The key factors contributing to the high mortality risk of ovarian cancer include late diagnosis, surgical recurrence, and treatment resistance. The new tumor biomarkers and drug approaches for ovarian cancer may be made easier by identifying molecular mechanisms causing the development and progression of ovarian cancer (Qin et al. 2023). Iatrogenic injury, inflammation, genetics, autoimmune, psychological, and environmental, variables are typical pathogenic factors involved in ovarian cancer. Reduced oestrogen secretion, ovulation problems, and infertility are the symptoms of premature ovarian insufficiency (Meng et al. 2023). About 70 80% of cases are of ovarian high-grade serious carcinoma among all types of ovarian cancer (Liberto et al. 2022). Few blood biomarkers are identified for ovarian cancer that is highly particular and adequately sensitive to identify early-stage diagnosis and have a great effect on patient mortality rates (Kim et al. 2022). Platinum resistance is caused by increased cisplatin-induced autophagy, upregulated de-ubiquitination of proteins targeted for proteasomal destruction, and reliance on mitochondrial oxidative phosphorylation. The hallmarks of cancer, metabolic reprogramming and angiogenesis provide the tumor with nutrition and oxygen, improve energy production, and promote cell survival, are also considered to have a role in platinum treatment resistance (Ortiz et al. 2022). Typically, platinum and taxane-based drugs are used to manage ovarian cancer; however, the choice of chemotherapeutic medications is dependent on the onset of the disease. Over 70% of patients decline subsequent treatment and sooner or later develop chemotherapeutic resistance as a result of serious side effects and risk of high-dose chemotherapy (Alatise et al. 2022). The phytochemicals and chemotherapeutic drugs are developed in combination to overcome cisplatin-resistance ovarian cancer as described in Table 17.7. Phytochemicals as chemosensitizers in combination are used as novel treatment strategies for patients suffering from ovarian cancer by overcome chemoresistance and increasing therapeutic efficacy of drug (Song et al. 2022). For instance, Jafarzadeh et al. has investigated the effect of the combination of cisplatin and fisetin in cisplatin-resistant cancer cells in the ovary (A2780 cell line) using a variety of techniques and concluded that A2780 cancer cells were destroyed by the combination in a concentration-dependent manner. The study revealed that the combination of these two drugs significantly reduces cell capability at concentrations below the IC50 of cisplatin and fisetin (Jafarzadeh et al. 2021). In another study, multi-purpose smart alginate/perfluorohexane nanodroplets of Curcumin and Doxorubicin were developed for combinatorial delivery by using the nanoemulsion process. The synergistic effect was observed on the andriamycin-resistant A2780 ovarian cancer cell, and the result showed enhanced cytotoxicity and efficient tumor regression with smart alginate/perfluorohexane nanodroplets of curcumin and doxorubicin as compared to doxorubicin nanodroplets alone (Baghbani and Moztarzadeh 2017).

Table 17.7 Nano formulation for combination of phytoconstituents and chemotherapeutic agents for the treatment of colon and colorectal cancer and their target cell lines

Phytochemical	Chemotherapeutic agent	Nano formulation	Target cell/cell line	References
Curcumin	Cisplatin	Nanoparticles	–	Sandhiutami et al. (2021)
Baicalein	Taxol	–	A2780 cells	Pan et al. (2016)
Berberine	Cisplatin	-	A2780/DDP cells	Chen et al. (2015)
Cardamonin	Cisplatin	–	SKOV3 and A2780 cell lines	Niu et al. (2018)
Curcumin	Doxorubicin	Alginate droplets	A2780 cells	Baghbani and Moztarzadeh (2017)

17.2 Conclusion

As the prevalence and mortality of cancer are rapidly rising, it is a matter of concern as conventional therapies have become less effective due to drug resistance and high doses. So, it has become important to evaluate the novel therapeutics by using the bioactive compounds with anti-cancer drugs or the combination of two or more bioactive compounds, which would be the most effective method for the treatment of various cancers due to their low toxicity, low doses, and cost effectiveness. As they have great anti-proliferative, anti-angiogenic, anti-inflammatory, and antioxidant properties, they showed excellent and best results in pre-clinical and clinical research when tested for various cancers, including breast cancer, lung cancer, colorectal cancer, liver cancer, prostate cancer, cervical cancer, and ovarian cancer. The combination regimen of phytochemicals with anti-cancer drugs and two or more phytochemicals together has piqued the interest of researchers because it reduces toxicity and side effects, lowers the dose, and overcomes resistance in various tissues and cells and proved to be very beneficial in the treatment of cancer.

References

Abd-Alhaseeb MM, Massoud SM, Elsayed F, Omran GA, Salahuddin A (2022) Evening primrose oil enhances tamoxifen's anticancer activity against breast cancer cells by inducing apoptosis, inhibiting angiogenesis, and arresting the cell cycle. Molecules 27(8):2391

Afzal M, Alharbi KS, Alruwaili NK, Al-Abassi FA, Al-Malki AAL, Kazmi I, Anwar F (2021) Nanomedicine in treatment of breast cancer—a challenge to conventional therapy. In: Seminars in cancer biology, vol 69. Academic Press, pp 279–292

Ahmed R, Samanta S, Banerjee J, Kar SS, Dash SK (2022) Modulatory role of miRNAs in thyroid and breast cancer progression and insights into their therapeutic manipulation. In: Current research in pharmacology and drug discovery, p 100131

Alam S, Mohammad T, Padder RA, Hassan MI, Husain M (2022) Thymoquinone and quercetin induce enhanced apoptosis in non-small cell lung cancer in combination through the Bax/Bcl2 cascade. J Cell Biochem 123(2):259–274

Alatise KL, Gardner S, Alexander-Bryant A (2022) Mechanisms of drug resistance in ovarian cancer and associated gene targets. Cancers 14(24):6246

Alemi A, Zavar Reza J, Haghiralsadat F, Zarei Jaliani H, Haghi Karamallah M, Hosseini SA, Haghi Karamallah S (2018) Paclitaxel and curcumin coadministration in novel cationic PEGylated niosomal formulations exhibit enhanced synergistic antitumor efficacy. J Nanobiotechnol 16(1):1–20

Al-Mutairi A, Rahman A, Rao MS (2021) Low doses of thymoquinone and ferulic acid in combination effectively inhibit proliferation of cultured MDA-MB 231 breast adenocarcinoma cells. Nutr Cancer 73(2):282–289

Alobaedi OH, Talib WH, Basheti IA (2017) Antitumor effect of thymoquinone combined with resveratrol on mice transplanted with breast cancer. Asian Pac J Trop Med 10(4):400–408

Alyami NM (2021) MicroRNAs role in breast cancer: theranostic application in Saudi Arabia. In: Frontiers in oncology, p 4313

Amirsaadat S, Jafari-Gharabaghlou D, Alijani S, Mousazadeh H, Dadashpour M, Zarghami N (2021) Metformin and Silibinin co-loaded PLGA-PEG nanoparticles for effective combination therapy against human breast cancer cells. J Drug Deliv Sci Technol 61:102107

Aromokeye R, Si H (2022) Combined curcumin and luteolin synergistically inhibit colon cancer associated with notch1 and TGF-β signaling pathways in cultured cells and xenograft mice. Cancers 14(12):3001

Askar MA, El Shawi OE, Mansour NA, Hanafy AM (2021) Breast cancer suppression by curcumin-naringenin-magnetic-nano-particles: in vitro and in vivo studies. Tumor Biol 43(1):225–247

Baek JS, Cho CW (2017) A multifunctional lipid nanoparticle for co-delivery of paclitaxel and curcumin for targeted delivery and enhanced cytotoxicity in multidrug resistant breast cancer cells. Oncotarget 8(18):30369

Baghbani F, Moztarzadeh F (2017) Bypassing multidrug resistant ovarian cancer using ultrasound responsive doxorubicin/curcumin co-deliver alginate nanodroplets. Colloids Surf B 153:132–140

Bhat A, Sharma A, Bharti AC (2018) Upstream Hedgehog signaling components are exported in exosomes of cervical cancer cell lines. Nanomedicine 13(17):2127–2138

Bhattacharjee R, Dey T, Kumar L, Kar S, Sarkar R, Ghorai M, Dey A et al (2022) Cellular landscaping of cisplatin resistance in cervical cancer. Biomed Pharmacother 153:113345

Bolat ZB, Islek Z, Demir BN, Yilmaz EN, Sahin F, Ucisik MH (2020) Curcumin-and piperine-loaded emulsomes as combinational treatment approach enhance the anticancer activity of curcumin on HCT-116 colorectal cancer model. Frontiers Bioeng Biotechnol 8:50

Bouvard V, Wentzensen N, Mackie A, Berkhof J, Brotherton J, Giorgi-Rossi P, Lauby-Secretan B et al (2022) The IARC perspective on cervical cancer screening. Obstet Gynecol Surv 77(3):154–156

Bracci L, Fabbri A, Del Cornò M, Conti L (2021) Dietary polyphenols: promising adjuvants for colorectal cancer therapies. Cancers 13(18):4499

Briguglio G, Costa C, Pollicino M, Giambo F, Catania S, Fenga C (2020) Polyphenols in cancer prevention: New insights. Int J Functional Nutr 1(2):1–1

Buhrmann C, Yazdi M, Popper B, Shayan P, Goel A, Aggarwal BB, Shakibaei M (2018) Resveratrol chemosensitizes TNF-β-induced survival of 5-FU-treated colorectal cancer cells. Nutrients 10(7):888

Burmeister CA, Khan SF, Schäfer G, Mbatani N, Adams T, Moodley J, Prince S (2022) Cervical cancer therapies: current challenges and future perspectives. In: Tumor virus research, p 200238

Cai Y, Zheng Y, Gu J, Wang S, Wang N, Yang B, Wang Z et al (2018) Betulinic acid chemosensitizes breast cancer by triggering ER stress-mediated apoptosis by directly targeting GRP78. Cell Death Dis 9(6):1–16

Cao LL, Peng M, Xie X, Chen GQ, Huang SY, Wang JY, Dietrich CF et al (2022) Artificial intelligence in liver ultrasound. World J Gastroenterol 28(27):3398

Cassell A, Yunusa B, Jalloh M, Ndoye M, Mbodji MM, Diallo A, Gueye SM et al (2019) Management of advanced and metastatic prostate cancer: a need for a Sub-Saharan guideline. J Oncol

Castañeda AM, Meléndez CM, Uribe D, Pedroza-Díaz J (2022) Synergistic effects of natural compounds and conventional chemotherapeutic agents: recent insights for the development of cancer treatment strategies. Heliyon e09519

Chen Q, Qin R, Fang Y, Li H (2015) Berberine sensitizes human ovarian cancer cells to cisplatin through miR-93/PTEN/Akt signaling pathway. Cell Physiol Biochem 36(3):956–965

Chen Y, Deng Y, Zhu C, Xiang C (2020) Anti prostate cancer therapy: aptamer-functionalized, curcumin and cabazitaxel co-delivered, tumor targeted lipid-polymer hybrid nanoparticles. Biomed Pharmacother 127:110181

Chen TY, Zhou J, Li PC, Tang CH, Xu K, Li T, Ren T (2022) SOX2 knockdown with siRNA reverses cisplatin resistance in NSCLC by regulating APE1 signaling. Med Oncol 39(3):1–10

Chen L, Lan J, Li Z, Zeng R, Wang Y, Zhen L, Zhang T et al (2022) A novel diosgenin-based liposome delivery system combined with doxorubicin for liver cancer therapy. Pharmaceutics 14(8):1685

Chen YH, Wu JX, Yang SF, Yang CK, Chen TH, Hsiao YH (2022) Anticancer effects and molecular mechanisms of apigenin in cervical cancer cells. Cancers 14(7):1824

Cheng T, Huang S (2021) Roles of non-coding RNAs in cervical cancer metastasis. Front Oncol 11:646192

Chung SS, Dutta P, Austin D, Wang P, Awad A, Vadgama JV (2018) Combination of resveratrol and 5-flurouracil enhanced anti-telomerase activity and apoptosis by inhibiting STAT3 and Akt signaling pathways in human colorectal cancer cells. Oncotarget 9(68):32943

Cree IA, Charlton P (2017) Molecular chess? Hallmarks of anti-cancer drug resistance. BMC Cancer 17:10

Dahiya S, Rani R, Dhingra D, Kumar S, Dilbaghi N (2018) Conjugation of epigallocatechin gallate and piperine into a zein nanocarrier: implication on antioxidant and anticancer potential. Adv Nat Sci Nanosci Nanotechnol 9(3):035011

de Oliveira Júnior RG, Adrielly AFC, da Silva Almeida JRG, Grougnet R, Thiery V, Picot L (2018) Sensitization of tumor cells to chemotherapy by natural products: a systematic review of preclinical data and molecular mechanisms. Fitoterapia 129:383–400

Dey DK, Chang SN, Vadlamudi Y, Park JG, Kang SC (2020) Synergistic therapy with tangeretin and 5-fluorouracil accelerates the ROS/JNK mediated apoptotic pathway in human colorectal cancer cell. Food Chem Toxicol 143:111529

Du L, Cheng Q, Zheng H, Liu J, Liu L, Chen Q (2022) Targeting stemness of cancer stem cells to fight colorectal cancers. In: Seminars in cancer biology, vol 82. Academic Press, pp 150–161

Eickelschulte S, Riediger AL, Angeles AK, Janke F, Duensing S, Sültmann H, Görtz M (2022) Biomarkers for the detection and risk stratification of aggressive prostate cancer. Cancers 14(24):6094

Elzoghby AO, El-Lakany SA, Helmy MW, Abu-Serie MM, Elgindy NA (2017) Shell-crosslinked zein nanocapsules for oral codelivery of exemestane and resveratrol in breast cancer therapy. Nanomedicine 12(24):2785–2805

Erdoğan MK, Ağca CA, Aşkın H (2022) Quercetin and luteolin improve the anticancer effects of 5-fluorouracil in human colorectal adenocarcinoma in vitro model: a mechanistic insight. Nutr Cancer 74(2):660–676

Fan X, Wang T, Ji Z, Li Q, Shen H, Wang J (2021) Synergistic combination therapy of lung cancer using lipid-layered cisplatin and oridonin co-encapsulated nanoparticles. Biomed Pharmacother 141:111830

Fathy M, Fawzy MA, Hintzsche H, Nikaido T, Dandekar T, Othman EM (2019) Eugenol exerts apoptotic effect and modulates the sensitivity of HeLa cells to cisplatin and radiation. Molecules 24(21):3979

Feng X, Pi C, Fu S, Yang H, Zheng X, Hou Y, Wang Y, Zhang X, Zhao L, Wei Y (2020) Combination of curcumin and paclitaxel liposomes exhibits enhanced cytotoxicity towards A549/A549-T cells and unaltered pharmacokinetics. J Biomed Nanotechnol 16(8):1304–1313. https://doi.org/10.1166/jbn.2020.2969

Feng B, Ma XH, Wang S, Cai W, Liu XB, Zhao XM (2021) Application of artificial intelligence in preoperative imaging of hepatocellular carcinoma: current status and future perspectives. World J Gastroenterol 27(32):5341

Gandaglia G, Leni R, Bray F, Fleshner N, Freedland SJ, Kibel A, La Vecchia C (2021) Epidemiology and prevention of prostate cancer. In: European urology oncology

Gao X, Wang B, Wu Q, Wei X, Zheng F, Men K, Gong C et al (2015) Combined delivery and anti-cancer activity of paclitaxel and curcumin using polymeric micelles. J Biomed Nanotechnol 11(4):578–589

Gogoi P, Kaur G, Singh NK (2022) Nanotechnology for colorectal cancer detection and treatment. World J Gastroenterol 28(46):6497

Gordan JD, Kennedy EB, Abou-Alfa GK, Beg MS, Brower ST, Gade TP, Rose MG et al (2020) Systemic therapy for advanced hepatocellular carcinoma: ASCO guideline. J Clin Oncol 38(36):4317–4345

Guo S, Zhang Y, Wu Z, Zhang L, He D, Li X, Wang Z (2019) Synergistic combination therapy of lung cancer: cetuximab functionalized nanostructured lipid carriers for the co-delivery of paclitaxel and 5-demethylnobiletin. Biomed Pharmacother 118:109225

Guo X, Zhao Z, Chen D, Qiao M, Wan F, Cun D, Yang M (2019) Co-delivery of resveratrol and docetaxel via polymeric micelles to improve the treatment of drug-resistant tumors. Asian J Pharm Sci 14(1):78–85

Guo Z, Xie J, Wan Y, Zhang M, Qiao L, Yu J, Yao Y et al (2022) A review of the current state of the computer-aided diagnosis (CAD) systems for breast cancer diagnosis. Open Life Sci 17(1):1600–1611

Guo J, Li W, Shi H, Xie X, Li L, Tang H, Xie X et al (2022) Correction to: synergistic effects of curcumin with emodin against the proliferation and invasion of breast cancer cells through upregulation of miR-34a. Mol Cell Biochem 477(3):979–980

Hashem S, Ali TA, Akhtar S, Nisar S, Sageena G, Ali S, Bhat AA et al (2022) Targeting cancer signaling pathways by natural products: exploring promising anti-cancer agents. Biomed Pharmacother 150:113054

Hassani N, Jafari-Gharabaghlou D, Dadashpour M, Zarghami N (2022) The effect of dual bioactive compounds artemisinin and metformin co-loaded in PLGA-PEG nano-particles on breast cancer cell lines: potential apoptotic and anti-proliferative action. In: Applied biochemistry and biotechnology, pp 1–16

Hodges N, Mackenzie H, D'Souza N, Brown G, Miskovic D (2022) Survival outcomes for right-versus left-sided colon cancer and rectal cancer in England: a propensity-score matched population-based cohort study. Eur J Surg Oncol 48(4):841–849

Hong Y, Che S, Hui B, Yang Y, Wang X, Zhang X, Ma H et al (2019) Lung cancer therapy using doxorubicin and curcumin combination: targeted prodrug based, pH sensitive nanomedicine. Biomed Pharmacother 112:108614

Hou Z, Liu J, Jin Z, Qiu G, Xie Q, Mi S, Huang J (2022) Use of chemotherapy to treat hepatocellular carcinoma. BioScience Trends

Huang YF, Zhu DJ, Chen XW, Chen QK, Luo ZT, Liu CC, Liao NZ et al (2017) Curcumin enhances the effects of irinotecan on colorectal cancer cells through the generation of reactive oxygen species and activation of the endoplasmic reticulum stress pathway. Oncotarget 8(25):40264

Huang Y, Wang X, Hu R, Pan G, Lin X (2022) SOX2 regulates paclitaxel resistance of A549 non-small cell lung cancer cells via promoting transcription of ClC-3. Oncol Rep 48(4):1–12

Hussain A, Priyani A, Sadrieh L, Brahmbhatt K, Ahmed M, Sharma C (2012) Concurrent sulforaphane and eugenol induces differential effects on human cervical cancer cells. Integr Cancer Ther 11(2):154–165

Imran M, Srivani G, Raju GSR (2022) Phytonanoformulations for hepatocellular carcinoma therapy. In: Theranostics and precision medicine for the management of hepatocellular carcinoma, vol 3. Academic Press, pp. 197–213

Ishola AA, La'ah AS, Le HD, Nguyen VQ, Yang YP, Chou SJ, Wang ML et al (2020) Non-coding RNA and lung cancer progression. J Chin Med Assoc 83(1):8–14

Jafarzadeh S, Baharara J, Tehranipour M (2021) Apoptosis induction with combined use of cisplatin and fisetin in cisplatin-resistant ovarian cancer cells (A2780). Avicenna J Med Biotechnol 13(4):176

Jang CH, Moon N, Lee J, Kwon MJ, Oh J, Kim JS (2022) Luteolin synergistically enhances antitumor activity of oxaliplatin in colorectal carcinoma via AMPK inhibition. Antioxidants 11(4):626

Javadi S, Rostamizadeh K, Hejazi J, Parsa M, Fathi M (2018) Curcumin mediated down-regulation of αVβ3 integrin and up-regulation of pyruvate dehydrogenase kinase 4 (PDK4) in Erlotinib resistant SW480 colon cancer cells. Phytother Res 32(2):355–364

Javan N, Khadem Ansari MH, Dadashpour M, Khojastehfard M, Bastami M, Rahmati-Yamchi M, Zarghami N (2019) Synergistic antiproliferative effects of co-nanoencapsulated curcumin and chrysin on mda-mb-231 breast cancer cells through upregulating mir-132 and mir-502c. Nutr Cancer 71(7):1201–1213

Jiang B, Tian Q, Shu C, Zhao J, Xue M, Zhu S (2022) Resveratrol enhances the anti-cancer effects of cis-platinum on human cervical cancer cell lines by activating the SIRT3 relative anti-oxidative pathway. Front Pharmacol 2273

Karami P, Othman G, Housein Z, Salihi A, Hosseinpour Feizi MA, Azeez HJ, Babaei E (2022) Nanoformulation of polyphenol curcumin enhances cisplatin-induced apoptosis in drug-resistant MDA-MB-231 breast cancer cells. Molecules 27(9):2917

Kashyap A, Umar SM, Mendiratta M, Prasad CP (2022) In vitro anticancer efficacy of a polyphenolic combination of Quercetin, Curcumin, and Berberine in triple negative breast cancer (TNBC) cells. Phytomedicine Plus 2(2):100265

Kashyap D, Tuli HS, Yerer MB, Sharma A, Sak K, Srivastava S, Bishayee A et al (2021) Natural product-based nanoformulations for cancer therapy: opportunities and challenges. In: Seminars in cancer biology, vol 69. Academic Press, pp 5–23

Kaur H, Kaur K, Singh A, Bedi N, Singh B, Alturki MS, Arora S et al (2022) Frankincense oil-loaded nanoemulsion formulation of paclitaxel and erucin: a synergistic combination for ameliorating drug resistance in breast cancer: in vitro and in vivo study. In: Frontiers in pharmacology, p 13

Khamis AA, Ali EM, Abd El-Moneim MA, Abd-Alhaseeb MM, El-Magd MA, Salim EI (2018) Hesperidin, piperine and bee venom synergistically potentiate the anticancer effect of tamoxifen against breast cancer cells. Biomed Pharmacother 105:1335–1343

Khan H, Ullah H, Martorell M, Valdes SE, Belwal T, Tejada S, Kamal MA et al (2021) Flavonoids nanoparticles in cancer: treatment, prevention and clinical prospects. In: Seminars in cancer biology, vol 69. Academic Press, pp 200–211

Khatoon E, Banik K, Harsha C, Sailo BL, Thakur KK, Khwairakpam AD, Kunnumakkara AB (2020) Phytochemicals in cancer cell chemosensitization: current knowledge and future perspectives. In: Seminars in cancer biology. Academic Press

Khoshravan Azar L, Dadashpour M, Hashemi M, Zarghami N (2022) Design and development of nanostructured co delivery of artemisinin and Chrysin for Targeting hTERT gene expression in breast cancer cell line: possible clinical application in cancer treatment. Asian Pac J Cancer Prev 23(3):919–927

Kilic U, Sahin K, Tuzcu M, Basak N, Orhan C, Elibol-Can B, Kucuk O et al (2015) Enhancement of cisplatin sensitivity in human cervical cancer: epigallocatechin-3-gallate. Front Nutr 1:28

Kim J, Lee H, Huang BW (2022) Lung cancer: diagnosis, treatment principles, and screening. Am Fam Physician 105(5):487–494

Kim M, Chen C, Wang P, Mulvey JJ, Yang Y, Wun C, Heller DA et al (2022) Detection of ovarian cancer via the spectral fingerprinting of quantum-defect-modified carbon nanotubes in serum by machine learning. Nature Biomed Eng 6(3):267–275

Kishiki T, Kuchta K, Matsuoka H, Kojima K, Asou N, Beniya A, Masaki T et al (2019) The impact of tumor location on the biological and oncological differences of colon cancer: multi-institutional propensity score-matched study. Am J Surg 217(1):46–52

Koraneekit A, Limpaiboon T, Sangka A, Boonsiri P, Daduang S, Daduang J (2018) Synergistic effects of cisplatin-caffeic acid induces apoptosis in human cervical cancer cells via the mitochondrial pathways. Oncol Lett 15(5):7397–7402

La X, Zhang L, Li Z, Li H, Yang Y (2019) (−)-Epigallocatechin Gallate (EGCG) enhances the sensitivity of colorectal cancer cells to 5-FU by inhibiting GRP78/NF-κB/miR-155-5p/MDR1 pathway. J Agric Food Chem 67(9):2510–2518

Lee GH, Malietzis G, Askari A, Bernardo D, Al-Hassi HO, Clark SK (2015) Is right-sided colon cancer different to left-sided colorectal cancer?—a systematic review. Eur J Surg Oncol (EJSO) 41(3):300–308

Li C, Ge X, Wang L (2017) Construction and comparison of different nanocarriers for co-delivery of cisplatin and curcumin: a synergistic combination nanotherapy for cervical cancer. Biomed Pharmacother 86:628–636

Li J, Zhong X, Zhao Y, Shen J, Pilapong C, Xiao Z (2022) Polyphenols as lung cancer chemopreventive agents by targeting microRNAs. Molecules 27(18):5903

Li B, Shao H, Gao L, Li H, Sheng H, Zhu L (2022) Nano-drug co-delivery system of natural active ingredients and chemotherapy drugs for cancer treatment: a review. Drug Deliv 29(1):2130–2161

Liberto JM, Chen SY, Shih IM, Wang TH, Wang TL, Pisanic TR (2022) Current and emerging methods for ovarian cancer screening and diagnostics: a comprehensive review. Cancers 14(12):2885

Lin TH, Hsu WH, Tsai PH, Huang YT, Lin CW, Chen KC, Cheng CH et al (2017) Dietary flavonoids, luteolin and quercetin, inhibit invasion of cervical cancer by reduction of UBE2S through epithelial–mesenchymal transition signaling. Food Funct 8(4):1558–1568

Liskova A, Samec M, Koklesova L, Brockmueller A, Zhai K, Abdellatif B, Kubatka P et al (2021) Flavonoids as an effective sensitizer for anti-cancer therapy: insights into multi-faceted mechanisms and applicability towards individualized patient profiles. EPMA J 12(2):155–176

Liu J, Zhong T, Yi P, Fan C, Zhang Z, Liang G, Fan Y et al (2020) A new epigallocatechin gallate derivative isolated from Anhua dark tea sensitizes the chemosensitivity of gefitinib via the suppression of PI3K/mTOR and epithelial-mesenchymal transition. Fitoterapia 143:104590

Lotfi-Attari J, Pilehvar-Soltanahmadi Y, Dadashpour M, Alipour S, Farajzadeh R, Javidfar S, Zarghami N (2017) Co-delivery of curcumin and chrysin by polymeric nanoparticles inhibit synergistically growth and hTERT gene expression in human colorectal cancer cells. Nutr Cancer 69(8):1290–1299

Lu C, Wang H, Chen S, Yang R, Li H, Zhang G (2018) Baicalein inhibits cell growth and increases cisplatin sensitivity of A549 and H460 cells via miR-424-3p and targeting PTEN/PI3K/Akt pathway. J Cell Mol Med 22(4):2478–2487

Lu Y, Zhang R, Zhang X, Zhang B, Yao Q (2020) Curcumin may reverse 5-fluorouracil resistance on colonic cancer cells by regulating TET1-NKD-Wnt signal pathway to inhibit the EMT progress. Biomed Pharmacother 129:110381

Ma Y, Yuwen D, Chen J, Zheng B, Gao J, Fan M, Shen Y et al (2019) Exosomal transfer of cisplatin-induced miR-425-3p confers cisplatin resistance in NSCLC through activating autophagy. Int J Nanomed 14:8121

Magura J, Moodley R, Mackraj I (2022) The effect of hesperidin and luteolin isolated from *Eriocephalus africanus* on apoptosis, cell cycle and miRNA expression in MCF-7. J Biomol Struct Dyn 40(4):1791–1800

Mahdy EM, Abdu SM, El Baseer MA, Mohamed WA (2020) Effect of thymoquinone and allicin on some antioxidant parameters in cancer prostate (PC3) and colon cancer ($CaCO_2$) cell lines. Sci J Al-Azhar Med Faculty Girls 4(2):85

Man YG, Mannion C, Jewett A, Hsiao YH, Liu A, Semczuk A, Cheng L et al (2022) The most effective but largely ignored target for prostate cancer early detection and intervention. J Cancer 13(13):3463–3475

Mandlik DS, Mandlik SK (2021) Herbal and natural dietary products: upcoming therapeutic approach for prevention and treatment of hepatocellular carcinoma. Nutr Cancer 73(11–12):2130–2154

Mawalizadeh F, Mohammadzadeh G, Khedri A, Rashidi M (2021) Quercetin potentiates the chemosensitivity of MCF-7 breast cancer cells to 5-fluorouracil. Mol Biol Rep 48(12):7733–7742

Meng L, Xia X, Yang Y, Ye J, Dong W, Ma P, Liu Y et al (2016) Co-encapsulation of paclitaxel and baicalein in nanoemulsions to overcome multidrug resistance via oxidative stress augmentation and P-glycoprotein inhibition. Int J Pharm 513(1–2):8–16

Meng X, Peng L, Wei X, Li S (2023) FOXO3 is a potential biomarker and therapeutic target for premature ovarian insufficiency. Mol Med Rep 27(2):1–8

Minaei A, Sabzichi M, Ramezani F, Hamishehkar H, Samadi N (2016) Co-delivery with nano-quercetin enhances doxorubicin-mediated cytotoxicity against MCF-7 cells. Mol Biol Rep 43(2):99–105

Mogheri F, Jokar E, Afshin R, Akbari AA, Dadashpour M, Firouzi-amandi A, Zarghami N et al (2021) Co-delivery of metformin and silibinin in dual-drug loaded nanoparticles synergistically improves chemotherapy in human non-small cell lung cancer A549 cells. J Drug Deliv Sci Technol 66:102752

Mohammadinejad S, Jafari-Gharabaghlou D, Zarghami N (2022) Development of PEGylated PLGA nanoparticles co-loaded with bioactive compounds: potential anticancer effect on breast cancer cell lines. Asian Pac J Cancer Prev 23(12):4063–4072. https://doi.org/10.31557/APJCP.2022.23.12.4063
Mohapatra P, Singh P, Singh D, Sahoo S, Sahoo SK (2022) Phytochemical based nanomedicine: a panacea for cancer treatment, present status and future prospective. OpenNano 100055
More MP, Pardeshi SR, Pardeshi CV, Sonawane GA, Shinde MN, Deshmukh PK, Kulkarni AD et al (2021) Recent advances in phytochemical-based Nano-formulation for drug-resistant cancer. Med Drug Discovery 10:100082
Moutabian H, Majdaeen M, Ghahramani-Asl R, Yadollahi M, Gharepapagh E, Ataei G, Farhood B et al (2022) A systematic review of the therapeutic effects of resveratrol in combination with 5-fluorouracil during colorectal cancer treatment: with a special focus on the oxidant, apoptotic, and anti-inflammatory activities. Cancer Cell Int 22(1):1–14
Muthusami S, Sabanayagam R, Periyasamy L, Muruganantham B, Park WY (2022) A review on the role of epidermal growth factor signaling in the development, progression and treatment of cervical cancer. Int J Biol Macromol 194:179–187
Nadile M, Retsidou MI, Gioti K, Beloukas A, Tsiani E (2022) Resveratrol against cervical cancer: evidence from in vitro and in vivo studies. Nutrients 14(24):5273
Neamtu AA, Maghiar TA, Alaya A, Olah NK, Turcus V, Pelea D, Mathe E et al (2022) A comprehensive view on the quercetin impact on colorectal cancer. Molecules 27(6):1873
Nguyen NT, Bui QA, Huynh PD, Nguyen QH, Tran NQ, Viet NT, Nguyen DT (2022) Curcumin and paclitaxel co-loaded heparin and poloxamer p403 hybrid nanocarrier for improved synergistic efficacy in breast cancer. Curr Drug Deliv 19(9):966–979
Niu P, Shi D, Zhang S, Zhu Y, Zhou J (2018) Cardamonin enhances the anti-proliferative effect of cisplatin on ovarian cancer. Oncol Lett 15(3):3991–3997
Noel B, Singh SK, Lillard JW Jr, Singh R (2020) Role of natural compounds in preventing and treating breast cancer. Front Biosci (Schol Ed) 12:137
Noh J, You C, Kang K, Kang K (2022) Activation of OSM-STAT3 epigenetically regulates tumor-promoting transcriptional programs in cervical cancer. Cancers 14(24):6090
Okamoto K, Saito Y, Narumi K, Furugen A, Iseki K, Kobayashi M (2020) Different mechanisms of cisplatin resistance development in human lung cancer cells. Biochem Biophys Res Commun 530(4):745–750
Opp S, Hurtado A, Pampeno C, Lin Z, Meruelo D (2023) Potent and targeted sindbis virus platform for immunotherapy of ovarian cancer. Cells 12(1):77
Ortiz M, Wabel E, Mitchell K, Horibata S (2022) Mechanisms of chemotherapy resistance in ovarian cancer. Cancer Drug Resistance 5(2):304
Palko-Labuz A, Sroda-Pomianek K, Uryga A, Kostrzewa-Suslow E, Michalak K (2017) Anticancer activity of baicalein and luteolin studied in colorectal adenocarcinoma LoVo cells and in drug-resistant LoVo/Dx cells. Biomed Pharmacother 88:232–241
Pan Q, Xue M, Xiao SS, Wan YJ, Xu DB (2016) A combination therapy with baicalein and taxol promotes mitochondria-mediated cell apoptosis: involving in Akt/β-catenin signaling pathway. DNA Cell Biol 35(11):646–656
Pan Y, Shao D, Zhao Y, Zhang F, Zheng X, Tan Y, Chen L et al (2017) Berberine reverses hypoxia-induced chemoresistance in breast cancer through the inhibition of AMPK-HIF-1α. Int J Biol Sci 13(6):794
Pandareesh MD, Kameshwar VH, Byrappa K (2021) Prostate carcinogenesis: insights in relation to epigenetics and inflammation. Endocrine Metabolic Immune Disorders-Drug Targets (Formerly Current Drug Targets-Immune, Endocrine & Metabolic Disorders) 21(2):253–267
Pandey N, Tyagi G, Kaur P, Pradhan S, Rajam MV, Srivastava T (2020) Allicin overcomes hypoxia mediated cisplatin resistance in lung cancer cells through ROS mediated cell death pathway and by suppressing hypoxia inducible factors. Cell Physiol Biochem 54(4):748–766

Torimura T, Iwamoto H (2022) Treatment and the prognosis of hepatocellular carcinoma in Asia. Liver Int 42(9):2042–2054

Trybus W, Trybus E, Król T (2022) Emodin sensitizes cervical cancer cells to vinblastine by inducing apoptosis and mitotic death. Int J Mol Sci 23(15):8510

Tsaur I, Thomas A, Monecke M, Zugelder M, Rutz J, Grein T, Blaheta RA et al (2022) Amygdalin exerts antitumor activity in taxane-resistant prostate cancer cells. Cancers 14(13):3111

Van Poppel H, Albreht T, Basu P, Hogenhout R, Collen S, Roobol M (2022) Serum PSA-based early detection of prostate cancer in Europe and globally: past, present and future. Nat Rev Urol 19(9):562–572

Vladu AF, Ficai D, Ene AG, Ficai A (2022) Combination therapy using polyphenols: an efficient way to improve antitumoral activity and reduce resistance. Int J Mol Sci 23(18):10244

Wang W, Xi M, Duan X, Wang Y, Kong F (2015) Delivery of baicalein and paclitaxel using self-assembled nanoparticles: synergistic antitumor effect in vitro and in vivo. Int J Nanomed 10:3737

Wang R, Yang M, Li G, Wang X, Zhang Z, Qiao H, Li J et al (2019) Paclitaxel-betulinic acid hybrid nanosuspensions for enhanced anti-breast cancer activity. Colloids Surf B 174:270–279

Wang ST, Cui WQ, Pan D, Jiang M, Chang B, Sang LX (2020) Tea polyphenols and their chemopreventive and therapeutic effects on colorectal cancer. World J Gastroenterol 26(6):562

Wang J, Xia S, Chen Y, Qin X, Liu S, Ren H (2023) Screening and validation of prognostic indicator genes in the progression of HBV related hepatocellular carcinoma. Biomed Technol 1:10–17

Wang R, Huang J, Chen J, Yang M, Wang H, Qiao H, Li J (2019) Enhanced anti-colon cancer efficacy of 5-fluorouracil by epigallocatechin-3-gallate co-loaded in wheat germ agglutinin-conjugated nanoparticles. In: Nanomedicine: nanotechnology, biology and medicine, vol 21, p 102068

Wei QY, He KM, Chen JL, Xu YM, Lau AT (2019) Phytofabrication of nanoparticles as novel drugs for anticancer applications. Molecules 24(23):4246

Wei Y, Zeng M, Pi C, Shen H, Yuan J, Zuo Y, Zhao L (2022) Novel curcumin derivative-decorated ultralong-circulating paclitaxel nanoparticles: a novel delivery system with superior anticancer efficacy and safety. Int J Nanomed 5265–5286

Wen C, Fu L, Huang J, Dai Y, Wang B, Xu G, Zhou H et al (2019) Curcumin reverses doxorubicin resistance via inhibition the efflux function of ABCB4 in doxorubicin-resistant breast cancer cells. Mol Med Rep 19(6):5162–5168

Wu J, Qi C, Wang H, Wang Q, Sun J, Dong J, Tian G et al (2022) Curcumin and berberine co-loaded liposomes for anti-hepatocellular carcinoma therapy by blocking the cross-talk between hepatic stellate cells and tumor cells. Frontiers Pharmacol 13

Xiao J, Wang F, Wong NK, He J, Zhang R, Sun R, Li C et al (2019) Global liver disease burdens and research trends: analysis from a Chinese perspective. J Hepatol 71(1):212–221

Xiong J, Sheng J, Wei Y, Sun Z, Xiao X, Zhang L (2022) Sesamol augments paclitaxel-induced apoptosis in human cervical cancer cell lines. Nutr Cancer 1–9

Yan J, Wang Y, Zhang X, Liu S, Tian C, Wang H (2016) Targeted nanomedicine for prostate cancer therapy: docetaxel and curcumin co-encapsulated lipid–polymer hybrid nanoparticles for the enhanced anti-tumor activity in vitro and in vivo. Drug Deliv 23(5):1757–1762

Yang Z, Sun N, Cheng R, Zhao C, Liu J, Tian Z (2017) Hybrid nanoparticles coated with hyaluronic acid lipoid for targeted co-delivery of paclitaxel and curcumin to synergistically eliminate breast cancer stem cells. J Mater Chem B 5(33):6762–6775

Yang JD, Hainaut P, Gores GJ, Amadou A, Plymoth A, Roberts LR (2019) A global view of hepatocellular carcinoma: trends, risk, prevention and management. Nat Rev Gastroenterol Hepatol 16(10):589–604

Yang MD, Sun Y, Zhou WJ, Xie XZ, Zhou QM, Lu YY, Su SB (2021) Resveratrol enhances inhibition effects of cisplatin on cell migration and invasion and tumor growth in breast cancer MDA-MB-231 cell models in vivo and in vitro. Molecules 26(8):2204

Yang Y, Li N, Wang TM, Di L (2021) Natural products with activity against lung cancer: a review focusing on the tumor microenvironment. Int J Mol Sci 22(19):10827

Yilmaz N, Yilmaz UE, Suer K, Goral V, Cakir N (2018) Screening for hepatocellular carcinoma: summary of current guidelines up to 2018. Hepatoma Res 4:46

Yu M, Qi B, Xiaoxiang W, Xu J, Liu X (2017) Baicalein increases cisplatin sensitivity of A549 lung adenocarcinoma cells via PI3K/Akt/NF-κB pathway. Biomed Pharmacother 90:677–685

Yu S, Gong LS, Li NF, Pan YF, Zhang L (2018) Galangin (GG) combined with cisplatin (DDP) to suppress human lung cancer by inhibition of STAT3-regulated NF-κB and Bcl-2/Bax signaling pathways. Biomed Pharmacother 97:213–224

Yuan JD, ZhuGe DL, Tong MQ, Lin MT, Xu XF, Tang X, Xu HL et al (2018) PH-sensitive polymeric nanoparticles of mPEG-PLGA-PGlu with hybrid core for simultaneous encapsulation of curcumin and doxorubicin to kill the heterogeneous tumor cells in breast cancer. Artif Cells Nanomed Biotechnol 46(sup1):302–313

Zhang Y, Wang X, Han L, Zhou Y, Sun S (2015) Green tea polyphenol EGCG reverse cisplatin resistance of A549/DDP cell line through candidate genes demethylation. Biomed Pharmacother 69:285–290

Zhang R, Qiao H, Chen S, Chen X, Dou K, Wei L, Zhang J (2016) Berberine reverses lapatinib resistance of HER2-positive breast cancer cells by increasing the level of ROS. Cancer Biol Ther 17(9):925–934

Zhang J, Li J, Shi Z, Yang Y, Xie X, Lee SM, Chen M et al (2017) PH-sensitive polymeric nanoparticles for co-delivery of doxorubicin and curcumin to treat cancer via enhanced pro-apoptotic and anti-angiogenic activities. Acta Biomater 58:349–364

Zhang L, Li H, Yuan M, Li M, Zhang S (2019) Cervical cancer cells-secreted exosomal microRNA-221-3p promotes invasion, migration and angiogenesis of microvascular endothelial cells in cervical cancer by down-regulating MAPK10 expression. Cancer Manage Res 11:10307

Zhang Y, Liu K, Yan C, Yin Y, He S, Qiu L, Li G (2022) Natural polyphenols for treatment of colorectal cancer. Molecules 27(24):8810

Zhao X, Chen Q, Li Y, Tang H, Liu W, Yang X (2015) Doxorubicin and curcumin co-delivery by lipid nanoparticles for enhanced treatment of diethylnitrosamine-induced hepatocellular carcinoma in mice. Eur J Pharm Biopharm 93:27–36

Zhao Y, Jing Z, Li Y, Mao W (2016) Berberine in combination with cisplatin suppresses breast cancer cell growth through induction of DNA breaks and caspase-3-dependent apoptosis. Oncol Rep 36(1):567–572

Zhao L, Zhao H, Zhao Y, Sui M, Liu J, Li P, Zhang K et al (2022) Role of ginseng, quercetin, and tea in enhancing chemotherapeutic efficacy of colorectal cancer. In: Frontiers in medicine, vol 9

Zhou H, Yuan Y, Wang Z, Ren Z, Hu M, Lu J, Zhu B et al (2023) Co-delivery of doxorubicin and quercetin by Janus hollow silica nanomotors for overcoming multidrug resistance in breast MCF-7/Adr cells. Colloids Surf A 658:130654

Chapter 18
Analysis on Usage of Sanitizers During Outbreak of COVID-19 and Its Impact on Indoor Air Quality and Health

Abhishek Nandan and Deepraj Sinha

18.1 Introduction

Indoor air quality (IAQ) refers to the condition of the air within and around buildings and structures, particularly as it relates to the health, comfort, and well-being of occupants. Poor IAQ has been associated with ailments such as Sick Building Syndrome (SBS), reduced work performance, and diminished academic achievement among students (Nandan et al. 2021a).

Common indoor air pollutants include gases (e.g., carbon monoxide, radon), volatile organic compounds (VOCs), particulate matter (PM), and microbial agents. These pollutants can lead to a range of adverse health effects depending on their concentration and duration of exposure (Nandan et al. 2021a).

Improving IAQ typically involves three main strategies:

1. **Source Control**—Identifying and eliminating or reducing emissions at the source.
2. **Ventilation**—Enhancing air exchange rates to dilute indoor contaminants.
3. **Air Filtration**—Using mechanical or electronic filters to capture airborne particles and pollutants.

Regular cleaning of carpets, rugs, and upholstery also contributes significantly to maintaining a healthy indoor environment. Assessment of IAQ involves multiple methods such as air sampling, surface contaminant testing, occupant health surveys, and computational modeling of airflow within the building (Demokritou et al. 2002).

In developing countries, indoor air pollution poses a critical health risk. The primary contributors are biomass fuels (wood, coal, crop residues, dung) used for cooking and heating in poorly ventilated spaces (Bruce et al. 2000). This results in

A. Nandan (✉) · D. Sinha
School of Advanced Engineering, UPES, Dehradun, India
e-mail: abhisheknandan24@gmail.com

N. A. Siddiqui et al. (eds.), *Advances in Disaster Management, Volume 2*, Springer Transactions in Civil and Environmental Engineering,
https://doi.org/10.1007/978-981-96-4047-8_19

high levels of particulate matter, contributing to an estimated 1.5–2 million premature deaths in the year 2000 alone.

Indoor workplaces such as warehouses, sales offices, clinics, libraries, schools, and childcare centers typically do not involve hazardous chemicals or high-noise environments (Akram 2014). Despite this, workers may still report symptoms associated with Sick Building Syndrome (SBS), including eye irritation, sore throat, nasal congestion, and headaches. These symptoms may not always be linked to measurable pollutants and must be thoroughly investigated alongside air quality monitoring.

Several other environmental factors contribute to the indoor occupational environment, including lighting conditions, ambient noise, thermal comfort, electromagnetic fields, ionizing radiation, and psychosocial stressors. A comprehensive approach is required to assess and manage these diverse influences on employee well-being.

From a policy standpoint, governmental efforts to address the health impacts of air pollution in occupational settings are still developing. However, significant contributions have been made by institutions such as the German Social Accident Insurance's Occupational Safety and Health Institutes. Over the past two decades, Indoor Air Quality (IAQ) has emerged as a growing public health concern due to its potential long-term effects on human health.

The term "indoor environment" in scientific literature encompasses a range of spaces, including residential areas, commercial workplaces, and recreational buildings. Commuting environments, such as vehicles, have also been studied for exposure to air pollutants. Studies indicate that individuals in developed countries spend up to 90% of their time indoors, with workplaces accounting for nearly 60% of this duration.

Modern energy conservation strategies such as tightly sealed buildings and restricted ventilation along with the use of synthetic building materials, have unintentionally worsened indoor air pollution. Numerous indoor contaminants, including allergens, carcinogens, neurotoxins, immunotoxins, and irritants, have been linked to SBS (Tsakas et al. 2011). Common health effects include nasal congestion, eye irritation, and generalized symptoms such as fatigue and headaches. A study by the National Institute for Occupational Health and Safety reported that over 70% of buildings surveyed exhibited at least three common SBS symptoms: dry skin, sore throat, and frequent headaches (Nag 2019).

Globally, the prevalence of indoor air contaminants has drawn significant attention. In Japan, since the 1990s, energy-efficient buildings have been constructed with low ventilation rates and new materials, contributing to increased reports of SBS and cases of Multiple Chemical Sensitivity (MCS). Consequently, the Japanese Ministry of Health, Labour, and Welfare has issued guidelines for indoor contaminants, including formaldehyde (HCHO) (Azuma et al. 2020).

Similarly, in Europe, the European Environment and Health Action Plan under the European Collaborative Action (ECA, 1998) acknowledged IAQ as a critical area for intervention (Settimo et al. 2023). The state of California, USA, has taken a proactive stance since the 1990s, implementing a wide range of policy instruments to tackle indoor air pollution, setting a benchmark in regulatory measures for IAQ management.

18.2 Literature Review

18.2.1 VOCs from Indoor Chemical Reactions and Health

Recent research has shown that chemical reactions between oxidizing agents particularly ozone and common indoor compounds such as terpenes, airborne organic materials, and carpet or textile surfaces can result in the formation of secondary pollutants (Weschler 2006). These reactions often involve the generation of hydroxyl radicals, which initiate further transformations. The byproducts of such indoor chemical interactions have been hypothesized to cause sensory irritation, manifesting as discomfort in the eyes, respiratory tract, or skin (Nielsen 1991).

Outdoor ozone infiltrating indoor spaces is a primary driver of these reactions. Epidemiological studies have linked outdoor ozone exposure to adverse respiratory health outcomes, raising concern about its indoor implications. While formaldehyde is a well-documented product of these reactions, its concentration indoors are often higher due to other sources such as pressed wood products, coatings, and adhesives, making it a significant pollutant in its own right.

Apart from infiltration from the outdoor environment, indoor ozone may originate from artificial sources such as (Hubbard et al. 2005):

- **Ozone generators** marketed as air purifiers.
- **Electronic air cleaners**, which unintentionally produce ozone as a byproduct.
- **Office equipment** (e.g., photocopiers and printers), which can emit trace amounts during operation.

In a pivotal study by Wolkoff et al., the health impact of five common terpene-derived byproducts was assessed through mouse bioassays. These compounds were evaluated for their potential to cause:

1. Acute sensory irritation in the upper respiratory tract.
2. Reduced pulmonary airflow.
3. General lung irritation.

The results suggested minimal sensory irritation or airflow limitations when extrapolated to realistic human exposure levels (Brüning et al. 2014). No significant pulmonary irritation was observed during prolonged exposure to these indoor pollutants. Nevertheless, due to relatively high estimated irritation potencies, Wolkoff and colleagues recommended precautionary measures to minimize exposure to two specific compounds:

- **Three-isopropenyl-6-oxo-heptanal (IPOH)**
- **Four-oxo-pentanal (4-OPA)**

Suggested preventive strategies include:

- Reducing indoor concentrations of ozone and reactive precursors like limonene.
- Improving indoor ventilation and limiting crowding.

- Regular cleaning to remove human and animal skin debris, which can participate in surface reactions.

Despite these insights, the current body of evidence remains insufficient to draw firm conclusions about the long-term health consequences of indoor chemical reaction byproducts. Emerging data indicate the possibility of other unrecognized reaction pathways triggering sensory irritation or inflammatory responses. Continued interdisciplinary research is essential to fully understand the mechanisms and health risks associated with indoor air chemistry.

18.3 VOCs in Cleaning/Sanitizing Products and Health

Consumers frequently use multiple cleaning and sanitizing products simultaneously, many of which contain complex chemical formulations. These combinations can lead to the emission of volatile organic compounds (VOCs), particularly when products are mixed during routine cleaning or disinfecting activities. Common household formulations such as sprays for dusting, polishing, and disinfecting often contain multiple VOC-producing ingredients, contributing to deteriorated Indoor Air Quality (Nandan 2020).

Numerous studies have established a correlation between the use of cleaning and sanitizing agents and respiratory health effects. While professional cleaners are at higher risk due to regular exposure, even individuals engaged in domestic cleaning can experience adverse symptoms (Nandan et al. 2021b). Scientific literature reviews consistently report a heightened risk of asthma exacerbation and other respiratory issues related to both professional and domestic use of such products, especially when applied in spray form. Although the exact biological mechanisms remain under investigation, the frequent use of chlorine-based disinfectants, aerosol sprays, and fragranced products is of particular concern.

Workplace studies have further substantiated this link. In occupational settings with stringent sanitation protocols, such as hospitals and commercial facilities, an increased prevalence of asthma symptoms has been observed among cleaning staff. Several cleaning agents have been implicated, including (Mwanga and Jeebhay 2020):

- Chlorine bleach.
- Ammonia.
- Toilet and bathroom cleaners.
- Abrasives and decalcifiers.
- Glutaraldehyde and orthophthalaldehyde.
- Chloramines.
- Ethylene oxide.

There is a dose-dependent relationship between the frequency and duration of exposure to these agents and the incidence of work-related asthma and respiratory distress. A review from Northern Europe identified the occupational group "cleaners and building caretakers" as one of the highest-risk populations for developing new cases of asthma.

In the domestic context, studies have demonstrated that:

- Frequent use of household spray cleaners—particularly glass cleaners, furniture polish, and air fresheners—is linked to a 40% increase in wheezing, a 50% rise in asthma-related symptoms or medication usage, and nearly a 100% increase in clinical asthma diagnoses.
- Women using spray cleaners regularly reported elevated levels of poorly controlled asthma and persistent respiratory irritation.
- During pregnancy, higher exposure to cleaning chemicals was associated with increased non-allergic respiratory symptoms in infants, indicating potential prenatal or early postnatal impacts on lung development.
- Among older adults, especially those with pre-existing pulmonary conditions, prolonged exposure to air fresheners and fragranced products has been linked to cardiovascular strain and chronic obstructive pulmonary symptoms.

18.4 Composition of Commercial Hand Sanitizers

18.4.1 Active Alcoholic Components

Ethanol (C_2H_5OH): Among ABHS constituents, ethanol is by far the most predominant due to its high antimicrobial effectiveness. A concentration of 60–90% v/v ethanol is effective against a large assortment of microorganisms that include Staphylococcus aureus, E. coli, and also various strains of influenza viruses (Birhanu et al. 2022).

Isopropyl Alcohol (C_3H_8O): Bactericidal and virucidal activity of isopropanol is exceptional at the 70–91.3% v/v range and so is its enveloped virus efficacy. It is better than ethanol in some formulations (Singh et al. 2020).

18.4.2 WHO-Recommended Formulations During COVID-19

The COVID-19 pandemic created a sudden and immense global need for hand sanitizers, which led to widespread shortages. To resolve this crisis, the World Health Organization (WHO) devised models for hand sanitizers that could be made locally or at home (Challenge 2009).

- WHO Formulation I (Ethanol-based):
  - Ethanol 80% v/v

Component	Typical Range (%)	Function
Ethanol / Isopropyl Alcohol	60–95% v/v	Primary antimicrobial agent
Glycerol / Propylene Glycol	0.5–2.0%	Moisturizer
Hydrogen Peroxide	0.1–0.2%	Sporicidal agent
Carbomer / Acrylates	0.1–0.3%	Gel-forming agents
Aminomethyl Propanol	0.1–0.5%	pH stabilizer
Fragrance & Essential Oils	0.01–0.2%	Aesthetic appeal

Fig. 18.1 Typical composition ranges in sanitizer formulations

- Glycerol 1.45% v/v (humectant)
- Hydrogen Peroxide 0.125% v/v (to inactivate bacterial spores)
- Sterile distilled or boiled cold water as diluent

- WHO Formulation II (Isopropyl alcohol-based):
  - Isopropyl alcohol 75% v/v (from 99.8%)
  - Glycerol 1.45% v/v
  - Hydrogen Peroxide 0.125% v/v
  - Sterile water

These formulations listed also in Fig. 18.1 were designed for small-scale, non-industrial settings and rapidly scaled in low-resource communities and healthcare centers worldwide.

Commercial Brands and Key Ingredients (Villa and Russo 2021)

Purell® Advanced Hand Sanitizer: 70% v/v ethanol, aminomethyl propanol, glycerin, and carbomer (for skin compatibility and gel formation) added to meet the requirements of E.U. cosmetics legislation.

Dettol® Hand Sanitizer: Alcohol Denat used, along with thickening agents and fragrances.

Lifebuoy® Hand Sanitizer: Alcohol-based (about 60% v/v ethanol), sometimes containing essential oils like thyme oil and herbal extracts.

Other Active Additives

Octenidine Dihydrochloride (Softasept®): Cationic antiseptic can be combined with ethanol in order to enhance antiseptic efficacy.

Hexamethylenebiguanide (Lavasept®): Functions as an additive antimicrobial agent with broad-spectrum activity.

Chlorhexidine Gluconate: Included in hospital-quality sanitizers for extended antimicrobial action.

Potential Health Hazards of Hand Sanitizers Containing Ethanol and Isopropyl Alcohol

Hand sanitizers have become a critical component of personal hygiene, particularly during global health emergencies like the COVID-19 pandemic. While generally effective and safe when used correctly, repeated or improper use of alcohol-based sanitizers, especially those containing ethanol and isopropyl alcohol (IPA), can pose health risks. These risks depend on several factors, including the concentration of alcohol, route of exposure (skin contact, inhalation, or accidental ingestion), frequency of use, and individual sensitivity.

18.4.3 Ethanol (Ethyl Alcohol) Toxicity

Ethanol is the most widely used alcohol in commercial hand sanitizers, typically present in concentrations ranging from 60 to 95% (v/v). Ethanol is a volatile, flammable compound that is also the active ingredient in alcoholic beverages. Though generally safe for topical use, frequent application, inhalation of vapors, or accidental ingestion, especially by children, can lead to adverse health effects (Kampf 2018).

Corresponding ethanol uptake was calculated as 1365 mg, 630 mg, and 358 mg, respectively. These levels are well below the toxic threshold of 200–300 mg/L, which is typically associated with central nervous system depression. However, the study also detected acetaldehyde, a known carcinogenic metabolite of ethanol, at peak levels of 0.57 mg/L in the blood—raising concern over long-term and cumulative exposure (Kramer et al. 2007).

Moreover, accidental ingestion of even small volumes (as little as 10 mL of a 70% ethanol solution) in children can lead to significant toxicity. Symptoms may include dizziness, nausea, vomiting, respiratory depression, and impaired motor coordination. In severe cases, such exposure could lead to coma or fatal outcomes (Schep et al. 2009).

18.5 Isopropyl Alcohol (IPA) Toxicity

Isopropyl alcohol, also referred to as isopropanol or 2-propanol, is another commonly used alcohol in hand sanitizers. Compared to ethanol, IPA has a higher molecular weight and different metabolic byproducts, resulting in greater toxicity, particularly when ingested or inhaled (Logsdon and Loke 2000).

IPA is rapidly absorbed through the gastrointestinal tract, reaching peak blood levels within 30 min to 2 h post-ingestion. It is primarily metabolized in the liver to acetone and then eliminated via the renal system. High concentrations can overwhelm metabolic detoxification mechanisms, leading to toxic effects. According to material safety data sheets (MSDS), ingestion of 20–30 mL of a 50% isopropanol solution may cause symptoms such as (Bradberry et al. 2014):

- Ataxia (loss of coordination).
- Hypotension (low blood pressure).
- Hypoglycemia.
- CNS depression leading to coma.

Children under the age of six are particularly vulnerable. Clinical evidence shows that ingestion of as little as 1 oz (~ 30 mL) of IPA can result in serious health outcomes in this age group. Additionally, dermal exposure to IPA is generally considered safe when skin is intact. However, frequent use or application on broken or damaged skin can enhance systemic absorption and increase the risk of toxicity. A dose of 0.5–1 mL/kg body weight is considered potentially toxic, though individual tolerance can vary based on health status, age, and metabolic function.

18.5.1 Clinical Observations and Risks in Children

Case studies and poison control reports have highlighted the vulnerability of children to IPA poisoning. Ingestion of as little as 20–30 mL of a 50% isopropanol solution can cause symptoms such as ataxia, hypotension, hypoglycemia, and in extreme cases, coma. For young children under 6 years of age, consuming just 1 oz (~ 30 mL) can result in serious clinical effects Fig. 18.2 highlights the same. Topical application of IPA-based hand sanitizers is generally safe for adults when used as directed. However, prolonged dermal exposure or use on broken skin can increase systemic absorption. A dose of 0.5–1 mL/kg body weight of IPA is considered toxic, though individual tolerance varies significantly.

Small children, including babies, are more likely to become ill than teenagers with alcohol poisoning. Young children have declined liver glycogen store to increase their hypoglycemia exposure or many pharmaceutical causes that render them more likely to be infected with alcohol. New studies of alcohol-based (alcoholic) medications that have been used in young children have reported serious problems including apnea, acidosis, and comas.

18.5.2 Increased Risk of Other Viral Diseases

Experts have begun warning against indirectly increasing the risk of skin disease infection through improper use for coronavirus protection of alcohol-based hand

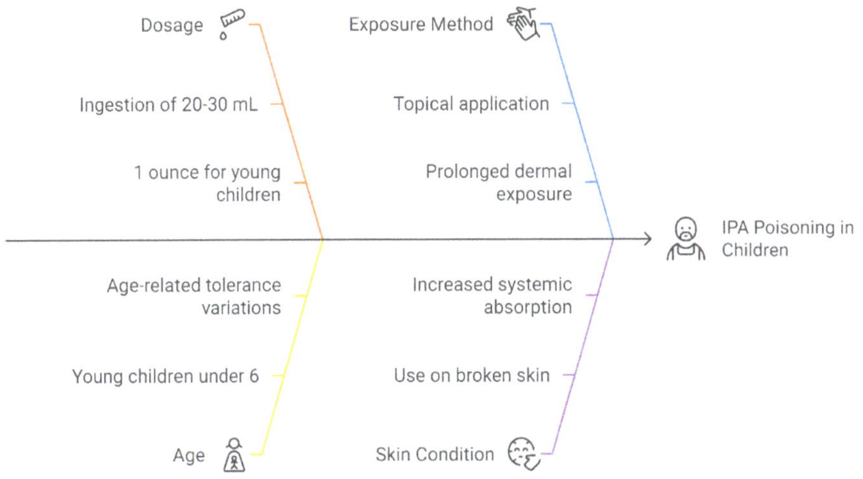

Fig. 18.2 IPA poisoning risks in children

sanitizers. Sanitis abuse for new viral pneumonia damages the skin and lowers the ability to protect them from other harmful viruses.

Sanitizers are also used worldwide to improve hand hygiene as a disinfectant. Excessive use of alcoholic sanitization enhances d-permeability and eliminates oil of the skin and water. Dry and damaged skin is hot for many conditions that are responsible for bacteria, raising the risk of infection with the virus. In some instances, excessive use of sanitizers can increase the risk of viral outbreaks. The previous study showed that the extensive use of alcoholic man-sanitizers leads to an increased risk of outbreak of norovirus. The survey of 160 care centers identified the correlation between the preferential consumption and norovirus outbreak of an alcohol-based sanitizer. In a survey with the results of 73 outbreaks, 91 out of the total facilities surveyed, 29 for norovirus were confirmed. Workers from facilities with norovirus were possibly six times as much as soap and water by hand sanitizers. Table 18.1 highlights the active Ingredients of the Sanitizer and their acute and chronic toxicity.

18.6 Data Analysis

18.6.1 Effect of COVID-19 on the Global Healthcare Industry

The new coronavirus pandemic started in the world, and holy grail became a hand sanitizer. Despite its decades of commercial availability of hand sanitizers, worldwide demand for disinfecting gel is growing due to the fact that it is capable of reducing germs and pathogens' transmission. In the end, many brands developed

Table 18.1 Active ingredients of the sanitizer and their acute and chronic toxicity

Active ingredient	Acute toxicity	Chronic toxicity
Ethanol (C_2H_5OH)	Central nervous system (CNS) depression, vomiting, hypotension, hypoglycemia, metabolic acidosis (lactic/ ketoacidosis)	Liver damage, electrolyte imbalances (hypokalemia, hypocalcemia), potential cardiopulmonary failure, and myoglobinuria (Bondy 1992)
Isopropanol (C_3H_8O)	CNS depression, nausea, vomiting, ataxia, and hypotension	Prolonged exposure may lead to rhabdomyolysis, ketosis, elevated osmolal gap, and renal dysfunction (Ashurst and Nappe 2018)
Hydrogen Peroxide (3%)	Oral irritation, vomiting, bloating, skin burns at high concentration	Rare but potential gastric and esophageal mucosal injury; long-term oral exposure may raise oxidative stress (Watt et al. 2004)
Benzalkonium Chloride	Irritation of skin, eyes, and mucous membranes; ingestion may cause vomiting or convulsions	Possible respiratory tract damage with repeated inhalation; allergic contact dermatitis (Johnson 2018)
Chlorhexidine Gluconate	Skin irritation, eye damage, oral mucosal damage with accidental ingestion	Repeated use may cause sensitization, allergic reactions, and anaphylaxis in rare cases (Basrani 2005)
Triclosan	Skin and eye irritation, potential endocrine-disrupting effects in high doses	Concerns over antimicrobial resistance and hormone disruption; bioaccumulative in aquatic organisms (Dann and Hontela 2011)
Glycerol (Glycerin)	Low toxicity; large oral doses may cause nausea, diarrhea, and headache	Rare cases of metabolic acidosis and renal stress after prolonged high-dose exposure (Becker et al. 2019)
Phenoxyethanol	Drowsiness, vomiting, and central nervous system effects in children when ingested	Liver and kidney effects noted in animal studies with prolonged exposure (Wang et al. 2020)
Aloe Vera Extract	Generally safe; ingestion of latex component may cause abdominal cramps or diarrhea	Potential for allergic skin reactions on repeated use; suspected genotoxicity in some non-purified forms (WHO 1999)
Carbomer	Eye and mucosal irritation in aerosol form	Considered inert, though long-term inhalation of powdered forms may cause respiratory irritation (Additives et al. 2021)

growth opportunities for plugging into the spring markets for the hand sanitizers, with panic purchases leading to over consumption of hand sanitizers, creating an enormous difference between supplies and demand. While COVID-19 has indiscriminately increased sales, a growing awareness about personal hygiene and well-being has led to a growth of hand sanitizers. The growth of the Global Hand Sanitary Market is likely driven by higher per capita revenues and by Westernization in developing countries. The number of cases of hand sanitation exposure to alcohol and alcohol reported by NPDS in 2011–2014 in children 12 years of age is highlighted in Fig. 18.3.

TechSci "Globe hand hygiene market by type of product (Gel, Liquid, Spray, Foam) by distribution channel, according to TechSci Research (Pharmacists, Supermarkets, Convenience Stores, Others) is projected to grow by more than 1 CAGR on the global market for hand sanitizers. The world market for hand sanitizers is projected to increase by over 1 CAGR by 2025 by regional, competitive, predictive and opportunities markets. In addition, the consumption of portable sanitizing and hand hygiene products is spluttering, in particular to avoid the occurrence of contagious diseases. Even various government initiatives emphasize the necessity of educating consumers on health issues and preventive measures that have driven growth of the global market for sanitization.

Global Hand Sanitizer Market: Timelines and Opportunities

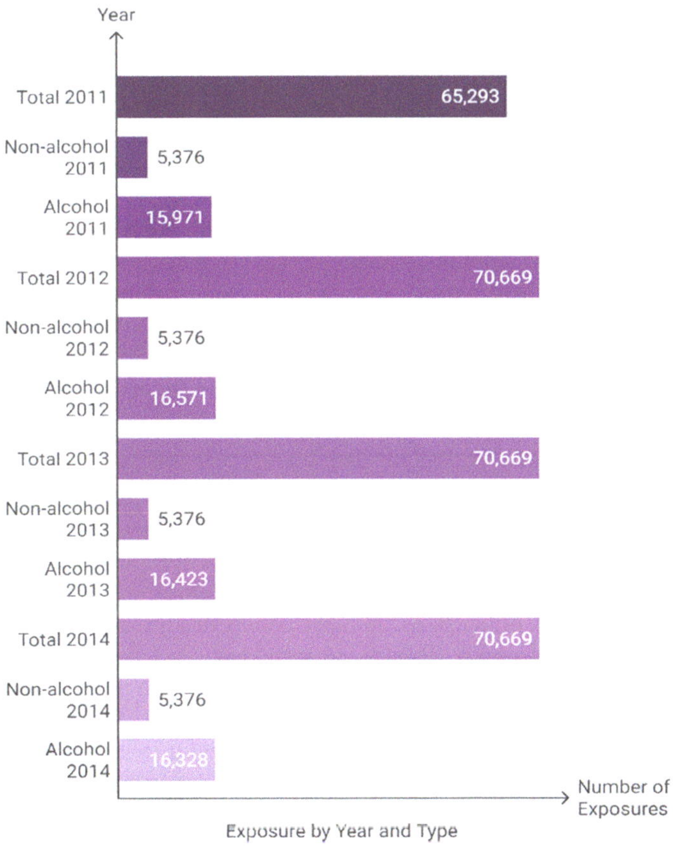

Fig. 18.3 Number of cases of hand sanitation exposure to alcohol and alcohol reported by NPDS in 2011–2014 in children 12 years of age (Santos 2017)

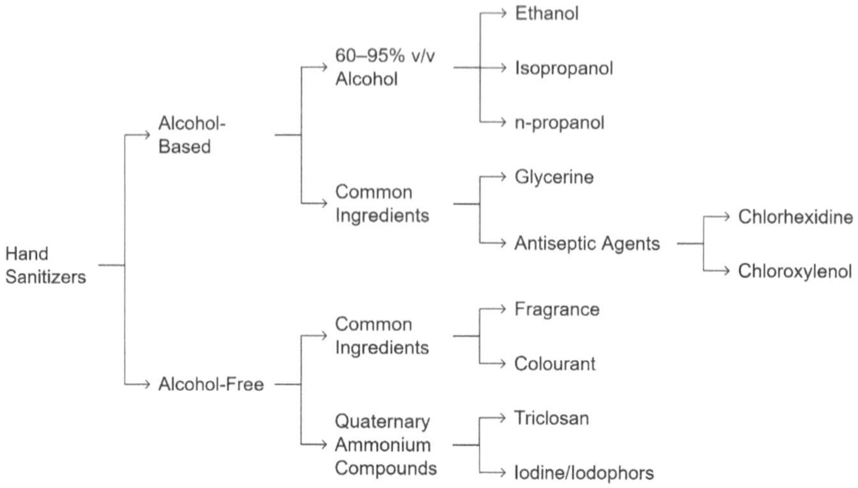

Fig. 18.4 Alcoholic, non-alcoholic, and frequently used manual hygienic excipient list

- **Improved awareness of hand hygiene for consumers**

Overall, there is increasing awareness of the importance of hygiene in hands which can avoid the spread of various diseases. Sensitizing of the possibility of nosocomial nutritional illness and infection by various organizations such as the WHO, FDA, and others. In addition, thousand years population is one of the largest purchasers of hand sanitizers since they tend to pursue a healthy lifestyle. Some of the factors that make handling sanitation products attractive for thousands of years, such as knowledge of germ prevention, ease of use, and portability. Moreover, a growing fear of contagious diseases like SRAS, swine flu H1N1, and COVID-19 has encouraged increased awareness of health. Figure 18.4 highlights the alcoholic, non-alcoholic, and frequently used manual hygienic excipient list.

- **Increased Internet influence in the buying behavior of end-users**

Customers' pick-up behavior was influenced by the web's web-based information and the increasing use of social media. Earlier, not so much money has been spent on personal care products as on other commodity products, particularly in developing countries. However, demand has increased over recent years for portable hand sanitizers due to a growing number of influencers who proactively report on how important it is to use hygiene and personal care products. Moreover, a growing number of e-commerce platforms continue to be crucial if toiletries, personal care, and hygiene products are to be sensitized, profitable reductions are to be offered and potential buyers can be informed, which eventually drives sales of hand sanitary supplies.

- **Growing demands for organic hand sanitizers and flavored products**

Many hand sanitizers in commercial applications contain big chemicals which contribute to the horrific smell of the product after application. Due to its long-term effects on the skin, some healthcare providers are reluctant to use alcohol-based sanitizers. Consequently, demand growth has emerged in recently as sustainable organic components range from niche to mainstream, for organic and chemical-free hand sanitizers. The demand for flavored and aromatic sanitation products has also risen, as people don't want to smell badly and keep your hands healthy. Many beautiful brands respond to consumer demands, who would like to use a clean product with lovely fragrances. There are also a new race of sanitizers with colorful rainbows, smelling pleasant, and extra care.

- **New players in the market of hand sanitizers**

Due to an increasing demand for pandemic coronavirus sanitizers, a number of brands from various industries, ranging from pharmaceutical to medical companies, major lifestyles, bravers, fragrances, and distillers, have entered the market. As many of the players in the market for hand sanitizers have entered the international market, increasing competition has resulted in the attraction of prospective customers by numerous different kinds, innovation and aggressive promotional activities. The growth of the national and international hand sanitizing market drives increasing quality and wide diversity of hand sanitizers.

Hand sanitizers can usually be categorized into groups without alcohol or alcohol (Fig. 18.4). ABHS may contain one or more types of alcohol that are applied to destroy microbes with or without the use of other excipients and humectants and that temporarily stop their growth.

Global market (Fig. 18.5) for sanitizers is expected to grow at 22.6% from 2020 to 2027 and is worth USD2.7 billion in 2019. The market is expected to shift consumer preference to convenient hygiene products. In addition, the recent COVID-19 pandemic at the start of 2020 stimulated the market for hand sanitizers. The demand for hand hygiene products has grown both online and in brick-and-mortar sales worldwide as a result of a global outbreak of the virus in a short time. The outbreak has strengthened the importance of regular manual cleaning and sanitization among consumers and is a key factor of market; Tables 18.2 and 18.3 highlight the same.

Hand sanitizers have a benefit over traditional handwashing products because they can be used directly without water. In addition, leading manufacturers such as Henkel Corporation, Unilever, and Procter and Gamble have conveniently packaged hand sanitizers, such as bags and minibottles, which can be easily carried in their bags or pockets. These factors expanded the spectrum of the market (Jing et al. 2020).

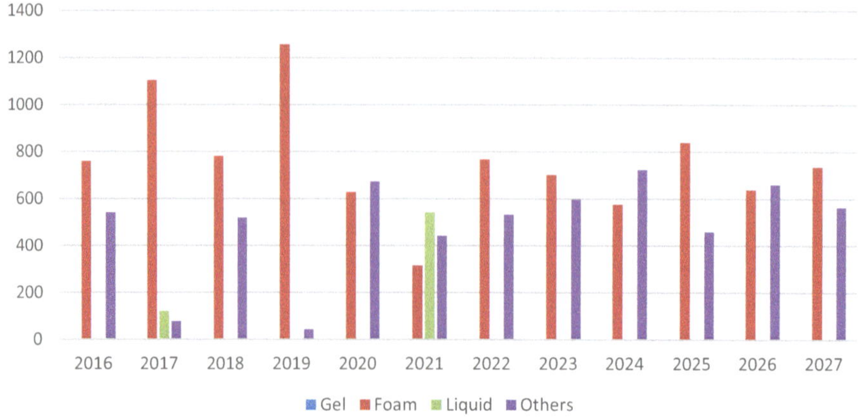

Fig. 18.5 US hand sanitizer market size (in USD Million) by product type (gel, foam, liquid, OTHERS) from 2016 to 2027

Table 18.2 Chemical grading of commonly used disinfectants in the hand sanitizer and antimicrobial mechanics of action (Jing et al. 2020)

Chemical group	Examples	Mechanism of action
Alcohol	• Ethanol (C_2H_6O)	Denaturation of proteins in the plasma membrane
	• Isopropanol (C_3H_8O)	
Chlorine compounds	• Hypochlorites (ClO^-) (e.g., sodium hypochlorite)	Halogenation/oxidation of cellular proteins
	• Chlorine dioxide (ClO_2)	
	• Chloramine-t trihydrate ($C_7H_7ClNNaO_2S$)	
Iodine compounds	• Povidone-iodine (polyvinylpyrrolidone with iodine)	Iodine can penetrate cell membranes of pathogens and attack proteins, nucleotides, and fatty acids
Quaternary ammonium compounds	• Benzalkonium chlorides (e.g., alkyl dimethyl benzyl ammonium chloride $C_{22}H4_{0}N^{+}$)	Lower surface tension
	• Benzyl dimethyl octyl ammonium chloride ($C_{17}H_{39}ClN$)	Inactivate enzymes
	• Didecyl dimethyl ammonium chloride ($C_{22}H_{48}ClN$)	Degrade cell proteins
Peroxygens	• Hydrogen peroxide (H_2O_2)	Free-radical oxidation of essential cell components
	• Peracetic acid (PAA) ($C_2H_4O_3$)	
(Bis) phenols	• Triclosan	Penetrate cytoplasmic bilayer
Biguanide	• Chlorhexidine	Ionic interaction
		Disrupt cell membrane

Table 18.3 Recommendations for formulation of alcoholic hand sanitizer compound United States Pharmacopeia (USP) and World Health Organization (Jing et al. 2020)

Components	Ethanol antiseptic 80% topical solution	Isopropyl alcohol antiseptic 75% topical solution	Isopropyl alcohol antiseptic 75% topical solution
Ethanol 96%	833.3 mL	–	–
Isopropyl alcohol 99%	–	757.6 mL	–
Isopropyl alcohol 91%	–	–	824.2 mL
Hydrogen peroxide 3%	41.7 mL	41.7 mL	41.7 mL
Glycerol 98%	14.5 mL	7.5 mL	7.5 mL
Water, sufficient quantity to make*	1000 mL	1000 mL	1000 mL

18.7 Conclusion

Proper manual hygiene is one of the most important infection control strategies, as it can undeniably reduce the probability of transmissions of direct or indirect microorganisms. As microorganisms are killed quickly and efficiently, ABHS becomes more common, especially when washing with water and soap. But in certain situations handwashing is preferable because ABHS is less effective when the hands are visibly dusty or bleeding and cannot cover certain types of pathogen. The right amount of alcoholic beverages should be selected, and the correct hand cleaning technique applied to ensure the efficacy of all microorganisms must be applied to ABHS. There is significant evidence of an increased risk of adversely affected respiratory health and asthma for individuals cleaning during working. Both of the research carried out in homes and colleges in all populations showed that PAPs decreased concentrations of PM2.5 between 22.6 and 92.0% in homes and 49% in schools. The nature of the study, the procedure duration, the internal airflow and user conformity are different because of this variability. In 2014–2019, the Indian market for the hand sanitizers grew by around 14% at a CAGR. Hand sanitizer is among several personal care products which, because of the new socio-economic scenario and growing occidentalizing and urbanization trends in India, have changed from being luxury to essential items. Moreover, as a result of the COVID-19 outbreak, demand for hand sanitizers has increased since 2020. This has led to a number of players entering this market—small and large.

References

Additives EPOF, Flavourings, Younes M, Aquilina G, Engel KH, Fowler P, Frutos Fernandez MJ, Fürst P, Gürtler R, Gundert-Remy U, Husøy T (2021) Safety evaluation of crosslinked polyacrylic acid polymers (carbomer) as a new food additive. EFSA J 19:e06693
Akram M (2014) Occupational disease and public health concerns of migrant construction workers: a social epidemiological study in western Uttar Pradesh. Soc Change 44:97–117
Ashurst JV, Nappe TM (2018) Isopropanol toxicity
Azuma K, Jinno H, Tanaka-Kagawa T, Sakai S (2020) Risk assessment concepts and approaches for indoor air chemicals in Japan. Int J Hyg Environ Health 225:113470
Basrani B (2005) Chlorhexidine gluconate. Aust Endod J 31:48–52
Becker LC, Bergfeld WF, Belsito DV, Hill RA, Klaassen CD, Lieblezr DC, Marks JG Jr, Shank RC, Slaga TJ, Snyder PW (2019) Safety assessment of glycerin as used in cosmetics. Int J Toxicol 38:6S-22S
BirhanuG, Selam MN, Habte BM, Marew T, Bitew M, Getachew T, Getachew S, Abate A, Mitiku M, Matsabisa M (2022) Evaluation of quality and antimicrobial efficacy of locally manufactured ethanol-based hand sanitizers marketed in Addis Ababa, Ethiopia in the era of COVID-19
Bondy SC (1992) Ethanol toxicity and oxidative stress. Toxicol Lett 63:231–241
Bradberry SM, Vale JA, Jefferson RD (2014) Common chemical. In: Oxford desk reference: toxicology, p 205
Bruce N, Perez-Padilla R, Albalak R (2000) Indoor air pollution in developing countries: a major environmental and public health challenge. Bull World Health Organ 78:1078–1092
Brüning T, Bartsch R, Bolt HM, Desel H, Drexler H, Gundert-Remy U, Harztwig A, Jäckh R, Leibold E, Pallapies D (2014) Sensory irritation as a basis for setting occupational exposure limits. Arch Toxicol 88:1855–1879
Challenge FGPS (2009) WHO guidelines on hand hygiene in health care. Retrieved from: whqlibdoc who int/publications/009 pdf
Dann AB, Hontela A (2011) Triclosan: environmental exposure, toxicity and mechanisms of action. J Appl Toxicol 31:285–311
Demokritou P, Yang C, Chen Q, Spengler JD (2002) An experimental method for contaminant dispersal characterization in large industrial buildings for indoor air quality (IAQ) applications. Build Environ 37:305–312
Hubbard H, Coleman B, Sarwar G, Corsi RL (2005) Effects of an ozone-generating air purifier on indoor secondary particles in three residential dwellings. Indoor Air 15
Jing JLJ, Pei Yi T, Bose RJ, Mccarthy JR, Tharmalingam N, Madheswaran T (2020) Hand sanitizers: a review on formulation aspects, adverse effects, and regulations. Int J Environ Res Public Health 17:3326
Johnson NF (2018) Pulmonary toxicity of benzalkonium chloride. J Aerosol Med Pulm Drug Deliv 31:1–17
Kampf G (2018) Efficacy of ethanol against viruses in hand disinfection. J Hosp Infect 98:331–338
Kramer A, Below H, Bieber N, Kampf G, Toma CD, Huebner N-O, Assadian O (2007) Quantity of ethanol absorption after excessive hand disinfection using three commercially available hand rubs is minimal and below toxic levels for humans. BMC Infect Dis 7:1–12
Logsdon JE, Loke RA (2000) Isopropyl alcohol. In: Kirk-Othmer encyclopedia of chemical technology
Mwanga HH, Jeebhay MF (2020) Work-related asthma and exposure to cleaning agents in healthcare settings–a review of the literature. Curr Allergy Clin Immunol 33:101–112
Nag PK (2019) Sick building syndrome and other building-related illnesses. In: Office buildings: health, safety and environment, pp 53–103
Nandan A (2020) Environmental and health impacts of printing and photocopying: measurement, evaluation and mitigation. UPES, Dehradun
Nandan A, Siddiqui N, Singh C, Aeri A (2021a) Occupational and environmental impacts of indoor air pollutant for different occupancy: a review. Toxicol Environ Heal Sci 13:303–322

Nandan A, Siddiqui N, Singh C, Aeri A, Gwenzi W, Ighalo JO, de Carvalho Nagliate P, Meili L, Singh P, Chaukura N (2021b) COVID-19 pandemic in Uttarakhand, India: environmental recovery or degradation? J Environ Chem Eng 9:106595

Nielsen GD (1991) Mechanisms of activation of the sensory irritant receptor by airborne chemicals. Crit Rev Toxicol 21:183–208

Santos C (2017) Reported adverse health effects in children from ingestion of alcohol-based hand sanitizers—United States, 2011–2014. MMWR Morb Mortal Wkly Rep 66

Schep LJ, Slaughter RJ, Temple WA, Beasley DMG (2009) Diethylene glycol poisoning. Clin Toxicol 47:525–535

Settimo G, Yu Y, Gola M, Buffoli M, Capolongo S (2023) Challenges in IAQ for indoor spaces: a comparison of the reference guideline values of indoor air pollutants from the governments and international institutions. Atmosphere 14:633

Singh D, Joshi K, Samuel A, Patra J, Mahindroo N (2020) Alcohol-based hand sanitisers as first line of defence against SARS-CoV-2: a review of biology, chemistry and formulations. Epidemiol Infect 148:e229

Tsakas MP, Siskos AP, Siskos PA (2011) Indoor air pollutants and the impact on human health. In: Chemistry, emission control, radioactive pollution and indoor air quality, pp 447–484

Villa C, Russo E (2021) Hydrogels in hand sanitizers. Materials 14:1577

Wang J, Liu Y, Kam WR, Li Y, Sullivan DA (2020) Toxicity of the cosmetic preservatives parabens, phenoxyethanol and chlorphenesin on human meibomian gland epithelial cells. Exp Eye Res 196:108057

Watt BE, Proudfoot AT, Vale JA (2004) Hydrogen peroxide poisoning. Toxicol Rev 23:51–57

Weschler CJ (2006) Ozone's impact on public health: contributions from indoor exposures to ozone and products of ozone-initiated chemistry. Environ Health Perspect 114:1489–1496

Chapter 19
Design of Effective Temporary Shelter for Post-disaster Settlement

Abhishek Nandan

19.1 Introduction

Many policies that have been designed as part of the disaster relief operation signify that the victims of the disaster are provided with attributes like temporary housing, food, and medical services but their dwellings are not made to be restored under most of the circumstances. To the youth who have been troubled due to the occurrence of a disaster, these serve as some of the short-term support but not as long-term support.

During the post-disaster management stage, all the challenges that arise in the reconstruction of the houses of the victims are pretty much the same as the challenges that arise in construction of many of the low-cost houses in countries which are in the developing phase (Taş and Taş 2015). However, in this situation there are many probable challenges that add to them which include the following;

- The supply of resources is scarce in nature and the scene which is present would likely be chaotic.
- Limited time is available for the process of recovery.
- Sustainability must be inculcated in the projects that are implemented (Hosseini Hosseini et al. 2016).

Even with a causal mode of observation, it is clearly evident that the current world which we live in is likely to be disaster-prone and this circumstance is increasing day by day with time.

It has been suggested by the researchers that the accessibility to the shelters designed and constructed for public utility is significantly less for women when compared to men (Gouda 2016). There is a necessity for services which take care

A. Nandan (✉)
School of Advance Engineering, UPES, Dehradun, India
e-mail: anandan@ddn.upes.ac.in

N. A. Siddiqui et al. (eds.), *Advances in Disaster Management, Volume 2*, Springer Transactions in Civil and Environmental Engineering,
https://doi.org/10.1007/978-981-96-4047-8_20

of the children, transportation to majority of the work sites, and legal assistance that can be affordable.

The neighbourhoods which are usually wider in nature and denser in networking implement recovery measures faster when there is any disaster. This is because of two reasons (Biswas 2019):

- Stronger social ties which account for informal insurance
- Victims are allowed to utilise the network support which is ready to use in terms of financial guidance, logistic approach, and physical support.

During any sort of emergency, people usually become traumatised and this will eventually lead to a state of confusion. Any kind of shelter will help them by providing the individuals, with an opportunity for a new beginning from the situation in which they are and also to identify all the necessary needs for their coming future. Space is usually used differently in each of the cultures. The people who will lose their homes after any disaster may try staying in the previous homes or can leave their homes like forced migrants.

The methodology of selecting the site and the planning strategy for these temporary forms of shelters is one of the important aspects which contribute to their success in the long run as these factors can affect the non-shelter issues like the administration of the site and the health of the citizens who live in that area.

If there arises a case where people tend to prepare the camps on their own then this tendency may cause several problems if the site selection made by the migrants by themselves is capable of being utilised for any durable purpose as a solution.

The process of constructing shelters for emergency management is an act to fulfil various needs corresponding to the ones who had lost shelters in relation to the appropriate mode of response (Xu 2006). The extent of necessity can vary from very small level communities to countries on the whole (Chang et al. 2010). In accordance to those who wish to fulfil these necessities will range from very small-level local agencies which focus on humanitarian aspects to governmental bodies which are internationally recognised in multilateral dimensions.

On the whole, there are three prominent kinds of programs which are used as an assistance to the families which are homeless.

They are:

- Shelters in case of emergency
- Delivery of services
- Programs towards transitional stage of living

Each of these plays a significant role in solution determination for these homeless people. The preliminary response to any disaster for the communities that consist of families with homelessness is the introduction of shelters for emergencies (Sumka 1987). Moreover, the provision of emergency shelter is a kind of relief for short period of time if the provision of permanent housing is absent.

It is always observed that the evacuees are made to move to shelters which are made for temporary purposes and they end up permanently being away for various reasons because of their institutional policies and institutional legislations which are restrictive in nature.

C is one of the software that is been found in addition to HAZUS-MH which is widely used to estimate the shelter needs with the consideration of all the diverse variables (El-Anwar et al. 2009). Therefore, a proper estimate which contains all the information of the population which is in the necessity of public shelter is required for the decision-making in the response to the disaster (Chowdhury et al. 1998). Factors which contribute toward socio-economy also play a significant role in the decision-making process of people to utilise the shelter provided as part of post-disaster management.

The main objective of the response teams during an emergency is the provision of shelter as well as assistance to various victims who have been affected by a disaster. But this action has to be done in the minimal time possible. This is because all these have to be delivered to them in the quickest manner possible so as to be more impactful to the victims who have been affected due to disasters.

Disasters result in worldwide global problems which ultimately is paving the way in resulting homelessness. The region which gets affected faces serious economic problems and social problems in addition to the damage caused to the environment on the whole (Fallis et al. 2013). This makes a way that shows the necessity for an integrated type of management so as to utilise the national resources effectively. Minimal amount of energy should be used throughout the reconstruction phase to let the development of the community grow in a sustainable manner (Johnson 2007). The total cost of a house which is built for temporary purposes consists of 10–15% of the total amount that constitutes to the expenses of the construction activity for a whole (Kulshrestha et al. 2011). Usage of recyclable material in this process always proves beneficial in the reconstruction process.

The location of these shelters which are built for temporary purposes should be away from all the vulnerable areas (Hosseini et al. 2016). If this consideration is failed then that means that the location to which the evacuees have been taken to is again a disaster-prone area and it lacks the main objective of providing support to those people who became homeless (Twigg 2011). This can even make things worse than that in the earlier case.

19.2 Classification of Shelters

Shelters are classified into different types as follows (Karaoglan and Alaçam 2018):

- Emergency shelters
- Temporary shelters
- Temporary housing
- Transitional shelters
- Progressive shelters
- Core Shelters

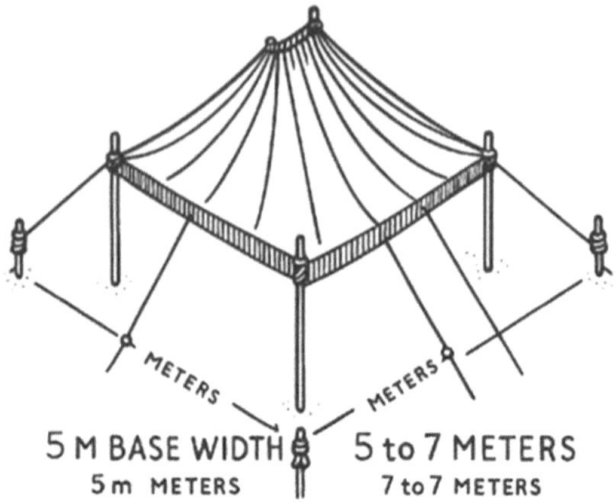

Fig. 19.1 Emergency shelter

- Permanent housing

19.2.1 Emergency Shelter

These kinds of shelters are widely utilised for very brief extent of time to give the support to save lives. This is one of the basic types of shelter used to provide support to the victims. This is used for a period of about a night to several days during the period of an emergency. This type of shelter neither allows the residents to prepare food extensively nor provides medical services for a prolonged duration of time.

Emergency shelter prototype and requirements (Schlaffer and Bonacci 2012) (Fig. 19.1):

Square sizing/occupancy: 5 m × 5 m, 6 m × 6 m, 7 m × 7 m
Ground conditions: concrete floor/ compacted ground
Material of construction: bamboo, timber, metal
Flooring: tarpaulin sheet, locally procured ground mat

19.2.2 Temporary Shelters

These kinds of shelters are widely utilised for a very short period of time (Hadafi 2010). A simple type of tent which is used for a mass of public as a shelter for a couple of weeks after the occurrence of a disaster is considered to be a temporary shelter. There would be a limited duration to stay in these kinds of shelters and hence there is a need to prioritise the pace and minimising the costs when thinking of constructing shelters of this type (Şener and Altun 2009) (Fig. 19.2).

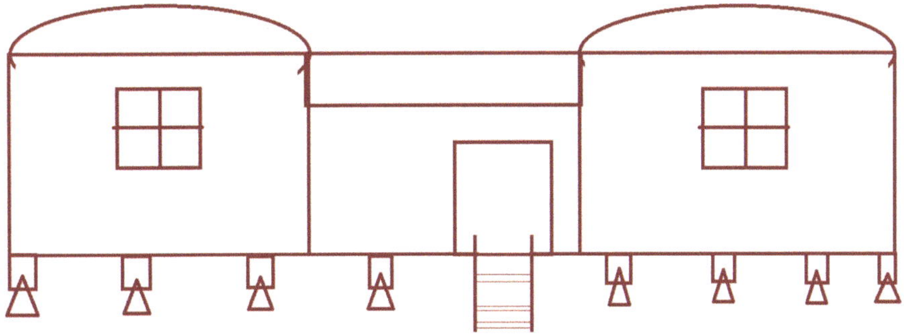

Fig. 19.2 Temporary shelter

19.2.3 Temporary Housing

These kinds of shelters are always distributed for a very long period of time from say six months to that of three years. Temporary housing consisting of houses which are on a rental basis and units which are prefabricated will allow the victims affected after the occurrence of a disaster to let them return back to normalised form of activities on a daily basis (Johnson et al. 2006). In most cases, the houses constructed on temporary basis are usually made to be installed on lands which are temporary (Hadafi 2010).

19.2.4 Transitional Shelters

These kinds of shelters are generally developed by individuals who have displaced themselves after the occurrence of any possible disaster. In this process there needs to be ample support for the self-management system inculcated by themselves and also their resourcefulness. These types of shelters are often relocated to a permanent site from a temporary site (Asefi and Sirus 2012). But here they try to use it as part of a permanent house later on. There is another opportunity here where there can be a sale of these parts for income generation in the course of recovery (Chou et al. 2013). Recycling and reusing these parts for reconstruction is also another way to utilise these shelters after their primary usage. These are on the whole expected to be utilised for a long time from several months to several years as well.

19.2.5 Progressive Shelters

The design and the construction of these types of shelters are with intension of using it permanently and modifying it as the time passes with alternative components (Johnson 2007).

19.2.6 Core Shelters

The main intension behind building these kinds of shelters is to transform them into a permanent form of the housing that can be usable for the future generations. This includes the important services and utilities along with the foundation. The main aim is to construct at least one room that will meet the housing as per the permanent utilisation standards. But these are very seldom recommended for permanent usage.

19.2.7 Permanent Housing

These are the houses that are prepared as a process of upgrading shelters from transitional kinds of shelters or in other cases progressive shelters or core shelters and also newly built houses. These kinds of houses must resist the hazards or any other disasters that are anticipated (Fayazi and Lizarralde 2013).

Out of all these shelters, the management authorities have to take decisions on which kind of shelter has to be used as per the necessities and the conditions that are prevailing (Bayram et al. 2015).

19.3 Design Criteria

The shelter's performance completely depends on the factors that are considered for the design. These factors are developed after consulting the people who have been affected due to the occurrence of any disaster, governmental and private departments and agencies that facilitate the disaster recovery (Melgarejo and Lakes 2014). This consultation is done just to avoid any kind of issues that may be encountered in terms of economy or sociocultural environment or technique (Tucker et al. 2014).

The following are the various types of factors that are to be considered for the design of temporary shelters at low cost:

- Environmental factors
- Economic factors
- Technical factors
- Socio-cultural factors

19.3.1 Environmental Factors

The following are the various environmental factors that are essential to be considered as part of the design criteria for temporary shelters:

- Climate variations

- Recycling, upgrading and disposal
- Hygienic

19.3.1.1 Climate Variations

Every season there will be a different experience of the weather that prevails in the locations where the disaster occurs. Temperature also plays a prominent role to make the people feel comfortable and hence different kinds of shelters are preferred in different regions (Lizarralde 2002). When it comes to the design detailing part aspects such as ceilings should be as high as possible and verandas also should be as wide as possible so as to keep the temperature warm in situations when the weather is too cold (Li et al. 2012).

19.3.1.2 Recycling, Upgrading and Disposal

The materials used in these shelters must be able to be recycled after its use or must be able to be resold or even must be able to relocate the entire shelter after it is made to be dismantled. A shelter is usually considered to be recycled if it is able to either partly or completely be able to reproduce itself from the materials that have been disassembled (Serpell and Alarcon 1998). The shelter should always have a wide range of applications and must be able to withstand any adverse conditions that prevail so as to make it environmentally friendly.

Hence the material that is usually selected for the making of these shelters must be able to be recycled or upgraded further or must be able to be reused rather than simply disposing it after its usage. Shelters must also clearly be ensured not to have any negative effects on the surrounding environment.

19.3.1.3 Hygiene

Health always improves by promoting personal hygiene which is one of the significant aspects. They must have adequate facilities for water and sanitation in the vicinity. This is always expensive to maintain. But the necessity of each will correspond to the improvement of health of the people to a greater extent. The standards help in the design of these programs including the facilities to wash their clothes and will also make sure the best possible usage of water and also disposal of waste water after using it for various purposes. Diseases can also be minimised by controlling the presence of mosquitos that carry diseases in and around the drainage systems. Nutrition is also considered to be playing a prominent role which includes their ability to properly store the raw food and also cook hygienic food.

19.3.2 *Economic Factors*

The following are the various economic factors that are essential to be considered as part of the design criteria of the temporary shelters.

- Type of shelter
- Lifetime
- Livelihood

19.3.2.1 Type of Shelter

Economy spent towards response activities and the recovery activities is a very vital aspect. It is always preferred to go for the process of up gradation and the improvement of older shelters rather than going for the shifting from phase to phase like from a stage of emergency response to that of temporary shelter to that of constructions which are intended to be for permanent usage (Chang and Liao 2015). Studies also suggest that cost-effective energy-reducing measures of shelter are impacted by increasing the overall base area of the shelter (Atmaca and Atmaca 2016). Price comparison should always be done for each of the possible options available.

19.3.2.2 Life Time

The life of these shelters should be well planned before the design phase as per the standards and the prevailing conditions (Silva 2007). Hence this aspect of life time has to be taken into account because this also includes the economy that needs to be invested in it correspondingly to make it last for a longer duration of time (Hendriks et al. 2017).

19.3.2.3 Livelihood

When it comes to the support in terms of provision of livelihood to the users who use the shelters has to be done for the long-term approach of displacement situations. Support should also be extended to the locals by the shelter-providing agencies to show them ways in which they can earn some money by setting up small-scale businesses (Xiang et al. 2018). One of the studies suggested four types of comprehensive and consolidated perspectives for livelihood support post-disaster (Joakim and Wismer 2015):

(a) Replacement of unnecessary assets
(b) Provision of monetary support for initiation of small business
(c) Providing effective and appropriate training for improving skill sets of people
(d) Effective advertisement support for sale of products

19.3.3 Technical Factors

The following are the various technical factors that are essential to be considered as part of the design criteria of the temporary shelters:

- Easy to erect and dismantle
- Materials and insulations
- Classification of hazards and performance
- Physical and Psychological effects

19.3.3.1 Ease to Erect and Dismantle

The process of both erection as well as dismantling of these kinds of shelters must be able to be done with ease so as to serve many crucial purposes. They should be very light in weight and must be very few in number. If the same is made with complexity then it requires a huge amount of training aids and resources which are required to build it which may even result in a lot of delays (Platt 2017).

19.3.3.2 Materials and Insulations

When it comes to the selection of materials in building the shelter one should consider the quality aspect, cost effectiveness, suitability and the availability of the material in the local vicinity. The properties of these materials should be eco-friendly. They should be in such a way that they can be reused for the next generations from time to time (Twigg 2006). The design of shelters should be done by the usage of such materials which can provide suitable insulation towards high noise levels, high temperature and adverse weather conditions.

19.3.3.3 Classification of Hazards and Performance

All the hazards that have the potential to cause serious threats to the occupants like earthquakes and other natural disasters must be taken into account so as to design a shelter which will withstand them in every possible scenario (Hadafi 2010). To withstand seismic waves during the earthquake one should prefer to use timber which may usually include bamboo to build the frames in the structures. This is because they hardly possess any weight and can hence cause least number of fatalities overall (Hendriks et al. 2017). While this can’t be the same to withstand the storms, they can prove to be more vulnerable.

19.3.3.4 Physical and Psychological Effects

The people to whom the damage has occurred to their houses will have a significant amount of stress. This is because of the situation of homelessness after their house gets damaged which accounts for psychological impact which eventually increases the stress level.

19.3.4 Socio-Cultural Factors

The following are the various socio-cultural factors that are essential to be considered as part of the design criteria of the temporary shelters (Félix et al. 2013).

- Cultural Differences
- Dignity and security
- Communication

19.3.4.1 Cultural Differences

In most of the countries, every region has a variety of design aspects and orientation patterns as per their own preferences (Hany Abulnour 2014). That is the reason why these parameters should also be taken into account based on the cultures that prevail in the communities of the local vicinity (Davidson et al. 2007). The agencies that provide these aids such as shelters must consider them. They must have a reflective approach towards the necessities of the one who uses them based on their values, genders, religions and styles of architecture (Tafti and Tomlinson 2013).

19.3.4.2 Dignity and Security

The two parameters out of which one is dignity and the other is security are very important parameters which vary from one region to the other based on the culture of the community. The development of these shelters should not only be as a structure alone but also should be made in such a way that the people who use it will feel that they are socially well integrated. Also, the provision of adequate space to have a happy living will ensure necessary dignity aspect as well as the security aspect of the users. There should be enough room for the users to promote extension of the shelter in accordance to better privacy. Doors and windows should possess suitable systems which can have proper locking systems to increase the degree of security (Twigg 2006).

19.3.4.3 Communication

Communication is always considered to have lot of importance to survivors during the initial stage of the recovery after the occurrence of any disaster. The negative impacts that can arise during the recovery stage can be minimised to the maximum extent with the participation of the survivors in making decisions during recovery. In this process, the main aspects that are going to be dealt with are the future requirements. Some of the various modes of communication within a society which are to be developed are the radio, mobile phone, leaflets, posters, newspapers, internet and television.

19.4 Tools and Techniques Used for Material Selection for Shelter

Types of shelter materials: usually the major shelter materials used are concrete/wood/bamboo/brick/steel-based structures. Depending on the availability of the material resource post-disaster.

Material and classification as per the usage and shelter model (Escamilla and Habert 2015):

Type of material	Shelter model
Timber	Transitional shelter
Concrete/brick	Core shelter
Concrete/timber	Core shelter
Steel	Transitional shelter
Bamboo	Transitional shelter

Criteria for selection of material.

- Environmental impact
- Physical planning
- Type of construction
- Life cycle cost assessment

19.4.1 Environmental Impact

Depending upon the structural material used in the construction of the shelter. Environmental impact varies from material grade selected and installed within the shelter (Audefroy 2010). Possible feasibility assessment is thoroughly conducted according to the shelter specifications. Increased disorder during the disaster should be considered as per the requirement. Based on the scale of the impact bamboo has significantly less impact (Escamilla and Habert 2015).

Environmental factors considered for assessment (Barenstein and Pittet 2007):

- Local availability of the material
- Environmental impact of the materials
- Pollutants in construction materials
- Impact of materials in case of demolition

19.4.2 Physical Planning

Develop contingency plans for all the required activities and the construction materials and further pursue eco-urbanisation process to support the orientation of the reconstruction process. Implement a structured approach towards post-disaster environmental responsibility and contingency and emergency plans (UNDP 2011). Possible steps to recycle and reuse the debris from the disaster.

19.4.3 Type of Construction

Several types of construction methods are utilised in post-disaster situations (Silva 2010).

- Concrete frame with masonry fill
- Reinforced blockwork
- Timber frame
- Prefabricated systems

19.4.4 Life Cycle Cost Assessment

This part covers significant project costs involved such as initial cost of materials, utility, labour, equipment, preventive maintenance, services and overall life cycle cost (Atmaca and Atmaca 2016). Life cycle assessment is completely dependent on the life span of the shelter and the usually temporary shelter are used for a span of 2–4 years depending on the resources and budget allocations. The energy efficiency of the shelter will be directly proportional to the cost endured and implementing solar panels and thermal efficient energy resources could benefit the settlement.

Various stages of life cycle:

- Construction phase
- Operational phase
- Demolition phase (in case)

19.4.5 Advances in Shelter Design Considerations

Recent developments in shelter design have been dependent on technical aspects and spatial considerations in initial development phase (Salman 2016). Parameters such as environmental considerations and cultural acceptance are emphasised (Schlaffer and Bonacci 2012).

Other design parameters considered (Krieger and Higgins 2002).

- Siting
- Square footage/occupancy requirements
- Shape and orientation
- Roof structure
- Utilizing environment

Legal aspects and regulations:

Disaster risk governance and reduction has been implemented by 168 UN members signing the accord of Hyogo framework for improving existing systems for better response towards disaster response and all aspects further emphasised to prioritise the action with accord of Sendai framework (2015–2030) (Briceño 2015). Thus, all UN members ensure improving appropriate legal compliance as well as enhancing the implementation and enforcement.

International Federation of Red Cross and Red Crescent Societies provides framework to all UN members regarding disaster governance (The Handbook on law and disaster risk reduction 2015).

- Implement dedicated laws for disaster risk management
- Establish clear roles and responsibilities related to disaster governance
- Allocate adequate budget and resources to disaster governance
- Government should provide clear procedures and responsibilities
- Provide guidelines for risk management with regard to disaster management
- Provide education, training and awareness raising within population
- Laws should adequately address gender considerations and special needs personal

19.5 Conclusion

It can be concluded on the whole that shelters play a crucial role as part of disaster response management and also disaster recovery management in many countries. They have various functions during the post-disaster management stage. There are many types of shelters. Each type of shelter is used for suitable purposes based on various factors. The design of temporary shelters depends on environmental factors, economic factors, technical factors and also the socio-cultural factors. The environmental factors include the climate variations from time to time, recycling capacity of the material used and hygiene in the location. The economic factors include the type of shelter to be constructed, the life span of the shelter and the livelihood provided to the people. The technical factors include the ease to either erect the shelter or dismantle the shelter, materials as well as the insulations used, types of hazards that prevail, effects that include physical as well as psychological aspects. The socio-cultural factors include the differences in various cultures, the dignity aspect and the security aspect and the modes of communication that are available to communicate.

References

Asefi M, Sirus FA (2012) Transformable shelter: evaluation and new architectural design proposals. Procedia—Soc Behav Sci 51:961–966

Atmaca A, Atmaca N (2016) Comparative life cycle energy and cost analysis of post-disaster temporary housings. Appl Energy 171:429–443

Audefroy JF (2010) Post-disaster emergency and reconstruction experiences in Asia and Latin America: an assessment. Dev Pract 20(6):664–677

Barenstein JD, Pittet D (2007) Post disaster housing reconstruction Current trends and sustainable alternatives for tsunami-affected communities in coastal Tamil Nadu. University of Applied Sciences of Southern Switzerland, Canobbio

Bayram V, BÇ Tansel, Yaman H (2015) Compromising system and user interests in shelter location and evacuation planning. Transp Res Part B: Methodol 72:146–163

Biswas A (2019) Exploring Indian post-disaster temporary housing strategy through a comparative review

Briceño S (2015) Looking back and beyond Sendai: 25 years of international policy experience on disaster risk reduction. Int J Disaster Risk Sci **6**(1):1–7

Chang Y et al (2010) Resourcing challenges for post-disaster housing reconstruction: a comparative analysis. Build Res Inf 38(3):247–264

Chang HS, Liao CH (2015) Planning emergency shelter locations based on evacuation behavior. Nat Hazards 76(3):1551–1571

Chou J-S et al (2013) Emergency shelter capacity estimation by earthquake damage analysis. Nat Hazards 65(3):2031–2061

Chowdhury JU et al (1998) Models for cyclone shelter planning in Bangladesh. Water Int 23(3):155–163

Da Silva JJSE (2007) Quality and standards in post-disaster shelter. Struct Eng 85(14):25

Da Silva J (2010) Key considerations in post-disaster reconstruction. Practical Action Publishing, UK

Davidson CH et al (2007) Truths and myths about community participation in post-disaster housing projects. Habitat Int 31(1):100–115

El-Anwar O, El-Rayes K, Elnashai A (2009) An automated system for optimizing post-disaster temporary housing allocation. Autom Constr 18(7):983–993

Escamilla EZ, Habert G (2015) Global or local construction materials for post-disaster reconstruction? Sustainability assessment of 20 post-disaster shelter designs. Build Environ 4:308–314

Fallis A et al (2013) No Title No Title

Fayazi M, Lizarralde G (2013) The role of low-cost housing in the path from vulnerability to resilience. Int J Architect Res 7(3):146

Félix D, Branco JM, Feio AJHI (2013) Temporary housing after disasters: a state of the art survey. Habitat Int 40:136–141

Gouda RK (2016) Post-disaster shelter in India: a study of the long-term outcomes of post-disaster shelter projects—Conclusions & recommendations. CARE India

Hadafi F, Fallahi A (2010) Temporary housing respond to disasters in developing countries-case study: Iran-Ardabil and Lorestan Province Earthquakes, vol 66, pp 1536–1542

Hany Abulnour AH (2014) The post-disaster temporary dwelling: fundamentals of provision, design and construction. HBRC J 10(1):10–24
Hendriks E et al (2017) Self-built housing as an alternative for post-disaster recovery. In: No-cost housing conference. ETH Zurich, Zurich, Switzerland
Hosseini SA et al (2016) Multicriteria decision-making method for sustainable site location of post-disaster temporary housing in urban areas. J Constr Eng Manage 142(9):04016036
Hosseini Hosseini SMA, de la Fuente A, Pons Valladares O (2016) Multi-criteria decision-making method for assessing the sustainability of post-disaster temporary housing units technologies: a case study in Bam, 2003
Joakim EP, Wismer SK (2015) Livelihood recovery after disaster. Dev Pract 25(3):401–418
Johnson CJD (2007) Strategic planning for post-disaster temporary housing. Disasters 31(4):435–458
Johnson C et al (2006) A systems view of temporary housing projects in post-disaster reconstruction. Constr Manage Econ 24(4):367–378
Johnson C (2007) Strategies for the reuse of temporary housing. In: Urban transformation—Holcim forum
Karaoglan FC, Alaçam S, Design of a post-disaster temporary living space through the use of shape evolution
Krieger J, Higgins DL (2002) Housing and health: time again for public health action. Am J Public Health 92(5):758–768
Kulshrestha A et al (2011) Robust shelter locations for evacuation planning with demand uncertainty. J Transp Saf Secur 3(4):272–288
Li AC et al (2012) Shelter location and transportation planning under hurricane conditions. Transp Res Part E: Logistics Transp Rev 48(4):715–729
Lizarralde G (2002) Organizational design, performance and evaluation of post-disaster reconstruction projects. In: Proceedings of the first international conference on post-disaster reconstruction: improving post-disaster reconstruction in developing countries
Melgarejo LF, Lakes T (2014) Urban adaptation planning and climate-related disasters: an integrated assessment of public infrastructure serving as temporary shelter during river floods in Colombia
Platt S (2017) Factors affecting the speed and quality of post-disaster recovery and resilience. In: International conference on earthquake engineering and structural dynamics. Springer
Salman M et al (2016) Construction materials from stainless steel slags: technical aspects, environmental benefits, and economic opportunities. J Ind Ecol 20(4):854–866
Santos C (2017) Reported adverse health effects in children from ingestion of alcohol-based hand sanitizers—United States, 2011–2014. MMWR. Morbidity and Mortality Weekly Report, pp 66
Schlaffer L, Bonacci P (2012) Shelter design, pp 21–35
Şener SM, Altun MC (2009) Architecture, design of a post disaster temporary shelter unit. J Fac Archit 6(2):58–72
Serpell A, Alarcon LF (1998) Construction process improvement methodology for construction projects. Int J Proj Manage 16(4):215–221
Sumka HJ (1987) Introduction in Shelter policy and planning in developing countries. 1987. J Am Plan Assoc 53(2):171–175
Tafti MT, Tomlinson RJHI (2013) The role of post-disaster public policy responses in housing recovery of tenants. Habitat Int 40:218–224
Taş N, Taş MJNH (2015) Determining of the local housing identity in urban transformation areas under disaster risk: Bursa, Turkey. Nat Hazards 75(1):119–139
The Handbook on law and disaster risk reduction (2015) International Federation of Red Cross and Red Crescent Societies
Tucker S, Gamage A, Wijeyesekera C (2014) Some design aspects of sustainable post-disaster housing
Twigg J (2006) Technology, post-disaster housing reconstruction and livelihood security
Twigg J et al (2011) Disability and public shelter in emergencies, vol 10, issue 3–4, pp 248–261

UNDP (2011) Methodological guide for post-disaster recovery planning processes—guidelines and actions for national, regional and local governments, pp 41
Xiang M, Zhao W, Chen JJS (2018) A comparison of different reconstruction modes and adaptive evaluation systems for community recovery following the Wenchuan earthquake. Sustainability 10(11):4115
Xu W et al (2006) Conceptual model of shelter planning based on the Vitae System. Annuals Disaster Prev Res Inst Kyoto Univ 49(B):181–187

Chapter 20
Multiple Factors and Its Effects on Construction Strategy for Post Disaster Shelter

Abhishek Nandan, Deesha Patel, and Prasenjit Mondal

20.1 Introduction

Natural disasters are the events that have demolished houses, schools, various infrastructures, it has also destroyed our crops and these events may also result in loss of lives. Disaster is the reason why people have been forced to displace to different locations and leave their own residence and belongings behind, to take shelter in some foreign and unknown place and be the refugee or migrants for a short time period or longer time period (Hoeppe 2016; Mallick et al. 2017). The data given by Internal Displacement Monitoring Centre around 26.4 million people or more all across the world were evacuated from their residencies each year due to Natural Disaster and Hazards in the years between 2008 to 2018, in this approximately 26.14% of those people were from South Asia (Ponserre and Ginnetti 2019; Torres and Casey 2017). Natural Disasters are the major cause of direct and indirect physical and mental damage to the people as they face various difficulties like loss of jobs, residency resulting in forced mitigation in an unfamiliar environment and social ties (Torres and Casey 2017; Park et al. 2017).

Ever since 2018 there have been various types of hazards and displacement or mitigation due to it, the highest mitigation or displacement of people that is 50.62% is caused by flood, after that comes 34.54% which is caused due to storm, followed by it next comes 12.23% which is caused by earthquake and Tsunami and 0.61% of displacement is caused by wildfire. In the year 2018 about 315 natural disasters happened and around 68 million people were affected and 11,804 casualties occurred, up-to 131.7 billion USD to re-establish and redeem the site to its near the

A. Nandan (✉) · D. Patel
School of Advance Engineering, UPES, Dehradun, India
e-mail: abhisheknandan24@gmail.com

P. Mondal
G. D Goenka University, Haryana, India

N. A. Siddiqui et al. (eds.), *Advances in Disaster Management, Volume 2*, Springer Transactions in Civil and Environmental Engineering,
https://doi.org/10.1007/978-981-96-4047-8_21

20.2.4 *Progressive Shelters*

Progressive shelters are the post disaster household that are planned and designed such that it could be to upgrade it to more permanent residence whenever needed. This could be achieved by including and considering future transformation and all the possible alteration that could occur in the structural unit at the very beginning of the designing phase (Transitional and Progressive Shelter Guidelines 2014).

20.2.5 *Core Shelter*

Core shelters are the type of post disaster household shelter which is planned and designed as a permanent residency; it is constructed while assuming it to be the part of the future permanent housing. This allows the design to facilitate the future process of extension of the household, by using all means and resources. The basic idea of constructing the core shelter is to create one or two rooms which ultimately provide protection as well as constructing the post disaster shelter that could meet the standards of permanent housing. This shelter is also known as *One Room Shelters*, it facilitates development and also meets all the standards of permanent housing but it is not a complete full permanent house but a shelter (Transitional and Progressive Shelter Guidelines 2014) (Table 20.1).

While analysing it could be easily understandable that facing any sort of crises it would be the most widely constructed shelters would be *emergency shelters.* It is very easy to establish any emergency shelters, although it is only constructed for a short-term period. But it is very easy to construct and does not really require highly skilled labour for construction as emergency shelters are cost effective and all the material required to build it is easily available. While preparing Transitional or core shelters it requires highly skilled labour, it is time consuming as well as it is costly. Thus, emergency shelters are most widely used.

20.3 Various Materials Used for the Constriction of Shelter

There are various types of materials required to construct different components of the shelter, like the structure of the shelter, covering, windows screen, seaming. The material for all the components must be selected based on the requirement (Walta Asfaw 2021).

Table 20.1 Comparative studies of various types of shelters

Sr. no.	Types of shelters	Material	Minimum area per person (m^2)	Minimum height (m)	Life span (months)	Capacity (persons)	Cost (INR)
1	Emergency shelter (https://emergency.unhcr.org/entry/36774/emergency-shelter-standard%20%20Transitional%20shelters)	Bamboo, Cardboard, Rubber, tarpaulin	3.5	2	3	5	26,000
2	Temporary shelter (https://www.shacc.ch/documents/files/library/Guideline/IFRC-Transitional-Shelters-8-designs.pdf)	Coco lumber, plywood, corrugated tin sheet	3.5	2	12–18	5	35,000
3	Transitional shelter (Transitional and Progressive Shelter Guidelines 2014)	Bamboo, Plywood, tin and steel sheets	3.5	2	24	5	74,000
4	Progressive shelter (Transitional and Progressive Shelter Guidelines 2014)	Reusable materials like tarpaulin sheets, wood, metals, bricks	3.5	2	120	5	74,000
5	Core shelter (Transitional and Progressive Shelter Guidelines 2014)	Bricks, cement, concrete, reinforcement	3.5	2	240	5	18,000

20.3.1 *Material for Structure of the Shelter*

There are various materials that could be used for construction of structure for the shelter but some of these materials are: (1) Wood; (2) Fibre-glass and Polymers.

20.3.1.1 Wood

In history the shelters were usually built with the help of wood as it is usually available all across the world. The only drawback was wood is very heavy and it has the strength sometimes more than what is needed. That was the reason for more open space design. Wood needs a lot of crafting and hard work to make the required structure. Using metal could have been an option because of its strength, but there has been an issue of corrosion. Even if there are many methods to prevent corrosion, it is not cost effective. There is one more reason why using metal is not a wise choice, as connecting pipes usually requires precise art of welding. To find a welder at a time of disaster is not an ideal situation (Walta Asfaw 2021).

20.3.1.2 Fibre-Glass and Polymers

Fibre-glass and polymers are the most effective material to be used for structure of the shelter, due to its high strength and flexibility. Of course, the strength of these materials cannot be compared to metal, but the problem of corrosion can be easily solved. Fibre-glass and Polymer are very easy to handle and are cost effective. Most commonly used polymer for structure construction is Acrylonitrile Butadiene Styrene (ABS) and Polyvinyl Chloride (PVC) as these are light in weight and rigid in nature. They are generally used for underground piping due to its resistance against rusting and corrosion. There is another useful quality of the polymers and that is even when deformed it will bend back, which is not the case with metal as it stays deformed (Walta Asfaw 2021; Al-Sherrawi 2021).

Fibre-glass is one of the best materials Known for its flexibility and its resistance to high Stress. This material is mostly used in the case of outdoor tents. Generally, Fibre-glass is used for the structure of the roof. Due to its flexibility, it could be arched for a very long time period without any sort of deformation (Al-Sherrawi 2021).

20.3.1.3 Using Matrix for Decision Making

Graph 20.1 shows the sample matrix that can be used for deciding the material to be used for structure of the shelter. The categories that are being used for comparing the materials are weight, cost and the ease of working. Here the scorning is done out of 5 according to the materials compatibility in the preceding criteria. At last, the total is done and the most suitable material can be chosen. The criteria can differ according to the requirement.

Using this matrix, it is very easy to decide on the material that should be used for constructing structure for the shelter, as the matrix shows the top two material based on the effective cost and the ease of working with the material are PVC (Polyvinyl Chloride) and ABS (Acrylonitrile Butadiene Styrene) (Walta Asfaw 2021).

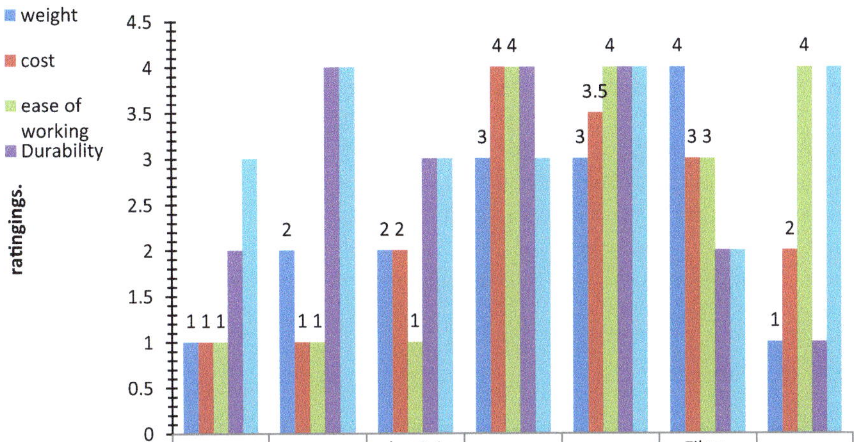

	copper	steel	aluminium	ABS	PVC	Fibre-glass	Wood
weight	1	2	2	3	3	4	1
cost	1	1	2	4	3.5	3	2
ease of working	1	1	1	4	4	3	4
Durability	2	4	3	4	4	2	1
Availability	3	4	3	3	4	2	4

Materials

Graph 20.1 Graphical representation of data of the matrix for selecting the material for structure (Walta Asfaw 2021; Al-Sherrawi 2021)

20.3.2 Material for the Fabric of the Shelter

The most common fabric materials for the cover material of the shelter are cotton, polyester, poly-cotton and nylon. There are various criteria on which the cover material has to be selected.

20.3.2.1 Cotton

Cotton is one of the cheaper and biodegradable materials that could be used in comparison to synthetic fabric that is more expensive to manufacture. Cotton threads are heavy in nature that is one of the main reasons why it is the most abrasion and durable resistant material. Cotton threads prevents any type of east tears in the fabric from growing any further, and if there is accidentally any tear occurring it is very easy to mend any tears by using sewing patches (Walta Asfaw 2021; https://expeditionportal.com/forum/threads/rooftop-tent-fabrics-and-care.50630/). It is a well-known quality of the cotton; its heat transfer is slower as not much ventilation can occur if closed completely. As the annual temperature of India is approximately 24.6 °C (Delhi) in India, using Cotton as the fabric for the cover of the shelter could help to maintain cooler temperature (https://en.climate-data.org/asia/india-129/). But there are some disadvantages of using Cotton as it does not possess any resistance towards UV which ultimately results in the fabric getting discoloured after a particular period of time while being in direct contact with sunlight. Another disadvantage is that cotton absorbs the moisture present in the atmosphere which then causes the cotton fabric to swell and become tightened. This will further cause the increase in weight of the material due to the swelled fibres. Due to the moisture absorbed by the cotton it will slowly start rotting and degrading, if not thoroughly dried before being stored away (https://outdoorcommand.com/canvas-tents-v-s-nylon-tents-which-are-better/).

20.3.2.2 Polyester

Polyester is the fabric which does not degrade that and it can also be stored even when it is wet unlike Cotton, as the fibres of Polyester do not absorb water instead the water easily slides off. Polyester is known to be lighter than cotton and it has very tightly bound fibres due to which the fabric is resistant to any form of stretching and tear. But there are few drawbacks, as the fibres are tightly bound it lacks the quality of breathable material. Similarly heat transfer occurs faster which could result in the inside of the shelter being warmer than the outside area. The most important one is that it is very hard to repair any tear in the fabric (Walta Asfaw 2021).

20.3.2.3 Nylon

Nylon is the type of fabric which has many similarities with polyester as it is also a synthetic material. It is known to be natural water resistant due to which it gets dry very quickly, it also allows water to condense the outer surface area before running off, it is a very light weighted fabric easy to handle. Once there is a tear it is very hard to mend it, but it is different from polyester as the presence of a net like pattern in the rip stop nylon stops the tear from increasing any further. There are few disadvantages as well nylon is not UV resistant and it slowly loses its abilities when exposed to extreme and harsh weather or when exposed to sunlight. Nylon will weaken if it gets wet, ultimately making it harder to be used as a fabric material for covering the shelter (https://outdoorcommand.com/canvas-tents-v-s-nylon-tents-which-are-better/; http://outdoorsewingsolutions.co.uk/tents-made-wide-range-fabrics/).

20.3.2.4 Poly-Cotton

Poly-cotton is the combination of 65% polyester and 35% cotton, as it is the combination of Polyester and Cotton it also has some of the combined advantages of both the materials. But it is still heavy in weight making it hard to handle and it will still absorb some of the moisture due to thick fibres of cotton which makes it susceptible for rotting and degrading (http://outdoorsewingsolutions.co.uk/tents-made-wide-range-fabrics/).

20.3.2.5 Using Matrix for Decision-Making

In this matrix, we are using eight different criteria for making decisions, and to choose the correct fabric that could be used as the cover material for the shelter (Graph 20.2).

This matrix indicates that polyester is the material that meets maximum criteria in this scenario and it should be selected as the cover material for the shelters.

20.3.3 Material Used for Window Screen

It is necessary to provide a source of ventilation in shelter, to increase the factor of breathability in shelter windows must be provided as in many cases the cover material used is Polyester or polyethylene which doesn't have the quality of ventilation. The most common materials used for the window screens are aluminium and fibre-glass.

Further, matrix is used to choose the better material for constructing the window screen six criteria were used (Graph 20.3).

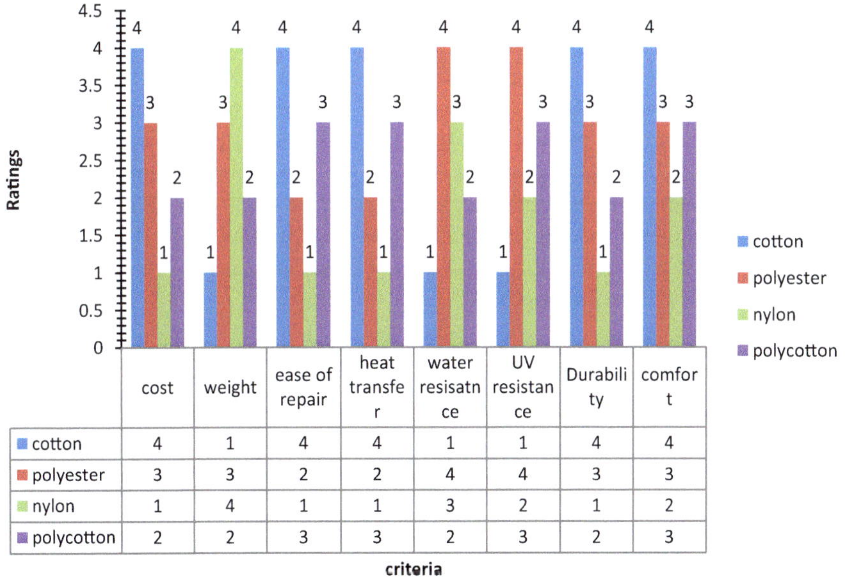

	cost	weight	ease of repair	heat transfer	water resisatnce	UV resistance	Durability	comfort
cotton	4	1	4	4	1	1	4	4
polyester	3	3	2	2	4	4	3	3
nylon	1	4	1	1	3	2	1	2
polycotton	2	2	3	3	2	3	2	3

Graph 20.2 Walta Asfaw (2021): Graphical representation of data of the matrix for selecting the material for cover fabric

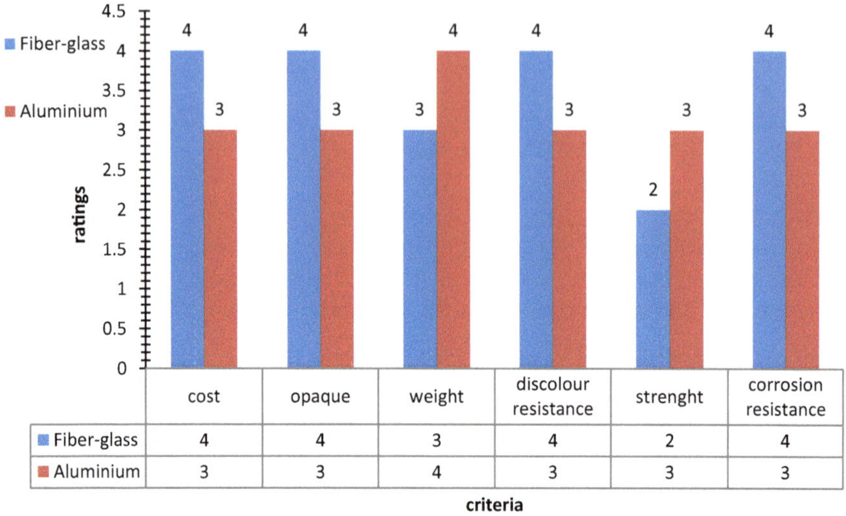

Graph 20.3 Graphical representation of data of the matrix for selecting the material for the window screen (Walta Asfaw 2021)

20.3.4 Material Used for Seaming

In case of seaming the material used is polyethylene tarp; it is a uni-body trap which means it is a single piece of fabric that covers the roof and all the four sides of the walls as well as its frame. There are two walls of three metres length with an arch which are directly sown with the single piece of five-metre trap cover.

All of this was done so that the trap could be stretched and it would decrease the caving of the trap that happens in between vertical members. Unibody design also provides some extra tarp material which could be sown at desired location for each vertical member. The sleeves are also shown on the house as well as on vertical members, which results with equal division of stress and energy between tarp and vertical members (Walta Asfaw 2021).

20.3.5 Water Supply

Safe water supply is a basic necessity for each and every type of shelter. The quality of water that is being supplied to the shelter should be met with the standards established by WHO (World Health Organization) (Hitchens 1972). The minimum requirement of water for drinking cooking and other basic hygiene are: For hospital and first aid the minimum water require is approximately 40–60 L per person per day, for cooking the water require is up to 20–30 L per person per day, and as for the shelter and hygiene purposes the water requires is up to 15–20 L per person per day (Assar 1971).

Various types of sources can be used for obtaining safe water such as: (1) Municipal Water, (2) Private water supply system, (3) springs and wells, (4) Surface water.

20.3.5.1 Municipal Water

If the multiple system is affected and damaged due to disaster it is the first priority to repair the system and restore it to its original value. Various new methods have been developed like coupling and plastic packing the pipe and repairing it in the fastest possible way. In case of disaster the water pressure should be raised so the water can be distributed to a maximum number of houses and shelters. But if in case the water pumping station and distribution system is damaged beyond repair while being affected by a disaster other method can be used (Assar 1971).

20.3.5.2 Private Water Supply System

At the time of disaster there are many cases when a private water supply system has been used by the community to meet the basic need of water. This type of system may belong to food plants, industrial establishments or other agriculture sectors. Usually in this case the source of water is a deep well, canal, river etc. Water collected from these must be treated and disinfected before distributing (Assar 1971).

20.3.5.3 Natural Springs and Wells

Springs and Wells can usually be found around the area where disaster has occurred, they are less contaminated as compared to the surface water. There is one more advantage and that is it could be used without any major treatment the only treatment which required would may be disinfection. But firstly, we must carefully check the geographic area around where the disaster occurred. As is their presence of limestone or certain types of rocks in case of earthquake they may contaminate the groundwater (Assar 1971).

20.3.5.4 Surface Water

This is the least used water supply system as there is a very high chance of surface water being contaminated with pollutants, it is usually turbid in nature thus, it must be avoided. Surface water must be only used after proper treatment, removal of turbidity and impurities, colour and odour must be removed, proper disinfection must be provided before consumption. Various types of techniques are used to reduce

Harvey P, Baghri S, Reed B (2002) Emergency sanitation: assessment and programme design, p 358. Accessed 09 Dec 2021 [Online]. Available: /articles/book/Emergency_sanitation_assessment_and_programme_design/9585065/1

Hitchens RAN (2021) International standards for drinking water, 3rd edn. Br J Ind Med 29(3):349, 1972. Accessed 09 Dec 2021 [Online]. Available: https://www.ncbi.nlm.nih.gov/pmc/articles/PMC1009447/

Hoeppe P (2016) Trends in weather related disasters—consequences for insurers and society. Weather Climate Extremes 11:70–79. https://doi.org/10.1016/J.WACE.2015.10.002

International Strategy for Disaster Reduction. Guideline Note on Recovery Shelter. https://www.recoveryplatform.org/assets/Guidance_Notes/Guidance%20Note%20on%20Recovery-Shelter.pdf. Accessed 09 Dec 2021

Indian Climate: Average Temperature, Weather by Month & Weather for India—Climate-Data.org. https://en.climate-data.org/asia/india-129/. Accessed 09 Dec 2021

IOM-QIM (2021) Transitional shelter guidelines. https://www.iom.int/sites/g/files/tmzbdl486/files/our_work/Shelter/documents/Transitional%20Shelter%20Guidelines%20-%202012.pdf. Accessed 09 Dec 2021

IStructE Temporary Demountable Structures—[PDF Document]. https://fdocuments.in/document/istructe-temporary-demountable-structures/. Accessed 09 Dec 2021

Johnson C (2007) Strategic planning for post-disaster temporary housing. Disasters 31(4):435–458. https://doi.org/10.1111/J.1467-7717.2007.01018.X

Kats GH (2003) Green building costs and financial benefits. Accessed 09 Dec 2021 [Online]. Available: www.masstech.org

Mallick B, Ahmed B, Vogt J (2017) Living with the risks of cyclone disasters in the South-Western Coastal Region of Bangladesh. Environments MDPI 4(1):1–17. https://doi.org/10.3390/environments4010013

Naylor A, Faure Walker J, Suppasri A (2018) Suitability of the early warning systems and temporary housing for the elderly population in the immediacy and transitional recovery phase of the 2011 Great East Japan Earthquake and Tsunami. Int J Disaster Risk Reduction 31:302–310. https://doi.org/10.1016/J.IJDRR.2018.05.022

Park JY, Son M, Park CK (2017) Natural disasters and deterrence of economic innovation: a case of temporary job losses by Hurricane Sandy. J Open Innov Technol Market Complexity 3(1):1–23. https://doi.org/10.1186/S40852-017-0055-2/TABLES/11

Peacock WG, Dash N, Zhang Y (2007) Sheltering and housing recovery following disaster, pp 258–274. https://doi.org/10.1007/978-0-387-32353-4_15

Pomponi F, Moghayedi A, Alshawawreh L, D'Amico B, Windapo A (2019) Sustainability of post-disaster and post-conflict sheltering in Africa: what matters? Sustain Prod Consumption 20:140–150. https://doi.org/10.1016/J.SPC.2019.06.007

Ponserre S, Ginnetti J (2021) Disaster displacement: a global review, 2008–2018, 2019. Accessed 09 Dec 2021 [Online]. Available: http://repo.floodalliance.net/jspui/handle/44111/3485

Quarantelli EL (1982) General and particular observations on sheltering and housing in American disasters. Disasters 6(4):277–281. https://doi.org/10.1111/J.1467-7717.1982.TB00550.X

Rooftop Tent Fabrics and Care|Expedition Portal. https://expeditionportal.com/forum/threads/rooftop-tent-fabrics-and-care.50630/. Accessed 09 Dec 2021

Shiozaki Y, Tanaka Y, Hokugo A, Bettencourt S (2013) Transitional shelter. Accessed 09 Dec 2021 [Online]. Available: https://openknowledge.worldbank.org/handle/10986/16148

Siriwardhana SD, Kulatunga U, Samaraweera A, Shanika VG (2021) Cultural issues of community resettlement in post-disaster reconstruction projects in Sri Lanka. Int J Disaster Risk Reduction 53:102017. https://doi.org/10.1016/J.IJDRR.2020.102017

Sukhwani V, Napitupulu H, Jingnan D, Yamaji M, Shaw R (2021) Enhancing cultural adequacy in post-disaster temporary housing. Progr Disaster Sci 11:100186. https://doi.org/10.1016/J.PDISAS.2021.100186

Sanitation in Emergencies Overview|SSWM—Find tools for sustainable sanitation and water management! https://sswm.info/water-nutrient-cycle/wastewater-treatment/hardwares/sanitation-emergencies/sanitation-in-emergencies-overview. Accessed 09 Dec 2021

Torres JM, Casey JA (2017) The centrality of social ties to climate migration and mental health. BMC Public Health 17(1):1–10. https://doi.org/10.1186/S12889-017-4508-0/FIGURES/1

Ten Great Emergency Shelter Designs|Engineering For Change. https://www.engineeringforchange.org/news/ten-great-emergency-shelter-designs/. Accessed 09 Dec 2021

Tents Are Made Out of a Wide Range of Fabrics—Outdoor Sewing Solutions. http://outdoorsewingsolutions.co.uk/tents-made-wide-range-fabrics/. Accessed 09 Dec 2021

Transitional and Progressive Shelter Guidelines (2014) Accessed 09 Dec 2021 [Online]. Available: www.haiyansheltercluster.org

Transitional shelter Eight Design. https://www.shacc.ch/documents/files/library/Guideline/IFRC-Transitional-Shelters-8-designs.pdf. Accessed 09 Dec 2021

Uddin K, Matin MA (2021) Potential flood hazard zonation and flood shelter suitability mapping for disaster risk mitigation in Bangladesh using geospatial technology. Progr Disaster Sci 11:100185. https://doi.org/10.1016/J.PDISAS.2021.100185

UNDRR (2021) UNDRR annual report 2020—World|ReliefWeb. https://reliefweb.int/report/world/undrr-annual-report-2020. Accessed 09 Dec 2021

UvA-DARE (Digital Academic Repository) (1997). https://doi.org/10.1023/A:1024842532117

Walta Asfaw DH (2021) Disaster relief shelter. https://www.humanitarianlibrary.org/sites/default/files/2014/02/DRS_FinalReport.pdf. Accessed 09 Dec 2021

Chapter 21
Study of Biological Hazard in Various Industries with Special Reference to COVID-19—A Review

Abhishek Nandan, Damini Rana, and Deepraj Sinha

21.1 Introduction

The major pandemic such as Antonine Plague, Plague of Justinian, The Black Death, Cholera Pandemic, Small pox, Flu Pandemic, Asian Flu, Poliomyelitis, HIV/AIDs Pandemic, Novel Coronavirus have made severe impact to the society (Hays 2005). These pandemics does not have any kind of pre-defined prevention measures. Hence, awareness on the biological agents is utmost necessary considering the current pandemic due to Novel Coronavirus (Shekhar et al. 2015). The problem when there is a pandemic is that it brings the entire world to a stop (Zizek 2020). Pandemic impact each and every sector starting with Industry, education, transportation, business, and many other. There are no specific defined measures for this biological hazards.

Industrial growth are one of the source which has various impact on the society. The growth of industry is ensured only when it operates following basic norms of safety from various hazards. The biological hazard is one such threat to industries which they have to deal in their workplace. The probability of occurrence of biological hazard is less but the severity factor and detectability factor is quite high.

Biological hazards are present everywhere in the surrounding. They are generally found in the form of various agents in various industrial sectors. The biological hazards are generally invisible the risk imposed by them are very critical. Exposure to biological hazard may result into various short term and long term disease including life threatening problems. Health impacts may cause various allergic reactions, infections, occupational asthma, multi-organ defects, zoonotic diseases, poisoning, sensitizations and various carcinogenic diseases. The Routes of Entry of these biological agents are inhalation during breathing activity, absorption through skin or eye, ingestion through hand while eating and injection due to puncture on skin.

A. Nandan (✉) · D. Rana · D. Sinha
UPES, Dehradun, India
e-mail: abhisheknandan24@gmail.com

N. A. Siddiqui et al. (eds.), *Advances in Disaster Management, Volume 2*, Springer Transactions in Civil and Environmental Engineering,
https://doi.org/10.1007/978-981-96-4047-8_22

Classification of Biological Agents (Lonergan 2013)

Group 1: This is the classification of biological agents which is not likely to cause any kind of diseases in human beings. Spread of such agents can be prevented by frequent handwashing.

Group 2: This is the classification of biological agents which generally cause mild symptoms in human beings. Spread of such agents can be prevented by frequent handwashing along with use of appropriate activity specific protective equipment.

Group 3: This is the classification of biological agents which generally cause very severe or fatal symptoms in human beings. Spread of such agents can be reduced by vaccines.

Group 4: This is the classification of biological agents which generally cause severe illness or even death in human beings. Currently there is not preventive measures for such illness.

The objective of this review is to create awareness among the workers working in Industry which are prone to biological hazard. There are very less information available on Biological Hazard. Hence such review will not only aware the people but will also protect them from the harmful biological agents by following basic precautionary measures.

21.2 Methodology

The review is done using Bio Safety Risk Assessment technique.

1. Identify the Biological Hazard—Specific Industry wise using Literature review

- Agricultural and animal Husbandry
- Forestry
- Hospital and medical industry
- Waste water treatment plant
- Engineering industry
- Process industry

    Food processing
    Wood processing
    Textile industry

The reason for selecting these industries as these industries play a major role in utilizing human being for completion of the activities. The outputs for these industries are important for day to day life usage. These industries play a pivotal role in maintaining financial, social and infrastructural growth of the country. Any disruption in any one of the industry sector will impact the other industrial sector in some or the other way. As it is witnessed how the impact of COVID-19 has impacted each one of this sector in different ways. It is the need of the hour for each and every sector to re-audit and find out the inherent biological hazard as well along with other known hazards. The priority should be given equally for biological hazard compared to what it is given for physical or chemical hazard.

2. Assess the Biological Hazard and Associated Risk in the Specified Industry—Consequences

For the first and second part of methodology, we have reviewed around 44 papers to understand the existing biological hazards and their associated risk in various industrial sector. For agriculture and animal husbandry sector, we have reviewed 9 papers to understand the biological agents that are causing diseases and infection in human beings. We also get the understanding on how the animal farming industry is contributing for the cutaneous anthrax among human beings due to the bacteria Bacillus Anthracis. And finally, the precautionary measures that could be adopted to protect people working in agriculture and animal husbandry from the agents. For the forestry sector, we have reviewed around 6 papers, to get to know the impactful virus in forestry industry. Hantavirus is one of the most critical biological agent impacting the forest workers and people residing near forest. The review of the control strategy on Hantavirus is also conducted. In the hospital sector, around 6 papers reviewed and the spread of coronavirus (MERS and SARS) was observed in health and medical facility workers including doctors and nurses. Studies on preventive measures are also reviewed on these virus in hospital sector. In the waste water treatment plant around 4 papers were reviewed to find out the biological hazards along with the source areas for the agents. The review is also done to find out the mechanism to control the spread of biological agents in waste water treatment plant. In the engineering industries sector, the review suggest the metal working fluid to be potential source for the biological agents. Also a review was done on the alternative metal fluid composition which will inhibit growth of biological agents. Similar reviews were conducted on three types of process industry—the food processing, wood processing and the textile industry to find out the biological hazards, risk associated and preventive measures.

3. Suggest a Strategy for Control Measures for Identified Hazards as Well as COVID-19 Hazard

The control strategy will be prepared in such a way that it will solve the purpose for any types of biological hazards. This strategy is prepared on 6 step barrier process. Every step has its own effectiveness and importance. It may be possible that for few cases all the steps may not be applicable. So the best applicable strategy has to be applied to ensure the precautionary measures from the biological agents in the workplace. This strategy is generally followed in industry for physical and chemical hazards. As biological hazard has the same potential to cause acute, chronic or fatal symptoms, such strategy to be adopted by industries to ensure a workplace free of biological hazard.

21.3 Literature Review

There is no boundary limit for biological hazard. Starting with agriculture industry, a research was conducted to find out the microorganisms present in the agricultural soil. As per the study there were various categories of bacteria and fungi identified. One of the identified bacteria species is Bacillus Anthracis (Jasuja et al. 2013). The researcher used the grain staining method to identify the bacteria content in the soil. The disease caused due to this specific bacteria is anthrax. In the study (Lalla 1992), the researcher showed how three members of a family who came in contact with the cow having Bacillus Anthracis were found out to be suffering from cutaneous anthrax. The study 25 cases of cutaneous anthrax was analyzed and it was found that 23 cases were linked to the animal farming (Demirdag, et al. 2003). Another study of 85 cases were done of cutaneous anthrax and the result revealed were 26 cases had contact with only sheep, 23 cases had contact with goats, 13 had contact with cows and 23 cases had contact with sheep, goats and cows (Karahocagil et al. 2008). Another study revealed 23 out 28 cases of cutaneous anthrax has direct contact relation with animal husbandry (Denk et al. 2016). A review conducted on 22 patients for 7 years and it was revealed that all the 22 people have suffered from cutaneous anthrax and all the patients have some kind of direct or indirect contact with animal husbandry (Doganay 2010). A different scale study was conducted for the count of cutaneous anthrax in human cases arising from the livestock disease cases for 3 years (Kracalik et al. 2014). The result says that in the year 2010, there were 9 livestock cases and 28 human cutaneous anthrax cases (Kracalik et al. 2014). The result for 2011 increased to 30 livestock cases and 81 human cutaneous anthrax cases (Kracalik et al. 2014). Similar increasing trend was observe for 2012 with 35 livestock cases and 142 human cutaneous anthrax cases (Kracalik et al. 2014). Based on these studies, it is utmost required to sensitized farmers and animal husbandry workers as there are lack of awareness on the anthrax. The affected people mainly were not having any knowledge on the anthrax and this is leading to the spread of the disease into the neighboring places (Kungu et al. 2020). One of the study suggested three important aspects in controlling the anthrax causing rate. Firstly awareness and spreading knowledge through community education forum on the agents causing anthrax and the symptoms in human beings. Secondly the reporting of anthrax related cases to the local health departments so as to ease the process on identification of the exact root causes of disease causation. And finally ensuring the effectiveness of animal immunization program (Kumar 2000). Various studies on the diagnosis, treatment procedure and the prevention measures including vaccination are conducted but awareness has to be created among the people to know the hazardous impact and associated consequence of the biological agents.

After agriculture, forestry was reviewed. Forestry workers are prone to various biological agents and many infectious diseases are resulting due to these agents. Most predominant biological agent found was Hantavirus. At a survey in Hungary 38 out of random 835 samples of forest workers, tested positive for hantavirus (Oldal 2014). A survey in Germany suggested similar trends. For female forest workers 64 samples were taken, 10 were found positive for hantavirus (Mertens et al. 2011). For male forest workers 499 samples were taken, 41 were found positive for hantavirus (Mertens et al. 2011). A study shows forest workers and people living in less than 100 m near forest are getting affected due to hantavirus (Sin et al. 2007). A study in Poland suggested that 25 samples tested positive, 76 were doubtful out of 820 samples from forest workers and people residing near forest areas (Wróblewska-Łuczka et al. 2017). For Hantavirus, there is no approved vaccine released yet. The basic treatment that is followed is antiviral immune therapy or providing antiviral drugs which can help to cure the person infected with Hantavirus (Krüger et al. 2001; Maes et al. 2004).

After forestry industry, the reviews are conducted on hospital and other health care facilities. Hospital is one such sector where we can find almost all types of disease causing micro-organisms virus, bacteria, protozoa, parasites, fungi and many others. In this review we have mainly focused on the coronavirus aspect in hospitals. As per the study conducted on 65 patients with MERS CoV infection, 42% of them were directly or indirectly related to hospital or medical facilities (Hunter 2016). Another interesting in depth study was conducted among 7 medical practitioners who were infected with MERS CoV (Memish et al. 2013). All the health care workers were female (Memish et al. 2013). The health care workers are checked for various natural symptoms to establish the potential symptoms of MERS CoV infection (Memish et al. 2013). The final conclusion of this study was also suggested for health care employees needs to be aware about the infections and take necessary precautionary measures (Memish et al. 2013). A study conducted on a single patient to check the evaluation of the MERS CoV infection transmission (Alfaraj et al. 2018). From 1 infected person it spread to 7 person within just 12 days (Alfaraj et al. 2018). All the contacts were traced as per the person tested positive (Alfaraj et al. 2018). The study also revealed that 4 out of those 7 patients were asymptomatic. This seriousness of the disease as it is not identifiable without testing (Alfaraj et al. 2018). A study conducted on 80 health care workers out of which 45 tested positive due to SARS CoV infection. The conclusion made was that even if the patients are asymptomatic they need to ensure the proper control measures (Wilder-Smith 2005). In the 2003 SARS-Cov outbreak, 21st of the cases globally were found in health care workers (Dowell 2004). Swab specimen were collected from various hospital surfaces and 26 were tested positive out of 94 swabs taken (Dowell 2004). Hence it reflects the severity of healthcare workers working in hospitals where SARS CoV patients are treated. On the basis of control measures for 4 systematic control measures were described for protecting from SARS CoV infection (Christian 2004). The 4 measures are admin control, environmental engineering based control, PPE and quality control were discussed as a precautionary measures for SARS CoV infection (Christian 2004).

After hospital sector, comes the waste water treatment industry. The sector which can have wide range of microorganisms starting with bacteria, fungi, virus, protozoa, roundworms and many others. Few of the diseases associated due to this are gastrointestinal diseases, infection related to Hepatitis A and B, viral disease, parasitic infection, leptospirosis and many others (Clark 1987). A study was conducted to check the limits of the endotoxin and bacteria counts present in the industrial effluents at waste water treatment plant. The effects were tested at various location of the waste water treatment like pumping zone, basin area, controlling room and other areas (Laitinen et al. 1994). The exposure rate for the waste water treatment plant workers mainly depends on the capacity and characteristics of the infrastructure. The movement of the sewage and the constituents also play a contributing factor for the worker exposure (Korzeniewska 2011). The impact due to the exposure of the microorganisms does not only lead to acute effect but also cause various chronic effect. A specific medical surveillance initiative was taken up to give focus on the chronic effects of the workers in waste treatment plant with the help of absentees' record (McCunney 1986).

After understanding the microorganisms' impacts on workers of waste water treatment plants, the review of the similar aspect is done in engineering industry. As it is known that metal working fluid contain various microorganisms which can cause various chronic diseases. One of the studies suggested 23 types of bacteria and 9 types of fungi is present in metal working fluid emulsion causing infectious diseases (Passman and Rossmore 2002). Respiratory related symptoms are mainly observed in the metal engineering industry workers. The workers also tend to develop various lung diseases due to exposure to aerosol present in the metal working fluids. This may lead to asthma or even severe pneumonia among the metal engineering workers. Hence use of water-based metal working fluid were tested on pilot scale. As these would result into reduction in growth of microorganisms in the fluid (Rosenman 2015).

Food Industry has lot of operations which are prone to biological hazard. Various studies have been conducted on the food industry to understand the biological hazard, its impact and its precautionary measures. The committee reviewed some 816 reports based on the biological hazard parameters. In these report there were total of 14 biological agents that has been identified from the 816 reports and these virus are getting spread in food industry as there are lot of hand contact operation are present (Greig et al. 2007). Another study was conducted on the 816 reports about mortality and morbidity rate of the people which was mainly dependent on the food industry. One of the major contributing factor in spreading the disease is the infected worker working in the food industry and also explaining the gravity of the outbreak (Todd et al. 2007). Further in depth study on the factors were carried out. A documented information were prepared on the characterization of the factors leading to the outbreak of the biological disease related to food industry (Todd et al. 2007). The infected workers were observed to find out the dosage of infection. The study also concluded the various mode of pathogen career in the infected workers (Todd et al. 2008). The background study for the route of the infection mode for the food workers is being considered so as to take necessary precaution to prevent the spread at the source (Todd et al. 2008). To reduce the illness in food industry due to transmission, the existence and transfer mode of pathogen is reviewed critically to so as to provide information and aware the workers about the biological hazards (Todd et al. 2009). Two types of barriers to prevent the transmission of pathogen is being studied so as to protect the people from various infections. The preventive measures are physical as well as chemical mode of barrier which were part of the studies (Todd et al.

2010). Specific study was conducted on one of the personal protective equipment and also mandated by legal jurisdiction. Critical study on the effectiveness and failure modes are discussed so as to aware the people using the gloves during food handling activities (Todd et al. 2010). The gloves may cause a kind of discomfort and hence it becomes necessary to keep the hands clean and free of germs. Based on this a study on the importance of washing and then drying the hand before handling the food was conducted to reduce the spread of the germs (Ewen et al. 2010). Washing hands with soap may not be effective in removal of germs from hand surface, hence study is conducted on alcoholic antiseptic hand wash shall be used for washing (Todd et al. 2010). If adequate water is not available, sanitizer may be used. Hand washing should be made as part of a legal compliance to ensure control over the spreading of the germs (Todd et al. 2010).

Wood-processing industry prone to various endotoxins. A study in Croatia was conducted in wood-processing industry by collecting samples airborne samples. The results revealed that the endotoxin concentration is higher than the proposed exposure level and this may result into various respiratory diseases among the industry workers (Pipinić et al. 2010). In a lumber mill in British Columbia, 9% of the collected were found to have endotoxin dosage more than permissible limit (Dennekamp et al. 1999). Such cases are leading to various occupational diseases among the workers in wood-processing industry. In a wood industry in Tanzania, a study revealed that the exposure level of endotoxin is observed higher in dry season compared to a wet or moist environment (Rongo et al. 2004). A study on the dosage level of the endotoxin is studied in various segments of the wood-processing industry in Italy and the difficulty in maintain the control measures is also addressed (Gioffrè et al. 2012).

Textile industry also possess various biological hazards. It may lead various acute and chronic effect. The long term exposure to endotoxin is leading to risk of lung cancer among textile industry workers (Astrakianakis et al. 2007). Lung defects are also being observed as the workers are not aware of the biological hazards in textile industries (Ghani et al. 2016). Hence, strict measures needs to be implemented to protect the workers from the same.

Most of the biological hazards are either leading to an acute effect or a chronic effect. In many cases it may even lead to death of the individual depending on the severity of the hazard and the exposure level.

COVID-19 virus is one of the biological agent which didn't restrict its spread to one sector. Starting with agriculture, forestry, hospital, waste water treatment, engineering industry, food processing, wood processing, textile and many others. Hence, the precautionary measures for individual sector will have to be taken considering the sector specific biological hazard and COVID-19 biological hazard.

21.4 Suggestion and Discussion

As per the review of various studies, for the control strategy the technique followed in Industries for other hazards is the hierarchy of control mechanism. A hierarchy of control for Biological Hazard is generally not followed but following this steps the measures can be taken (Figs. 21.1 and 21.2).

As per the control measure strategy, the classification is done on 6 categories:

1st Category—Elimination

This is the most effective control measures. In this measures, the biological hazard can be totally removed.

2nd Category—Substitution

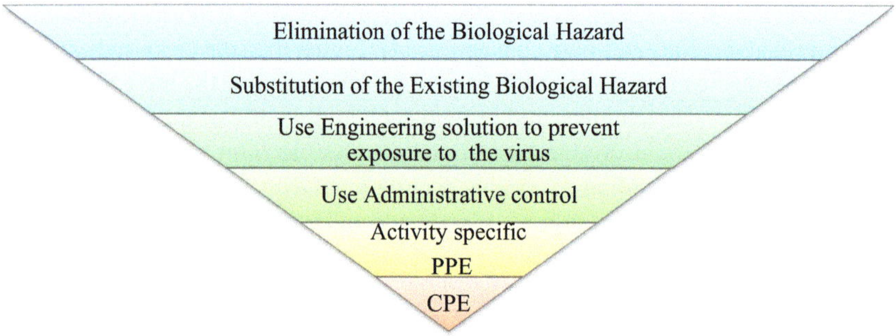

Fig. 21.1 Hierarchy of control for biological hazard

Fig. 21.2 Working from home technique prevents from getting infected from virus

Fig. 21.3 Engineering shield to prevent contact with virus source

In this measures, the process in which the biological hazard cannot be totally removed. In place of the existing process, a process to be applied which will have less impactful hazard.

3st Category—Engineering Solution

This is the one of the effective control measures. In this measures, the biological hazard can be controlled using engineering controls (Fig. 21.3).

(a) Isolation of the biological hazard source carrier.
(b) Separation of the biological agent carrier source from other source.
(c) Partial Enclosure by ensuring adequate ventilation
(d) Safety Interlock device systems

4th Category—Administrative approach

This is the one of the effective control measures. In this measures, the biological hazard can be controlled using administration controls (Fig. 21.4).

(a) Safe personal hygiene procedure.
(b) Reducing the time span of exposure.
(c) Maintaining a safe distance from the hazard
(d) Creating awareness, training and observing
(e) Health Surveillance and biological hazard monitoring

5th Category—Activity specific Personal Protective Equipment

This is the one of the least effective control measures. In this measures, the biological hazard can be controlled using a personal protective equipment during the activity (Fig. 21.5).

6th Category—Community Protective Equipment

Fig. 21.4 Training to aware people about the risk associated with biological hazard

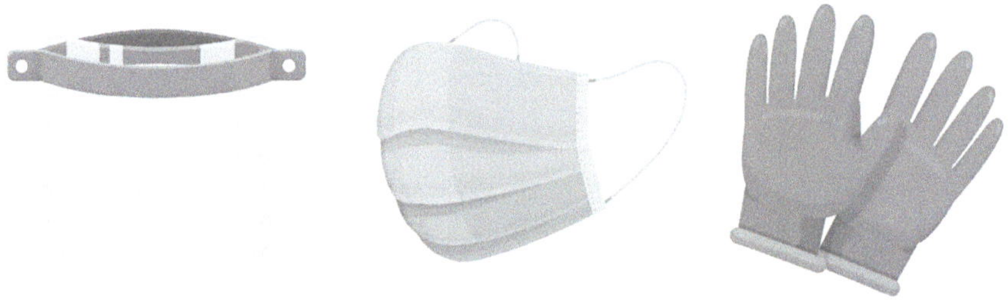

Fig. 21.5 Activity-specific PPE like face shield, approved mask and gloves

This is the one of the least effective control measures. In this measures, the biological hazard can be controlled using a protective equipment to reduce the transmission of biological agents from one person to another (Fig. 21.6).

Fig. 21.6 Wearing mask on public places to avoid spreading of germs and to protect from germs

21.5 Conclusion

The most important factor about the severity of the biological hazard factor is the lack of awareness among the people. The biological hazard known to people in the current scenario is the COVID-19 coronavirus. But, the review revealed that there are lot of other biological hazard that do exist and we ignore just because of the lack of awareness. Proper imparting of training and supervision will increase the awareness of people in terms of biological hazard. Health monitoring and surveillance is a handy source in order to keep check on the impact on human being due to biological hazard. All biological hazard are not air-borne and all are not transferred through human beings. All biological agents as a preventive measure may not require a vaccine. But, all biological hazard requires a control measures to avoid any acute or chronic or fatal effect. Control measures need not to be a kind of expensive measures. It can be as simple as washing hands frequently with a soap for at least 20 s. Wearing mask and gloves is not only protecting the person but it is also protecting the community spread of a biological agent through human being. It is difficult to find out the preventive strategy for all biological hazard but the basic control strategy suggested will be helpful in reducing the exposure level. The control strategy suggested is applicable not only for COVID-19 coronavirus measures but it will also protect from the other associated biological hazards in respective workplace.

References

Alfaraj SH et al (2018) Middle East respiratory syndrome coronavirus transmission among health care workers: implication for infection control 46(2):165–168

Astrakianakis G et al (2007) Lung cancer risk among female textile workers exposed to endotoxin 99(5):357–364

Christian MD et al (2004) Possible SARS coronavirus transmission during cardiopulmonary resuscitation 10(2):287

Clark CSJJ (1987) Potential and actual biological related health risks of wastewater industry employment 59(12):999–1008

De Lalla F et al (1992) Familial outbreak of agricultural anthrax in an area of northern Italy 11(9):839–842

Demirdag K et al (2003) Cutaneous anthrax in adults: a review of 25 cases in the eastern Anatolian region of Turkey 31(5): 327–330

Denk A et al (2016) Cutaneous anthrax: evaluation of 28 cases in the Eastern Anatolian region of Turkey 35(3):177–180

Dennekamp M et al (1999) Endotoxin exposure among softwood lumber mill workers in the Canadian Province of British Columbia 6(2):141–146

Doganay M et al (2010) A review of cutaneous anthrax and its outcome 3(3):98–105

Dowell SF et al (2004) Severe acute respiratory syndrome coronavirus on hospital surfaces 39(5):652–657

Ewen C et al (2010) Outbreaks where food workers have been implicated in the spread of foodborne disease. Part 9. Washing and drying of hands to reduce microbial contamination

Ghani N, Khalid A, Tahir A (2016) Cross-sectional study on the endotoxin exposure and lung function impairment in the workers of textile industry near Lahore, Pakistan 66(7):803–814

Chapter 22
Nature-Based Solutions for Drought Mitigation

Sanayanbi Hodam, Richa Srivastava, and Anil Kumar Gupta

22.1 Introduction

Drought, a recurrent climatic phenomenon, poses substantial challenges to agriculture, water resources, and socio-economic stability. Drought happens when there's not enough water because it hasn't rained enough for a short or long time (Naumann et al. 2015). It mostly occurs naturally. Water scarcity is when there's not enough water for a more m extended period, like months or even years, because people need more water than what the natural system can provide in rivers. This happens because of how people use water for different things like industries and drinking. Water stress is when there's not enough water, and it includes issues like water quality and accessibility. This problem usually gets worse during droughts (Duel et al. 2022). The impact of drought on a specific sector can vary between countries due to differences in inherent exposure and vulnerability, as well as the presence of existing mitigation measures and the adaptive capacity of each country (Duel et al., 2022; Naumann et al., 2015; Mens et al. 2022).The global rise in drought-related risks is attributed to both climate change and demographic shifts. Particularly for marginalized communities, the consequences of droughts are substantial, resulting in heightened poverty, shortages of food, and limited access to clean and adequate

S. Hodam
North Eastern Regional Institute of Water and Land Management, DoWR, RD & GR, Ministry of Jal Shakti, Tezpur, Assam, India
e-mail: sanayanbi.hodam@gov.in

R. Srivastava · A. K. Gupta (✉)
National Institute of Disaster Management, Ministry of Home Affairs, Delhi, India
e-mail: anilg.gov.in@gmail.com

R. Srivastava
e-mail: richa_roli@yahoo.com

N. A. Siddiqui et al. (eds.), *Advances in Disaster Management, Volume 2*, Springer Transactions in Civil and Environmental Engineering,
https://doi.org/10.1007/978-981-96-4047-8_23

drinking water (Deltares 2022). In India, a country heavily reliant on rain-fed agriculture, droughts have profound implications for food security and rural livelihoods. According to the report "Droughts in a Changing Climate: A Regional Perspective" published by the Intergovernmental Panel on Climate Change (IPCC), India is expected to experience an increase in the frequency and intensity of droughts due to climate change. The report underscores the urgency of implementing adaptive strategies and populations. As climate patterns evolve, addressing the multifaceted aspects of drought in India requires a comprehensive and interdisciplinary approach, emphasizing the integration of scientific insights, community engagement, and innovative solutions to build resilience against the growing threat of water scarcity (IPCC 2018).

The impact of drought on a specific sector can vary between countries due to differences in inherent exposure and vulnerability, as well as the presence of existing mitigation measures and the adaptive capacity of each country. The significance of mitigating drought lies in its ability to lessen the negative effects of prolonged water scarcity on communities, ecosystems, and economies. Drought, marked by enduring water shortage, presents substantial threats that go beyond immediate concerns related to water. Drought prevention involves many actions, which we can put into three groups: making more water available, using less water, and minimizing the effects of drought. Each group helps reduce how drought affects the environment, economy, and people in different ways (Rossi 2000). Several other researchers have identified different methods for drought mitigation like establishing early warning systems, more utilization of groundwater, recycling and reuse of waste water, etc. (Ndiritu 2019; Mugabe et al. 2019; Wilhite 2016). One of the most notable recent advances for drought mitigation is the nature-based approach. The rise of nature-based solutions for drought mitigation signifies a growing recognition of the effectiveness of natural processes and ecosystems in addressing water scarcity challenges. Nature-based solutions utilize the inherent capabilities of the environment to manage and mitigate the impacts of drought. These solutions include practices such as afforestation, watershed management, sustainable land use, and the restoration of natural ecosystems. By harnessing the power of nature, these approaches contribute to enhancing water retention, reducing soil erosion, and promoting overall resilience in the face of water scarcity. The emergence of nature-based solutions represents a shift toward more sustainable and environmentally friendly strategies for mitigating the impacts of drought, emphasizing the importance of working with nature to build resilience in water management. Many researchers have already conducted different studies to evaluate the effectiveness of NbS to mitigate drought risk (Sintayehu et al. 2023; EEA 2021).

22.2 Drought and Its Impact Chain

22.2.1 Key Drivers of Drought

Drought is a big problem caused by deficit rainfall for a long time. It comes in different types, each with its own effects on nature and people. First, there's meteorological drought, where there's not enough rain compared to what's normal for a long time. It occurs when there is insufficient precipitation compared to long-term averages (Dai 2011). Then, there's agricultural drought, which happens when the soil doesn't have enough water for crops to grow properly. This can lead to less food being produced. Further there is hydrological drought which refers to deficiency in surface water and groundwater availability, affecting how much water we have for things like farming, industry, and our homes (Stagge et al., 2015). Another type is socioeconomic drought which extends beyond physical water scarcity to encompass the social and economic impacts of prolonged water shortages causing problems for communities, jobs, and the economy (Wilhite and Glanthz 1985). So, from the definitions of drought, it can be understood that the key drivers which are crucial for predicting, monitoring, and mitigating the impacts of drought are climate change, temperature extremes, precipitation deficits, land-use changes, soil conditions, population growth, mismanagement of water, etc. (Fig. 22.1).

22.2.2 The Impact Chain of Drought

Drought significantly affects agricultural productivity, leading to reduced crop yields, diminished pasture quality, and livestock losses. The resulting food scarcity and increased prices can exacerbate food insecurity, particularly in regions heavily reliant on rain-fed agriculture (Lesk et al. 2016). The long-term consequences of agricultural drought can also include soil degradation and decreased land productivity. One of the most palpable impacts of drought is the strain it places on water resources. Reduced precipitation and streamflow contribute to water scarcity, affecting drinking water supplies, irrigation for agriculture, and industrial processes. Moreover, prolonged droughts can lead to the depletion of reservoirs and groundwater aquifers, posing long-term challenges for water availability (Wilhite 2000). Ecosystems are highly sensitive to drought, with impacts extending to biodiversity loss, deforestation, and desertification. Drought-induced stress on plant and animal species can disrupt ecological balance and lead to habitat degradation (Trenberth et al. 2014). Ecosystem resilience is further compromised as drought exacerbates the frequency and intensity of wildfires, further altering landscapes and biodiversity. Drought impacts human health through various channels, including compromised access to clean water, heightened air pollution due to dust storms, and increased prevalence of vector-borne diseases (Watts et al. 2019). Mental health is also affected as communities grapple with economic losses, displacement, and uncertainty about the future. The impacts

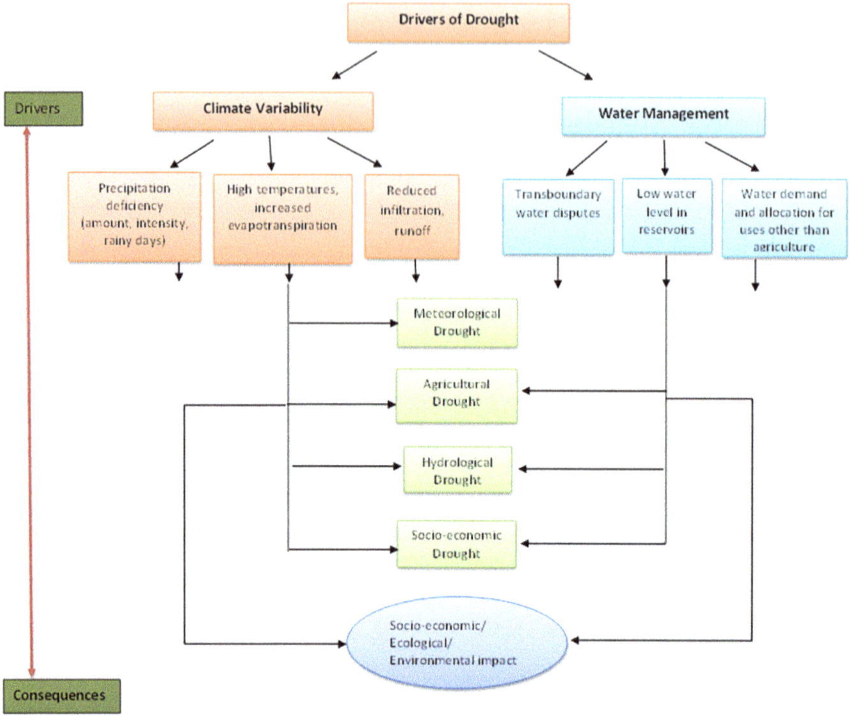

Fig. 22.1 Drivers and impact of drought

of drought are far-reaching and multifaceted, affecting agriculture, water resources, ecosystems, and human well-being. Understanding these impacts is crucial for developing effective mitigation and adaptation strategies to build resilience in the face of drought. The impact chain of drought is shown in Fig. 22.1. From the figure it can be seen that the final impact is on the socio-economy, ecology and environment as a whole namely food insecurity, health, water availability, livelihoods, poverty, infrastructure loss, energy crisis, disruption in ecosystem services, short- and long-term ecosystem changes cause potential loss, etc.

22.3 Drought Mitigation

With the global increase in vulnerability to drought, there is a growing focus on minimizing the associated risks by implementing proactive measures. This involves enhancing operational capabilities, such as monitoring climate and water supply, and building institutional capacity. There has been a paradigm shift in Disaster Management. Initially, it was just response. But with time, efforts are being directed toward mitigation measures aimed at lessening the impact of any disaster. This shift in

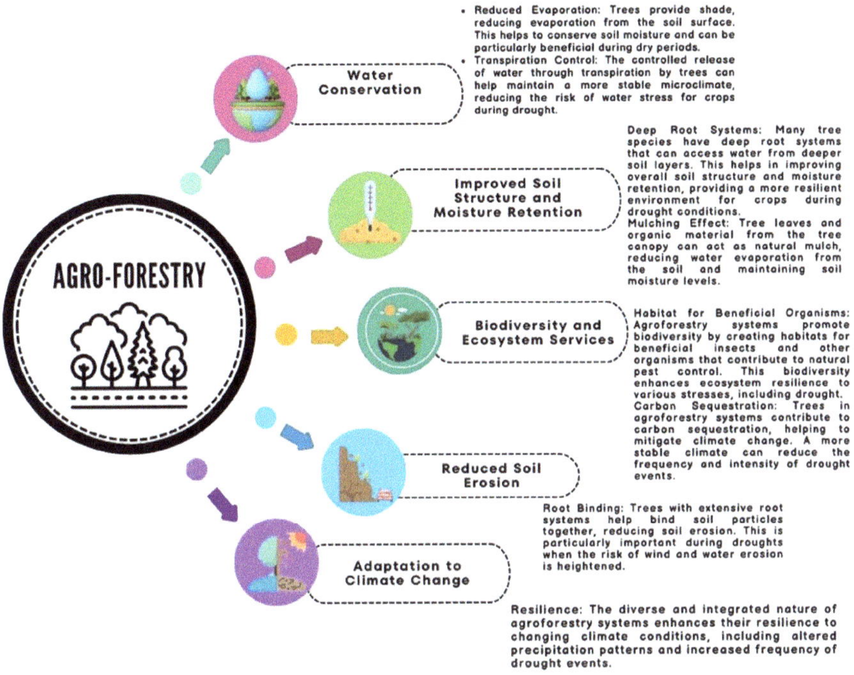

Fig. 22.2 Role of agroforestry in drought mitigation (author's own)

emphasis is long overdue. Traditionally, after a natural hazard event and resulting disaster, governments and donors have primarily focused on assessing impacts, responding to emergencies, facilitating recovery, and reconstructing affected regions to their pre-disaster state. Unfortunately, there has been limited attention given to preparedness, mitigation, and prediction/early warning actions, which fall under the umbrella of risk management. This lack of emphasis on proactive measures has resulted in a recurring cycle of moving from one disaster to another, with minimal reduction in overall risk (Wilhite et al. 2014). The shift in disaster management is shown in Fig. 22.2. In any disaster management, the first step is mitigation and hence there should be proper mitigation measures for all types of disaster. For drought mitigation, some of the core components are efficient water use, rainwater harvesting, water recycling, diversification of water sources, nature based solutions, etc.

22.4 Nature-Based Solutions for Drought Mitigation

Out of all the measures given in Table 1, Nature-based solutions is one such measure which offers a holistic and sustainable approach to addressing environmental challenges like drought, recognizing the interconnectedness of ecosystems and their role in supporting life on Earth. Integrating NbS into policies and practices is crucial for promoting resilience and sustainability across various sectors. The various effective Nature based solutions are explained in detail as follows.

22.4.1 Agroforestry

Agroforestry is a land-use management system that integrates trees and shrubs into agricultural practices, combining elements of forestry and agriculture in a sustainable manner. This approach seeks to optimize the benefits derived from both trees and crops, fostering ecological, economic, and social resilience. In their research, Zhang et al. (2001) employed hydrological modeling to demonstrate the positive effects of afforestation on mean annual evapotranspiration and water balance. This strategy plays a vital role in enhancing water availability, soil conservation, and overall resilience in the face of water scarcity. Trees contribute to soil moisture retention by acting as natural sponges, reducing water runoff and preventing soil erosion. The shade provided by the tree canopy helps decrease evaporation, conserving water in the soil. Additionally, tree roots help bind the soil, promoting its ability to retain water. Reforestation/Agroforestry also contributes to the overall health of ecosystems, fostering biodiversity and creating a more sustainable environment. By harnessing the power of trees and restoring green cover, afforestation proves to be an effective and sustainable approach to mitigating the impacts of drought while promoting long-term environmental benefits. Agroforestry has the ability to establish microenvironments characterized by lower average air temperatures. This practice can decrease the rates of transpiration in crops by providing shade, enabling crops to access water from deeper soil layers, facilitating root water uptake, preventing significant soil erosion during intense rainfall or downstream flooding, and forming windbreaks to shield crops from storms (Quandt et al. 2023). Agroforestry also contributes to many of the SDGs which will ultimately help in drought mitigation (Agroforestry Network 2018). Agroforestry can help in achieving the following SDG targets 2.4 (sustainable food production and resilient agricultural), 15.2 (end deforestation and restore degraded forests) and 15.3 (end desertification and restore degraded land). Figure 22.2 shows the role of agroforestry in drought mitigation.

22.4.2 Wetland Restoration

Wetland restoration can play a significant role in drought mitigation by enhancing water availability, improving water quality, and supporting ecosystem resilience. Wetland restoration is a critical component of comprehensive drought management strategies, providing multifaceted benefits that contribute to water resource resilience. According to Mitsch and Gosselink (2015), restored wetlands function as natural reservoirs, storing excess water during periods of ample rainfall and gradually releasing it during dry spells, thus playing a vital role in maintaining water availability. This process supports groundwater recharge and helps sustain consistent water flow in rivers, mitigating the impacts of drought (Mitsch and Gosselink 2015). Additionally, wetlands act as biodiversity hotspots, providing habitat for diverse plant and animal species, contributing to ecosystem resilience during periods of water scarcity. According to the Ramsar Convention, wetlands are recognized as essential for water purification and flood control, enhancing water quality and minimizing the risk of downstream flooding during heavy rainfall or storm events (Ramsar Convention 1971). Furthermore, wetland restoration projects offer socio-economic benefits, including opportunities for recreation, tourism, and sustainable resource use, contributing to community resilience in the face of drought-related challenges (Mitsch and Gosselink 2015; Ramsar Convention 1971). Wetlands play a significant role in regulating water flow, acting as a natural solution to enhance the resilience of watersheds against extreme hydrological events and to aid the functionality of downstream hydrological infrastructure (Wu et al. 2023). The functions of the wetland are highlighted in Fig. 22.3. Wetland restoration emerges as a holistic and scientifically supported approach for effective drought management, addressing ecological, hydrological, and socio-economic dimensions. Some common activities undertaken during wetland restoration:

- ***Hydrological Restoration***: Reestablish natural water flow patterns by modifying drainage systems, restoring natural water levels, or implementing water control structures to enhance the wetland's hydrological function.
- ***Invasive Species Management***: Identify and control invasive plant species that may have colonized the wetland. Implement strategies such as manual removal, herbicide application, or biological control to mitigate the impact of invasives.
- ***Vegetation Planting***: Select and plant native vegetation that is adapted to the wetland environment. This may involve establishing a diverse mix of aquatic plants, wetland grasses, and other native species to enhance habitat and ecosystem functions.
- ***Soil Conservation Measures***: Implement erosion control measures, such as stabilizing banks, installing silt fences, or using erosion control blankets, to prevent soil erosion and protect the wetland's soil structure.
- ***Bank Stabilization***: Address erosion along the wetland's edges by stabilizing banks using natural materials, vegetation, or structural elements to prevent further degradation.
- ***Riparian Buffer Establishment***: Create or enhance riparian buffers by planting native vegetation along watercourses adjacent to the wetland. Riparian buffers contribute to water quality improvement and prevent sedimentation.
- ***Wetland Excavation or Dredging***: Undertake controlled excavation or dredging activities to restore natural wetland features, such as ponds or channels, and to improve water circulation within the wetland.
- ***Water Quality Monitoring***: Establish a water quality monitoring program to assess parameters such as nutrient levels, pH, and contaminants. Monitoring helps track changes in water quality and guides management decisions.

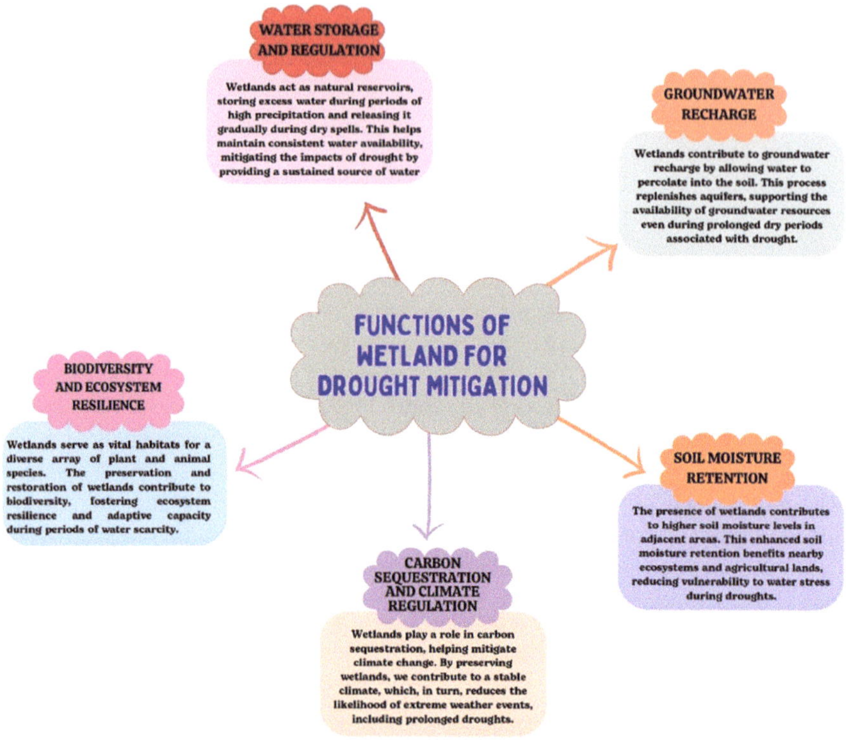

Fig. 22.3 Functions of wetland for drought mitigation (author's own)

22.4.3 Agricultural Best Management Practices

Implementing agricultural measures for drought mitigation is critical to ensuring food security and sustainable farming practices in water-scarce environments. According to studies such as the one conducted by Lobell et al. (2014), selecting drought-resistant crop varieties is a fundamental strategy in adapting agriculture to changing climatic conditions. Farmers can further enhance water-use efficiency by adopting precision irrigation methods, as recommended by Zhang et al. (2019). Soil moisture conservation practices, including cover cropping and mulching, contribute to water retention in the soil, as outlined by Lal (2015). Agroforestry, integrating trees into agricultural systems, has been recognized for its role in improving soil structure and providing shade, enhancing overall resilience to drought (Jose 2009). Precision farming technologies, discussed by O'Connell et al. (2016), enable farmers to optimize resource use by precisely tailoring water and nutrient applications to crop requirements. Crop rotation, diversification, and resilient crop planning based on climate projections contribute to building agricultural resilience, as emphasized by Porter et al. (2014). Educational outreach programs are essential to disseminate knowledge and encourage the adoption of these practices among farmers

(Gornall et al. 2010). By incorporating these measures, agriculture can better adapt to drought conditions, ensuring sustainable and productive farming systems. Watershed management stands as a crucial nature-based solution for drought mitigation, addressing water scarcity challenges at a broader scale. Watershed management involves the holistic and sustainable planning of land and water resources within a specific drainage area. This approach aims to optimize the use of water resources, reduce soil erosion, and enhance overall resilience to drought conditions. Key components of watershed management include natural resources management, soil conservation, reforestation, and the creation of water retention structures. By strategically managing the entire watershed, from its upper reaches to downstream areas, water flow can be regulated, ensuring a steady supply during dry periods. This nature-based solution not only helps mitigate the immediate impacts of drought but also contributes to the long-term health of ecosystems, promoting sustainable water use for both human and environmental needs. Some of the best agricultural best management practices are explained below and shown in Fig. 22.4.

- ***Drought-Resistant Crop Varieties***: Planting crop varieties that are specifically bred or adapted to withstand water stress. These drought-resistant varieties can better cope with limited water availability while maintaining productivity.
- ***Conservation Tillage***: Practicing conservation tillage to reduce soil disturbance and improve water retention. Reduced tillage minimizes soil erosion, enhances soil structure, and conserves soil moisture during dry periods.
- ***Rainwater Harvesting***: Implementing rainwater harvesting systems to capture and store rainwater during wet periods. Stored water can be used during dry spells, supplementing irrigation and reducing reliance on external water sources.

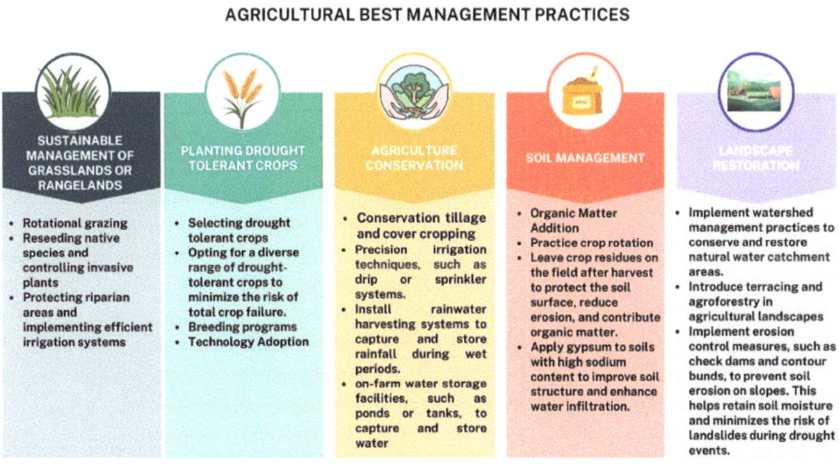

Fig. 22.4 Agricultural best management practices (author's own)

- ***Cover Cropping***: Using cover crops during fallow periods to protect the soil from erosion, increase organic matter, and improve water retention. Cover crops contribute to soil health and moisture conservation.
- ***Mulching***: Apply organic or synthetic mulch on the soil surface to reduce evaporation, suppress weed growth, and conserve soil moisture. Mulching is particularly effective in retaining moisture around plant roots.
- ***Diversified Cropping Systems***: Implementing diversified cropping systems to spread the risk of crop failure during drought. A mix of crops with varying water requirements can improve overall resilience in the face of water scarcity.
- ***Agroforestry:*** Integrating trees and shrubs into agricultural landscapes through agroforestry practices. Trees provide shade, reduce evaporation, and improve soil structure, contributing to enhanced water retention.
- ***Improved Livestock Management***: Practicing sustainable grazing and rotational livestock management to prevent overgrazing and soil compaction. Well-managed pastures are more resilient to drought conditions.
- ***Smart Crop Rotation***: Planning crop rotations strategically, considering water requirements and nutrient needs. Smart crop rotation can enhance soil fertility and reduce the risk of pest and disease outbreaks, contributing to drought resilience.

22.5 Community Engagement

Community engagement plays a pivotal role as a nature-based solution for effective drought mitigation. This approach involves actively involving local communities in the planning, decision-making, and implementation of strategies to address water scarcity. By fostering a collaborative relationship with residents, community engagement ensures that nature-based solutions are culturally appropriate, locally relevant, and sustainable. Communities are often intimately connected to their natural surroundings and possess valuable traditional knowledge about water management. Engaging community members in initiatives such as watershed protection, tree planting and sustainable agricultural practices not only enhance the effectiveness of drought mitigation efforts but also promote a sense of ownership and responsibility. Community engagement is a holistic and inclusive approach that recognizes the importance of working with people to build resilience, protect ecosystems, and address the complex challenges posed by drought conditions.

22.6 Monitoring and Evaluation

Establishing monitoring mechanisms is essential to gauge the effectiveness of nature-based approaches for drought mitigation. To begin, clearly define the objectives of the nature-based interventions, specifying desired outcomes related to water retention, soil conservation, and overall resilience to drought. Identify key indicators such as changes in soil moisture levels, vegetation health, and water table depth. Conduct a baseline assessment before implementation to establish a reference point for comparison. Utilize technology like remote sensing and GIS for real-time data, complemented by regular field surveys and community involvement to gather on-the-ground insights. Develop clear protocols for data collection, ensuring consistency, and determine the frequency of monitoring based on intervention dynamics. Regularly analyze data to assess trends and adjust strategies as needed, applying an adaptive management approach. Seek stakeholder feedback and communicate monitoring results transparently to build trust and garner support. This comprehensive monitoring framework ensures a thorough evaluation of nature-based solutions, facilitating informed decision-making and continuous improvement in drought mitigation efforts.

22.7 Adaptive Management

After thorough monitoring and assessment of the status of water resources, vegetation, and ecosystem health, adaptation and adjustment strategies can be made for changing environmental conditions. The adaptive management can be made in such a way to foster ecosystems that are resilient to changing climatic conditions, enhancing their ability to withstand and recover from drought events.

22.8 Conclusions and Next Steps

Drought presents complicated difficulties that call for a holistic and multifaceted approach that incorporates technology, community interaction, and nature-based solutions. Nature-based solutions (NbS) play a crucial role in mitigating the impacts of drought by leveraging the inherent resilience of natural ecosystems. These solutions encompass a range of strategies that work with, rather than against, nature to enhance water availability, support ecosystems, and build community resilience. It is urgent to put comprehensive and long-lasting drought mitigation measures such as nature based solutions into place as climate change increases the frequency and intensity of drought episodes. Nature-based Solutions (NbS) alone may not be the singular remedy for all water security challenges. However, they can play a pivotal role as components within a comprehensive set of solutions aimed at addressing hazards and diminishing risks within communities. Implementing effective nature-based solutions (NbS) for drought mitigation requires a strategic and collaborative approach. Some of the recommended key strategies are given below:

- ***International Collaboration***: Encouraging global cooperation for the exchange of knowledge, experiences, and effective practices regarding Nature-based Solutions (NbS) for mitigating drought. Establish partnerships that advance initiatives spanning international borders and tackle mutual water challenges collectively.
- ***Financial Incentives and Funding Mechanisms***: Creating financial incentives for implementing NbS, such as grants, subsidies, or tax benefits. Establishing funding mechanisms that support NbS projects at various scales, encouraging widespread adoption.
- ***Capacity Building***: Building the capacity of local communities, government agencies, and practitioners in the planning, implementation, and maintenance of NbS. Provide training on sustainable practices, ecological restoration, and adaptive management.
- ***Policy Integration and Coordination***: Integrating NbS into national and regional policies related to water management, agriculture, and land use. Ensure coordination among various government agencies, NGOs, and local communities to optimize the impact of NbS.
- ***Integrated Water Resource Management***: Promoting integrated water resource management that incorporates NbS as a key component. Develop and implement policies that prioritize the sustainable use and protection of water resources, considering both natural and human-made systems.

References

Dai A (2011) Drought under global warming: a review. Wiley Interdisc Rev Climate Change 2(1):45–65

Deltares (2022) Nature-based solutions to mitigate impacts of droughts. Policy Brief published by Deltares, Delft. Chromeextension://efaidnbmnnnibpcajpcglclefindmkaj/https://atuk.com.ec/documentos/deltares-2022-policy-brief-based-solutions-and-droughts.pdf. Last assessed on 30th Dec 2023

Duel H, Henk W, Ingrid T, Maat TJ, Matthews J (2022) HELP guiding principles for drought risk management under a changing climate. Catalysing actions for enhancing climate resilience. HELP report

EEA (2021) Nature-based solutions in Europe: policy, knowledge and practice for climate change adaptation and disaster risk reduction. EEA Report 1/2021

Gornall J et al (2010) Impacts of climate change on the agricultural and aquatic systems: the development of a collaborative framework for integrated assessment. Environ Manage 45(3):492–505

IPCC (2018) The IPCC report will have profound effects on climate governance. Climate change, intergovenrmental panel on climate change. https://www.cigionline.org/articles/ipcc-report-will-have-profound-effects-climate-governance/?utm_source=google_ads&utm_medium=grant&gad_source=1&gclid=EAIaIQobChMIrpPn9pejjAMV-KVmAh3OKjrJEAAYASAAEgJIXPD_BwE. Last accessed on 24th March 2025

Jose S (2009) Agroforestry for ecosystem services and environmental benefits: an overview. Agrofor Syst 76(1):1–10

Lal R (2015) Restoring soil quality to mitigate soil degradation. Sustainability 7(5):5875–5895

Lesk C, Rowhani P, Ramankutty N (2016) Influence of extreme weather disasters on global crop production. Nature 529(7584):84–87

Lobell DB, Field CB (2014) Global scale climate–crop yield relationships and the impacts of recent warming. Environ Res Lett 9(3):034002

Mens M, van Rhee G, Schasfoort F, Kielen N (2022) Integrated drought risk assessment to support adaptive policy making in the Netherlands. Nat Hazards Earth Syst Sci Discuss [preprint]. https://doi.org/10.5194/nhess-2021-407. in review

Mitsch WJ, Gosselink JG (2015) Wetlands. John Wiley & Sons

Mugabe AA, Mwaniki F, Mamary KA, Ngibuini HM (2019) An assessment of drought monitoring and early warning systems in Tanzania, Kenya, and Mali. Mapedza E, Tsegai D, Bruntrup M, Mcleman R (eds) Current directions in water scarcity research. Published by Elsevier, vol 2, pp 211–219. ISSN 2542-7946, ISBN 9780128148204. https://doi.org/10.1016/B978-0-12-814820-4.00014-6

Naumann G, Spinoni J, Vogt JV, Barbosa P (2015) Assessment of drought damages and their uncertainties in Europe. Environ Res Lett 10:124013. https://doi.org/10.1088/1748-9326/10/12/124013

Ndiritu SW (2019) Drought preparedness and livestock management strategies by pastoralists in semi-arid lands: Laikipia North, Kenya. Mapedza E, Tsegai D, Bruntrup M, Mcleman R (eds) Current directions in water scarcity research. Published by Elsevier, vol 2, pp 247–263. ISSN 2542-7946, ISBN 9780128148204. https://doi.org/10.1016/B978-0-12-814820-4.00017-1

O'Connell MG, Grove IG, Rippke U, Thornley JH (2016) Strategies for crop development and water use efficiency under climate change. Glob Change Biol 22(5):1888–1903

Porter JR (2014) Food security and food production systems. In: Climate change 2014: impacts, adaptation, and vulnerability. Contribution of Working Group II to the Fifth Assessment Report of the Intergovernmental Panel on Climate Change. Cambridge University Press, pp 485–533

Quandt A, Larsen AE, Bartel G, Okamura K, Sousa D (2023) Sustainable groundwater management and its implications for agricultural 405 land repurposing, Reg Environ Change 23:120

Ramsar Convention (1971) The Ramsar convention on Wetlands. Retrieved from https://www.ramsar.org/

Rossi G (2000) Drought mitigation measures: a comprehensive framework. In: Vogt JV, Somma F (eds) Drought and drought mitigation in Europe. Advances in natural and technological hazards research, vol 14. Springer, Dordrecht. https://doi.org/10.1007/978-94-015-9472-1_18

Sintayehu DW, Kassa AK, Tessema N, Girma B, Alemayehu S, Hassen JY (2023) Drought characterization and potential of nature-based solutions for drought risk mitigation in Eastern Ethiopia. Sustainability 15(5):11613. https://doi.org/10.3390/su151511613

Stagge JH, Tallaksen LM, Gudmundsson L, Van Loon AF, Stahl K (2015) Candidate distributions for climatological drought indices (SPI and SPEI). Int J Climatol 35(14):4027–4040

Trenberth KE, Dai A, van der Schrier G, Jones PD, Barichivich J, Briffa KR, Sheffield J (2014) Global warming and changes in drought. Nat Clim Chang 4(1):17–22

Verchot LV, Van Noordwijk M, Kandji S, Tomich T, Ong C, Albrecht A, Mackensen J, Bantilan C, Anupama KV, Palm C (2007) Climate change: linking adaptation and mitigation through agroforestry. Mitigation Adaptation Strategies Global Change 12:901–918. https://doi.org/10.1007/s11027-007-9105-6

Wardlow BD, Anderson MC (2009) Remote sensing of drought: global trends. Photogramm Eng Remote Sens 75(4):377–386

Watts N, Amann M, Arnell N, Ayeb-Karlsson S, Beagley J, Belesova K, Costello A (2019) The 2019 report of The Lancet Countdown on health and climate change: ensuring that the health of a child born today is not defined by a changing climate. Lancet 394(10211):1836–1878

Wilhite DA (2016) Introduction: managing drought risk in a changing climate. Clim Res 70:99–102. https://doi.org/10.3354/cr01430

Wilhite DA (2000) Drought as a natural hazard: concepts and definitions. In: Drought: a global assessment. Routledge, pp 3–18

Wilhite DA, Glantz MH (1985) Understanding: the drought phenomenon: the role of definitions. Water Int 10(3):111–120

Wilhite DA, Sivakumar MVK, Pulwarty RS (2014) Managing drought risk in a changing climate: the role of nationaldrought policy. Weather Clim Extrem 3:4–13

Wu Y, Sun J, Hu B, Zhang G, Rousseau AN (2023) Wetland-based solutions against extreme flood and severe drought: efficiency evaluation of risk mitigation. In: Climate risk management, vol 40, 100505. ISSN 2212-0963. https://doi.org/10.1016/j.crm.2023.100505

Zhang H, Wang D, Li Z, Huang Y, Huang B (2019) Precision irrigation management in agriculture: a review. J Clean Prod 234:635–644

Zhang L, Dawes WR, Walker GR (2001) Response of mean annual evapotranspiration to vegetation changes and catchment scale. Water Resour Res 37(3):701–708

Chapter 23
Blending Indigenous and Scientific Knowledge for Effective Risk Reduction

Sapna Tiwari, Fatima Amin, and Anil K. Gupta

23.1 Introduction

The integration of Indigenous and Scientific Knowledge for Effective Risk Reduction emerges as a transformative paradigm in the realm of disaster management. Indigenous knowledge offers unique insights into local environments, weather patterns, and sustainable practices (Berkes 1999). The ancestral knowledge, when synergistically combined with empirical insights derived from scientific methodologies, presents an opportunity to cultivate a holistic understanding of risks and resilience strategies.

The collaboration between these two knowledge systems is increasingly recognized as essential for creating more robust and culturally sensitive risk reduction approaches. Indigenous observations, grounded in lived experiences, can be integrated with scientific data to enhance the accuracy and timeliness of early warning systems (Fernandez-Gimenez 2000). The dynamic interaction of both knowledge not only acknowledges the importance of preserving cultural heritage but also enriches the scientific understanding of risk through the incorporation of traditional practices that have sustained communities in harmony with their environments for generations (Turner et al. 2008).

The collaborative approach shaped by mutual respect and recognition of the value each knowledge system brings holds the promise of fostering resilience and enhancing the effectiveness of risk-reduction strategies (Diaz et al. 2011). The fusion of community experiences and scientific progress forms a robust foundation, enabling adaptable plans and inclusive actions across diverse scenarios.

S. Tiwari · F. Amin · A. K. Gupta (✉)
National Institute of Disaster Management, Ministry of Home Affairs, GoI, Delhi, India
e-mail: anilg.gov.in@gmail.com

N. A. Siddiqui et al. (eds.), *Advances in Disaster Management, Volume 2*, Springer Transactions in Civil and Environmental Engineering,
https://doi.org/10.1007/978-981-96-4047-8_24

23.2 Integrating Indigenous and Scientific Knowledge

The integration of Indigenous knowledge and scientific knowledge represents a holistic approach to addressing global challenges, drawing on distinctive perspectives, methodologies, and insights (Berkes 2012; Johnson and Murton 2007). This fusion enriches our understanding and offers comprehensive solutions to contemporary issues. Rooted in generations of lived experience, Indigenous knowledge provides profound insights into local ecosystems, biodiversity, and sustainable resource management (Berkes 1999; Armitage et al., 2011). When combined with scientific understanding, it offers a nuanced perspective on environmental issues. Collaboration with Indigenous communities, especially in vulnerable regions, brings unique insights into adapting to environmental changes and addressing climate change impacts (Ford and Cameron 2016; Nakashima et al. 2012). This collaborative and adaptive approach extends to cultural preservation, fostering inclusivity and equity in the research environment (Battiste 2013; Cajete 2000). The integration of Indigenous and scientific knowledge enhances problem-solving and innovation, drawing on the creativity inherent in diverse knowledge systems (Agrawal 1995; Turner et al. 2008). Moreover, collaborative research promotes community engagement, empowerment, and ethical research practices, respecting Indigenous knowledge and ensuring fair benefit-sharing (United Nations 2007; Battiste 2008). Indigenous and scientific knowledge, like using two sets of eyes, contribute to finding sustainable ways that benefit both people and the environment, combining practices rooted in local experiences with rigorous scientific approaches. The figure given below shows the amalgamation of integration of indigenous and scientific knowledge (Fig. 23.1).

23.3 Historical Perspectives on Traditional and Scientific Approaches to Risk Reduction

Historical perspectives on risk reduction reveal a nuanced interplay between traditional and scientific approaches. Traditional methods, deeply embedded in cultural practices, have been integral to communities for centuries. For instance, indigenous knowledge in managing natural disasters is documented in works like "Indigenous Knowledge for Disaster Risk Reduction" by Rajib Shaw and Yukiko Takeuchi.

However, the synergy between traditional and scientific approaches is exemplified in disaster risk reduction. The Sendai Framework for Disaster Risk Reduction (2015–2030) emphasizes the incorporation of indigenous knowledge alongside scientific insights, acknowledging the importance of local perspectives in building resilient communities. The given figure shows Historical perspectives on both traditional and scientific approaches to reducing risks in their communities (Fig. 23.2).

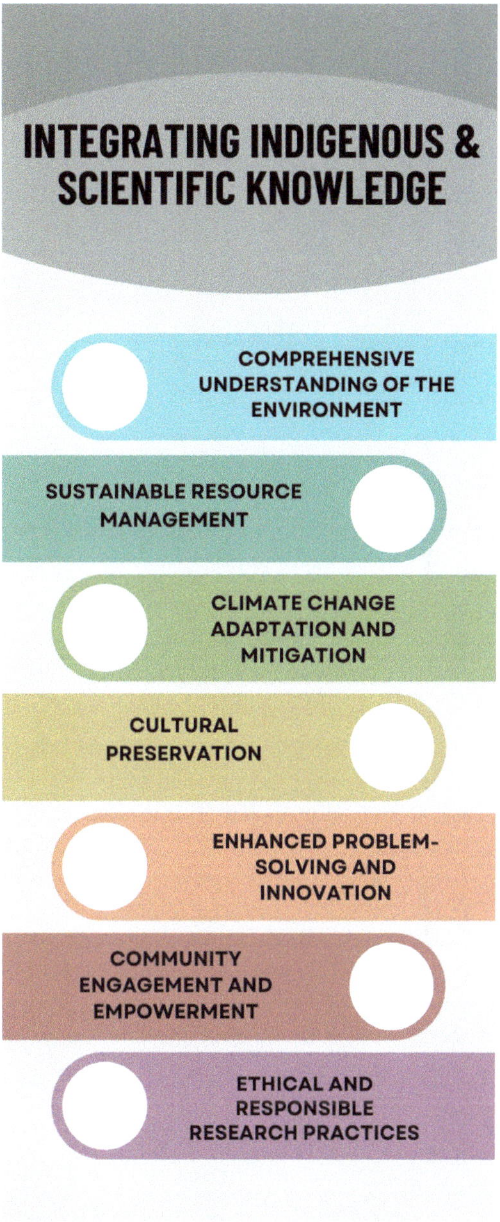

Fig. 23.1 Integration of indigenous and scientific knowledge

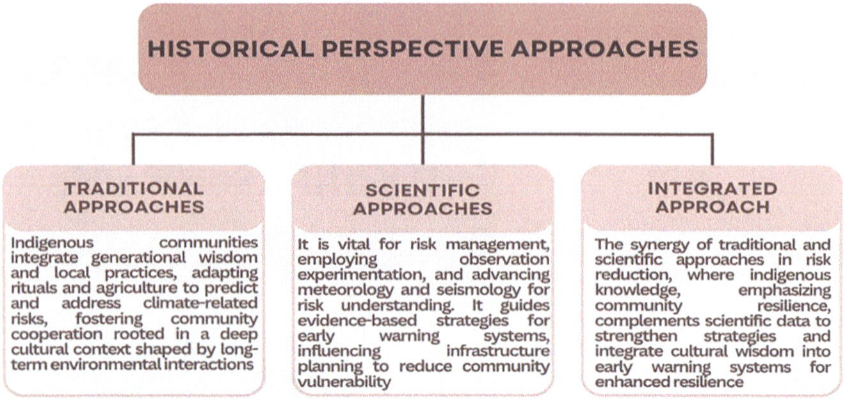

Fig. 23.2 Historical approaches

23.4 Understanding Indigenous Knowledge Systems

Understanding Indigenous Knowledge Systems means appreciating how indigenous communities see, interact with, and learn from their surroundings. These knowledge systems are rooted in the cultural, spiritual, and historical experiences of indigenous people. Important Parts of Indigenous Knowledge Systems:

- *Holistic Perspective*: Indigenous knowledge looks at everything together which includes not just the physical parts of the environment but also the spiritual, social, and cultural sides. This recognizes that nature and human communities are connected.
- *Oral Tradition and Wisdom Transmission*: A lot of indigenous knowledge is passed down orally, not in writing. Elders, or older community members, are important because they share wisdom, traditions, and stories. This talking tradition helps in creating a strong sense of community and keeps important knowledge alive.
- *Land Stewardship and Environmental Understanding*: Indigenous knowledge is closely linked to the land and know a lot about their environment. Traditional practices involve taking care of the land sustainably, using it for farming, and understanding local plants, animals, and seasons.
- *Cultural Harmony and Responsibility*: Indigenous knowledge includes ceremonies, dances, and rituals that have meaning for the environment. These practices promote harmony with nature and remind people of their responsibility to take care of the environment.
- *Adapting and Resilience*: Indigenous knowledge shows how communities have adapted to different and sometimes tough environments. They have learned to deal with changes in climate, weather, and other natural events. This ability to adapt shows how effective indigenous knowledge is in helping communities survive.

23.5 Exploration of Traditional Ecological Knowledge (TEK)

Traditional Ecological Knowledge (TEK) is a dynamic system deeply rooted in indigenous and local communities, passed down through generations, shaping their identity through a holistic understanding of ecosystems and the interconnectedness of all living beings (Vinyeta 2013). This practical, cultural, and spiritual knowledge reflects centuries of observation, adaptation, and sustainable practices, spanning domains such as ethnoecology, ethnozoology, and ethnoastronomy (Finn 2017). Recognized for its value in sustainable resource management and biodiversity conservation, TEK is acknowledged for complementing scientific approaches, providing context-specific insights and practical solutions to global environmental challenges. Integration of TEK into environmental policies holds the potential to foster resilience and collaboration between indigenous communities and external stakeholders (Hosen 2020).

In addressing climate change impacts, TEK proves instrumental, as evidenced by its application in Sarawak, Malaysian Borneo, where indigenous communities utilize it for local climate change adaptation (Hosen 2020). Beyond environmental considerations, TEK contributes significantly to environmental health sciences and biomedical research, offering culturally relevant perspectives on health and disease (Finn 2017). However, documenting and preserving this vast knowledge, especially in regions like North East India, is crucial to prevent the loss of this invaluable resource (Tynsong 2020) (Fig. 23.3).

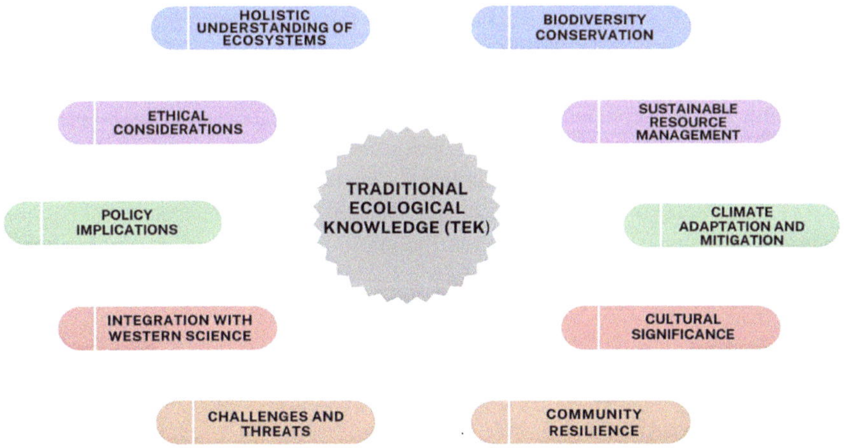

Fig. 23.3 Traditional ecological knowledge

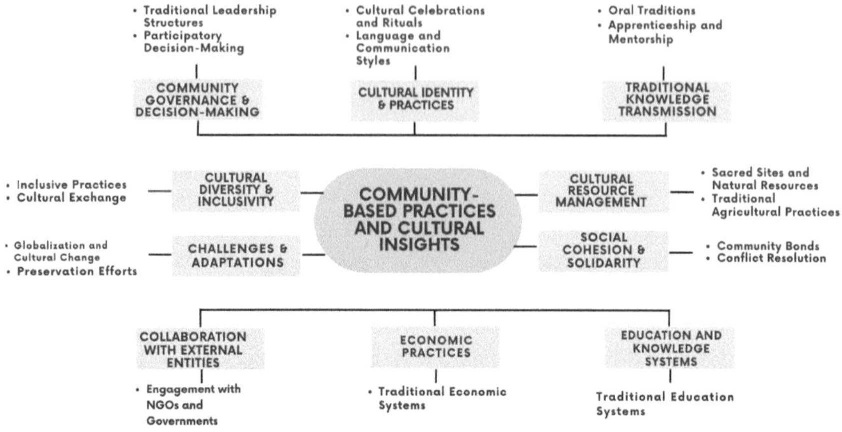

Fig. 23.4 Community-based practices and cultural insights

23.6 Examination of Community-Based Practices and Cultural Insights

Community-based practices are vital for promoting the well-being of diverse populations, and cultural insights play a crucial role in tailoring interventions to meet the unique needs of communities. By engaging with local communities, practitioners can gain a deeper understanding of cultural norms, values, and social dynamics, allowing for the development of more effective and culturally sensitive programs. This approach is consistent with the principles of community-based participatory research (CBPR), which emphasizes collaboration between researchers and community members to address health disparities and improve overall community health. Cultural competence in community-based practices involves recognizing and respecting the diversity of cultural backgrounds, adapting interventions accordingly, and fostering a sense of community ownership. As highlighted by Trickett (2011) and Israel et al. (2003), such an approach enhances the relevance and sustainability of interventions, ultimately contributing to the empowerment and resilience of communities (Fig. 23.4).

23.7 Role of Indigenous Wisdom in Risk Perception and Adaptation

Indigenous wisdom plays a vital role in how communities perceive and adapt to risks, especially when it comes to resisting disasters. The knowledge passed down through generations in indigenous communities helps them understand the potential dangers posed by disasters. This wisdom includes insights into the environment, weather patterns, and local ecosystems. Indigenous wisdom guides communities in developing strategies to resist and cope with disasters. This might involve traditional practices for building resilient structures, cultivating crops that are more resistant to extreme weather, or establishing early warning systems based on local signs and signals. It considers not only the physical aspects of the environment but also incorporates cultural, spiritual, and social dimensions. By blending traditional knowledge with modern approaches, indigenous communities can enhance their resilience and better withstand the challenges posed by disasters.

23.8 Technological Advancements in Disaster Monitoring and Early Warning Systems

Technological advancements have revolutionized disaster monitoring and early warning systems, contributing to more effective and timely responses to potential hazards. Remote sensing technologies, such as satellites and unmanned aerial vehicles (UAVs), provide real-time and high-resolution data for monitoring natural disasters like hurricanes, floods, and wildfires (Li et al. 2018). Advanced geospatial information systems (GIS) enable the integration of diverse data sources, facilitating better analysis and visualization of disaster-prone areas (Zhang et al. 2019). Additionally, the development of sophisticated sensors and monitoring devices has allowed for more accurate and timely detection of seismic activities, weather patterns, and environmental changes (Bates et al. 2020). Artificial intelligence (AI) and machine learning algorithms contribute to the prediction and interpretation of complex patterns in large datasets, improving the precision of early warnings (Yun et al. 2018). Mobile communication technologies play a crucial role in disseminating timely alerts to at-risk populations, while social media platforms provide valuable real-time information and facilitate community engagement (Bruns et al. 2018). These technological innovations collectively enhance our preparedness and resilience in the face of natural disasters, potentially minimizing their impact on human lives and infrastructure.

Recent advancements in disaster monitoring and early warning systems have been driven by the integration of Internet of Things (IoT) solutions, which have improved the collection and dissemination of environmental data (Esposito 2022). This has been further enhanced by the use of Information and Communication Technology (ICT) in early warning dissemination systems, which has enabled more reliable and timely data input from remote sensors (Rajendran 2021). The development of self-organizing sensor networks, high-resolution satellite imagery, and multi-sensor observational capacities has also significantly improved early warning technologies for geological disasters (Wenzel 2014). Additionally, the use of Unmanned Aerial Vehicles (UAVs) and low-altitude remote sensing technology has provided a new technical approach for remote sensing monitoring in dangerous areas, leading to the development of a comprehensive natural disaster early warning system (Wu 2021).

23.9 Case Studies Highlighting Successful Applications of Scientific Knowledge in Risk Reduction

Weichselgartner (2015) emphasizes the critical role of knowledge in disaster risk reduction, particularly in integrating knowledge systems and improving understanding of disaster risk. Facin (2014) and Islam (2012) both highlight the importance of knowledge management in reducing risks, with Facin focusing on the software development process and Islam on open innovation. Kelman (2012) underscores the potential of indigenous knowledge in disaster risk reduction, emphasizing the need for a balance between different knowledge forms. These studies collectively demonstrate the successful application of scientific knowledge in risk reduction, particularly in the areas of disaster risk, software development, and open innovation.

Box: 1 International Case Studies

- *Earthquake Early Warning Systems in Japan*:
  Japan has implemented an advanced earthquake early warning system, utilizing scientific knowledge to mitigate seismic impacts. This system, using seismometers, detects initial seismic waves, providing crucial warning time during events like the 2011 Great East Japan Earthquake (Japan Meteorological Agency 2011).
- *Flood Risk Management in the Netherlands*:
  In the Netherlands, flood-prone due to low-lying topography, flood risk is managed with the Delta Works. This construction series, based on hydraulic engineering and scientific understanding, strategically places dams, sluices, locks, dikes, and storm surge barriers to prevent catastrophic flooding (Rijkswaterstaat, n.d.).
- *Tsunami Warning System in the Pacific*:
  The Pacific Tsunami Warning Center (PTWC) by NOAA has implemented a robust tsunami warning system, integrating scientific knowledge into algorithms. Using seismographic and sea level monitoring stations, this system rapidly assesses undersea earthquakes, significantly improving preparedness in the Pacific region (NOAA, n.d.).

Box: 2 National Case Studies

- *Cyclone Early Warning Systems in Odisha*:
  In Odisha, India, vulnerable to tropical cyclones, the Indian Meteorological Department (IMD) implemented a Cyclone Early Warning System after the 1999 Super Cyclone. This system, grounded in scientific knowledge

of meteorology, oceanography, and satellite technology, accurately predicts cyclone paths and intensities. Notably effective during Cyclone Phailin in 2013 and Cyclone Fani in 2019, it facilitates timely evacuations, showcasing the impact of scientific interventions (IMD, n.d.).

- *Flood Management in Bihar*:
  Bihar, a state in northern India, faces recurrent flooding due to its geography and the annual monsoon rains. Scientific knowledge has been applied to develop a Flood Early Warning System in collaboration with the Bihar State Disaster Management Authority (BSDMA). This system integrates meteorological data, river gauging, and hydrological modeling to predict potential flood scenarios. The timely dissemination of warnings to vulnerable communities enables better evacuation planning and resource allocation, contributing to reduced flood-related casualties and damages.
- *Earthquake-Resilient Building Practices in Gujarat*:
  After the devastating earthquake in Gujarat in 2001, scientific research on seismic engineering and building construction played a crucial role in developing earthquake-resistant building practices. The Gujarat State Disaster Management Authority (GSDMA) implemented building codes based on seismic vulnerability assessments, and public awareness campaigns promoted the adoption of earthquake-resistant construction techniques. This application of scientific knowledge has significantly improved the resilience of buildings and infrastructure in earthquake-prone regions.
- *Drought Management in Maharashtra*:
  Maharashtra has faced recurring drought conditions, impacting agriculture and water resources. Scientific knowledge has been employed to develop Drought Early Warning Systems that incorporate satellite-based remote sensing, meteorological data, and hydrological modeling. These systems provide timely information on drought conditions, allowing for proactive water management strategies, crop planning, and the implementation of relief measures. By integrating scientific insights, Maharashtra has enhanced its capacity to mitigate the impacts of drought and support affected communities.
- *Landslide Risk Reduction in Uttarakhand*:
  The hilly terrain of Uttarakhand is prone to landslides, particularly during the monsoon season. Scientific research on geological and topographical factors influencing landslides has informed risk reduction strategies. The implementation of early warning systems, slope stabilization measures, and land-use planning based on scientific assessments has contributed to minimizing the risk of landslides and protecting vulnerable communities in the region. The application of scientific knowledge in Uttarakhand showcases a proactive approach to landslides.

23.10 Overlapping Concepts in Risk Assessment, Mitigation, and Preparedness

The concepts of risk assessment, mitigation, and preparedness are closely tangled, with overlapping elements. Christoplos (2001) highlights the need to address risk and improve disaster mitigation and preparedness, positioning them within poverty alleviation agendas. Schwab (2006) emphasizes the role of the federal government, state, and local governments, as well as the private sector in building resilient communities through risk assessment and preparedness activities. Abkowitz (2012) introduces the all-hazards risk management approach, which considers multiple disaster risks and emphasizes the need for a systematic and holistic approach to risk management. These studies collectively underscore the interconnectedness of risk assessment, mitigation, and preparedness, and the need for a comprehensive and integrated approach to address them.

Risk assessment, mitigation, and preparedness are interconnected concepts within the broader framework of disaster management. While each term represents a distinct phase of the disaster management cycle, there are overlapping elements that demonstrate the holistic nature of effective risk reduction efforts. The seamless integration of these concepts is essential for building a comprehensive and effective disaster risk reduction framework. Recognizing the interconnection of risk assessment, mitigation, and preparedness leads to more resilient and adaptive communities better equipped to face and recover from disasters (Fig. 23.5).

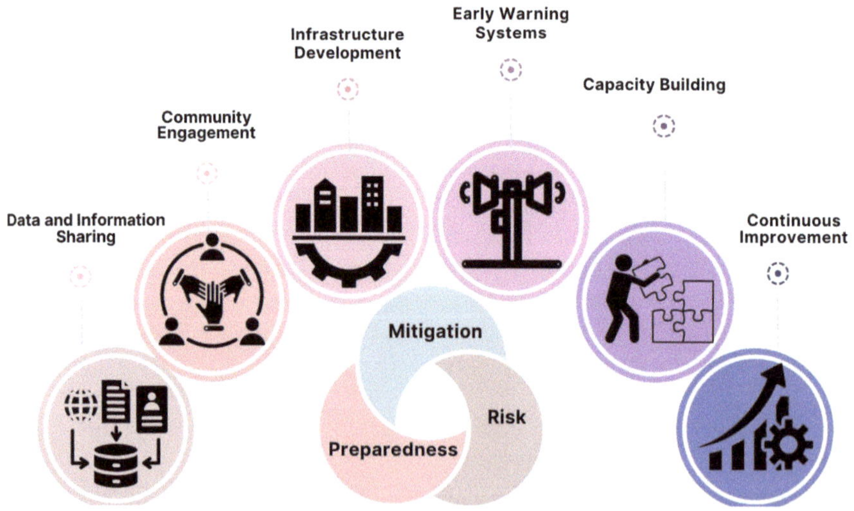

Fig. 23.5 Risk assessment, mitigation, and preparedness

23.11 Collaborative Governance and Policy

The integration of Indigenous perspectives into disaster management policies is crucial for effective and inclusive risk reduction strategies. Lambert (2019) and Radel (2023) both emphasize the need for culturally appropriate incentives, accurate data collection, and collaboration with Indigenous knowledge. Ali (2021) and Molina (2016) highlight the role of governmental and non-governmental organizations in fostering collaboration, with Ali specifically recommending interdependent and complementary partnerships. Policy recommendations include better consultation with local communities, the inclusion of Indigenous knowledge in decision-making processes, and the implementation of national laws on disaster risk reduction and indigenous peoples.

Governments and non-governmental organizations (NGOs) can play a pivotal role in fostering collaboration by establishing platforms for meaningful participation of Indigenous representatives in policy-making processes. This includes the inclusion of Indigenous voices in risk assessments, mitigation planning, and preparedness strategies. Governments should recognize and support Indigenous-led initiatives, ensuring that policies are informed by traditional ecological knowledge and community-specific needs. NGOs can contribute by facilitating cultural competency training for emergency responders, promoting community-led resilience projects, and advocating for policies that uphold Indigenous rights and self-determination. Policy recommendations should emphasize the co-creation of inclusive risk reduction strategies, respecting Indigenous governance structures, and acknowledging the interconnectedness of environmental stewardship and disaster resilience within Indigenous worldviews. Through these collaborative efforts, disaster management policies can be more holistic, culturally sensitive, and effective in building resilience within Indigenous communities.

23.12 Empowerment and Engagement

A range of studies have highlighted the importance of recognizing and integrating indigenous knowledge in community-based disaster risk reduction (Ali 2021; Rajeev 2014). This approach not only strengthens the resilience of indigenous communities but also contributes to sustainable development (Pandey 2005). The need for inclusive community-based disaster planning, which involves the active participation of local communities, has been emphasized as a key strategy for sustainable risk reduction efforts (Rajeev 2014; Pandey 2005). Building partnerships between development agencies and indigenous communities is also crucial for the success of these efforts (Ali 2021).

Empowering communities through the recognition of indigenous knowledge is a transformative approach to disaster resilience. Indigenous communities possess rich traditional knowledge deeply rooted in their connection to the environment and ecosystems. Acknowledging and incorporating this wisdom into community-based disaster planning fosters inclusivity and enhances the effectiveness of preparedness strategies. Inclusive community-based disaster planning ensures that the unique needs, vulnerabilities, and strengths of indigenous populations are considered. This approach encourages active participation from community members, promoting a sense of ownership and cultural relevance in risk reduction initiatives. To further strengthen resilience, building partnerships becomes essential. Collaborative efforts involving governmental bodies, non-governmental organizations, and indigenous leaders can pave the way for sustainable risk reduction endeavors, aligning cultural practices with modern strategies. This inclusive and collaborative approach not only enhances community empowerment but also contributes to the creation of more robust, culturally sensitive disaster management frameworks.

23.13 Future Prospects: Toward a Unified Framework

The future of disaster management requires a unified approach that bridges the gap between research and practice (Trainor 2018). This can be achieved through fostering collaboration and knowledge exchange among diverse stakeholders (Sitas 2016). Information flow and collaboration patterns can be enhanced through the integration of information and communication technologies (ICT) (Sagun 2009). An interdisciplinary, industry-based paradigm can further support efficient disaster response (Noran 2014). To achieve these goals, continuous collaboration and the use of ICT tools are recommended (Sagun 2009).

The Sustainable Development Goals (SDGs), established by the United Nations in 2015, acknowledge the invaluable role of indigenous knowledge in global sustainability. Indigenous practices, wisdom, and insights align with key SDGs, such as poverty reduction (SDG 1), sustainable agriculture (SDG 2), holistic healthcare (SDG 3), quality education (SDG 4), clean water access (SDG 6), climate action (SDG 13), and biodiversity conservation (SDG 15). Recognizing and integrating indigenous knowledge into development efforts is vital for fostering cultural diversity, enhancing effectiveness, and promoting a more inclusive, equitable, and culturally sensitive approach to global development. Embracing this wealth of knowledge ensures that no community is left behind in the pursuit of a sustainable and prosperous future.

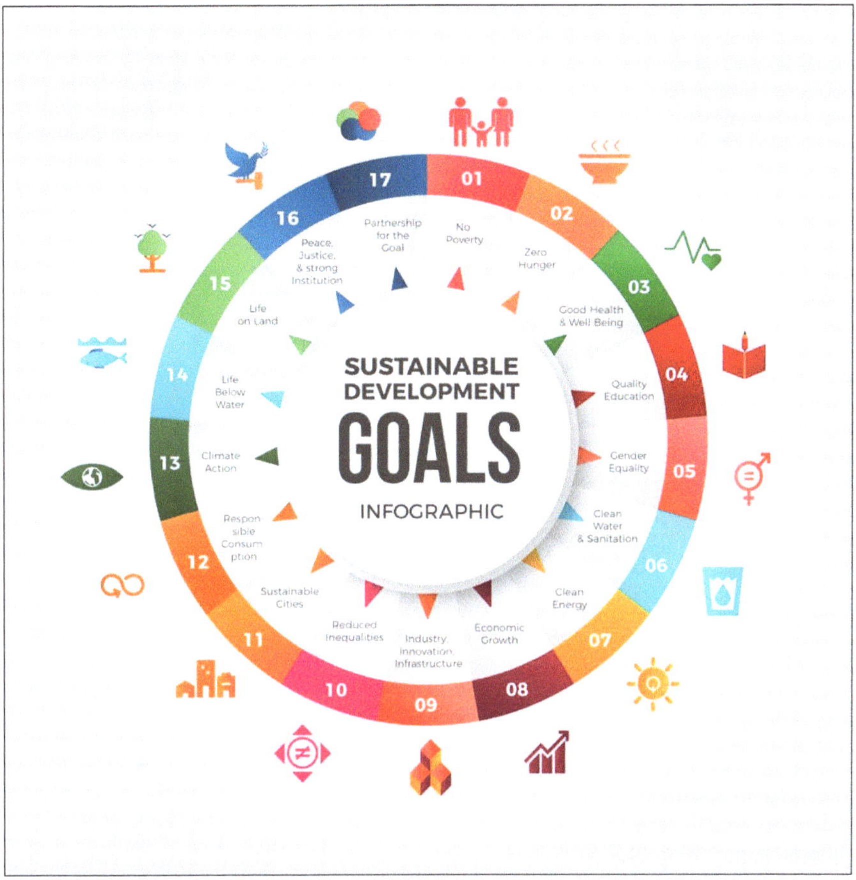

Source https://in.images.search.yahoo.com/images/view

The integration of technological advancements, community engagement, and interdisciplinary research can revolutionize our understanding and response to disasters. A holistic perspective, blending insights from environmental science, social sciences, and technological innovation, will be crucial for developing comprehensive strategies that address the complex and interconnected nature of contemporary disaster challenges. This unified approach should prioritize inclusivity, cultural sensitivity, and the empowerment of communities, laying the foundation for a resilient and adaptive disaster management framework.

The Sendai Framework (2015–2030) promotes a people-centered approach, emphasizing direct engagement with indigenous communities in disaster risk reduction. It encourages governments to involve them in decision-making, incorporating traditional knowledge alongside scientific data for more culturally sensitive strategies. This collaborative approach aims to address the specific vulnerabilities and strengths of diverse communities, respecting and integrating the wisdom embedded in indigenous cultures to foster resilience (Fig. 23.6).

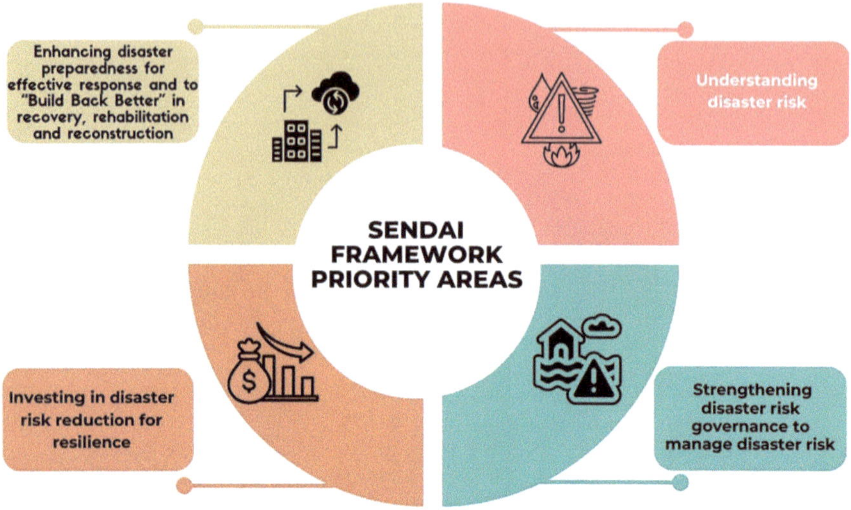

Fig. 23.6 Sendai framework priority areas

Identifying research gaps and areas for further exploration is imperative to advance the field of disaster management. Research should focus on refining predictive models, understanding the cascading effects of multiple hazards, and assessing the long-term social and environmental impacts of disasters. Additionally, there is a need for studies that explore the psychological aspects of disaster response, the role of emerging technologies like artificial intelligence, and the integration of indigenous knowledge into mainstream disaster management practices. By addressing these research gaps, the field can evolve to meet the evolving challenges of the twenty-first century and provide evidence-based insights for the development of effective policies and interventions.

Continuous collaboration is fundamental for the success of future disaster management efforts. Recommendations include fostering multi-disciplinary partnerships between academic institutions, governmental agencies, non-profit organizations, and local communities.

23.14 Conclusion

The chapter highlights the integration of Indigenous and Scientific Knowledge for Effective Risk Reduction in the field of disaster management, emphasizing the transformative potential of combining ancestral wisdom with empirical insights. The collaborative paradigm is deemed essential for developing resilient and culturally sensitive approaches to risk reduction. The rationale for integrating these knowledge systems lies in the comprehensive understanding they offer of the environment, combining local insights with scientific understanding. Indigenous knowledge is recognized for its valuable contributions to sustainable resource management, adaptation to environmental changes, and addressing the challenges posed by climate change. Historical perspectives acknowledge the enduring role of traditional methods deeply rooted in cultural practices, with frameworks like the Sendai Framework for Disaster Risk Reduction exemplifying the increasing recognition of the importance of incorporating Indigenous knowledge.

It encompasses understanding Indigenous Knowledge Systems, highlighting their holistic nature, oral transmission, and connection to the land, emphasizing their role in community adaptation. Traditional Ecological Knowledge (TEK) is introduced as a dynamic system deeply embedded in indigenous communities, contributing significantly to sustainable resource management and biodiversity conservation. Additionally, community-based practices and cultural insights are underscored as vital components for tailoring interventions to meet the unique needs of communities, with cultural competence deemed crucial for the effectiveness and sustainability of programs. The second paragraph emphasizes the role of Indigenous wisdom in risk perception and adaptation, illustrating how it guides communities in developing strategies to resist and cope with disasters. The integration of traditional knowledge with modern approaches is seen as a key factor in enhancing resilience. The chapter explores the scientific foundations of risk reduction, focusing on technological advancements in disaster monitoring and early warning systems, showcasing case studies from different countries to illustrate successful applications. The focus is on bringing together disaster management, teamwork in governance, and blending policies, highlighting the need for community strength, and aiming for a strong and united strategy that connects research and practical actions for a resilient future. The chapter concludes by urging continuous collaboration, addressing research gaps, and fostering inclusivity and cultural sensitivity in disaster management frameworks.

References

Abkowitz MD, Chatterjee S (2012) Regional disaster risk: assessment and mitigation concepts in an all-hazards context. J Homeland Security Emergency Manage 9(1)

Ali T, Paton D, Buergelt PT, Smith JA, Jehan N, Siddique A (2021) Integrating Indigenous perspectives and community-based disaster risk reduction: a pathway for sustainable Indigenous development in Northern Pakistan. Int J Disaster Risk Reduction 59:102263

Anderson R (2014) Early warning for geological disasters: scientific methods and current practice. Environ Eng Geosci 20(4):404–404

Armitage D, Berkes F, Dale A, Kocho-Schellenberg E, Patton E (2011) Co-management and the co-production of knowledge: learning to adapt in Canada's Arctic. Glob Environ Chang 21(3):995–1004

Bates PD, Quinn N, Smith A, Sampson CC, Neal JC, Alfieri L, Freer JE (2020) The need for better understanding of uncertainty in flood early warning systems: an example of the River Severn in the UK. Environ Sci Policy 112:108–117

Battiste M (2017) Decolonizing education: nourishing the learning spirit. UBC Press

Berkes F, Colding J, Folke C (2000) Rediscovery of traditional ecological knowledge as adaptive management. Ecol Appl 10(5):1251–1262
Bruns A, Highfield T, Burgess J (2018) The Arab Spring and the role of ICTs. Oxford Research Encyclopedia of Communication
Castleden H, Morgan VS, Lamb C (2012) "I spent the first year drinking tea": exploring Canadian university researchers' perspectives on community-based participatory research involving Indigenous peoples. Canadian Geographer/le Géographe Canadien 56(2):160–179
Christoplos I, Mitchell J, Liljelund A (2001) Re-framing risk: the changing context of disaster mitigation and preparedness. Disasters 25(3):185–198
Das A, Gujre N, Devi RJ, Mitra S (2021) A review on traditional ecological knowledge and its role in natural resources management: North East India, a cultural paradise. In: Environmental management, pp 1–22
Dkhar M, Tiwari BK (2020) Traditional ecological knowledge of tribal communities of North East India. Biodiversitas J Biol Diversity 21(7)
Esposito M, Palma L, Belli A, Sabbatini L, Pierleoni P (2022) Recent advances in internet of things solutions for early warning systems: a review. Sensors 22(6):2124
Facin ALF, Laurindo FJB, de Mesquita Spinola M (2014) Knowledge management to deal with risk in the process of software development: a case study. In: Proceedings of PICMET' 14 conference: Portland International Center for Management of Engineering and Technology; Infrastructure and Service Integration. IEEE, pp 2673–2681
Finn S, Herne M, Castille D (2017) The value of traditional ecological knowledge for the environmental health sciences and biomedical research. Environ Health Perspect 125(8):085006
Ford JD, Cameron L, Rubis J, Maillet M, Nakashima D, Willox AC, Pearce T (2016) Including indigenous knowledge and experience in IPCC assessment reports. Nat Clim Chang 6(4):349–353
Hosen N, Nakamura H, Hamzah A (2020) Adaptation to climate change: does traditional ecological knowledge hold the key? Sustainability 12(2):676
Islam AM (2012) Methods of open innovation knowledge sharing risk reduction: a case study. Int J e-Education e-Business e-Management e-Learning 2(4):294
Israel BA, Schulz AJ, Parker EA, Becker AB (2003) Review of community-based research: assessing partnership approaches to improve public health. Annu Rev Public Health 19:173–202
Johnson JT, Louis RP, Kliskey A (2014) Weaving indigenous and sustainability sciences: diversifying our methods. National Science Foundation, Arlington, pp 1–116
Kelman I, Mercer J, Gaillard JC (2012) Indigenous knowledge and disaster risk reduction. Geography 97(1):12–21
Lambert S, Scott J (2019) International disaster risk reduction strategies and indigenous peoples. Int Indigenous Policy J 10(2):1–21
Li Z, Wu L, Zhang Y, Zhang Y (2018) Unmanned aerial vehicle remote sensing for field-based crop phenotyping: current status and perspectives. Front Plant Sci 9:1389
Molina JGJ, Neef A (2016) Integration of indigenous knowledge into disaster risk reduction and management (DRRM) policies for sustainable development: the case of the Agta in Casiguran, Philippines. In: Sustainable development and disaster risk reduction, pp 247–264
Nakashima D, Galloway McLean K, Thulstrup H, Ramos Castillo A, Rubis J (2012) Traditional knowledge initiative. Weathering uncertainty: traditional knowledge for climate change assessment and adaptation
Noran O (2014) Collaborative disaster management: an interdisciplinary approach. Comput Ind 65(6):1032–1040
Pandey BH, Okazaki K (2005) Community-based disaster management: empowering communities to cope with disaster risks. Regional Dev Dialogue 26(2):52
Radel K, Sukumaran A, Daniels C (2023) Incorporating First Nations knowledges into disaster management plans: an analysis. Austr J Emergency Manage 38(2):36–41

Rajendran L, Shankaran S (2021) ICT enabled early warning dissemination system for disaster management. In: 2021 6th International Conference on Inventive Computation Technologies (ICICT). IEEE, pp 443–448

Sagun A, Bouchlaghem D, Anumba CJ (2009) A scenario-based study on information flow and collaboration patterns in disaster management. Disasters 33(2):214–238

Schwab AK, Brower DJ, Eschelbac K (2007) Hazard mitigation and preparedness: building resilient communities (No Title)

Sitas N, Reyers B, Cundill G, Prozesky HE, Nel JL, Esler KJ (2016) Fostering collaboration for knowledge and action in disaster management in South Africa. Curr Opinion Environ Sustain 19:94–102

Smith LT (2021) Decolonizing methodologies: research and indigenous peoples. Bloomsbury Publishing

Trainor JE, Stern EK, Subbio T (2018) On bridging research and practice in disaster science and management: unified system or impossible mission? In: Handbook of disaster research, pp 161–178

Trickett EJ (2011) Community-based participatory research as a worldview or instrumental strategy: is it lost in translation(al) research? Am J Public Health 101(8):1353–1355

Turner NJ, Ignace MB, Ignace R (2000) Traditional ecological knowledge and wisdom of aboriginal peoples in British Columbia. Ecol Appl 10(5):1275–1287

Turner NJ, Ignace MB, Ignace R (2008) Traditional ecological knowledge and wisdom of aboriginal peoples in British Columbia. Ecol Appl 10(5):1275–1287

United Nations (2007) United Nations Declaration on the Rights of Indigenous Peoples

Vinyeta K, Lynn K (2013) Exploring the role of traditional ecological knowledge in climate change initiatives. US Department of Agriculture, Forest Service, Pacific Northwest Research Station, Portland, OR

Weichselgartner J, Pigeon P (2015) The role of knowledge in disaster risk reduction. Int J Disaster Risk Sci 6:107–116

Yun Y, Xi C, Hong J, Guo W (2018) Machine learning applications for geospatial big data: a review. ISPRS J Photogramm Remote Sens 146:49–61

Zhang Y, Qian H, Wang Y, Li Y (2019) Applications of geographic information systems (GIS) data and methods in energy-related environmental research. Energy Rep 5:169–181

Chapter 24
Physical, Socio-Cultural, Economic, and Ecological Studies on Disasters for Planning Risk Reduction in Uttarkashi Region, Uttarakhand, India

Damini Rana and Surya Parkash

24.1 Introduction

24.1.1 Disaster

Disaster, when anyone hears about that utterance, the first trounce of consideration is, Where it strikes? When it strikes? What is the intensity level of disaster? How much fatality, damage, pain occurs? What are the transportation and accessibility echelon? How far are the medical facilities from the disaster-prone zone? What are the conditions of local perception and participation? What are the disaster management and government plans? And other negative reflections on the mind thought process. All of these above queries come in mind because directly or indirectly they are related to community. Community is always in the nucleus of disaster management process. (Pearce 2003) highlighted community involvement makes disaster management more eminent. There are mainly natural and manmade disasters and different disasters have diverse impact upon the people, their livelihoods, environment and economy and infrastructure. Communities are the first who are affected by any disaster and the first who respond, so it is the important to give the local community involvement adequate place in the disaster management programs.

D. Rana (✉)
UPES, Dehradun, Uttarakhand, India
e-mail: daminiranaofficial@gmail.com

S. Parkash
Head Geo-Meteorological Risks Management Division & CBRN, Industrial and Cyber DRR Division at NIDM, New Delhi, India

N. A. Siddiqui et al. (eds.), *Advances in Disaster Management, Volume 2*, Springer Transactions in Civil and Environmental Engineering,
https://doi.org/10.1007/978-981-96-4047-8_25

24.1.2 Vulnerability of Hazards

India has been traditionally vulnerable to natural disasters on account of its unique geo-climatic conditions. Floods, droughts, cyclones, earthquakes, and landslides have been recurrent phenomena. According to (Kumar KJA 2013) about 60% of the landmass is prone to earthquakes of various intensities; more than 40 million hectares are at risk of flooding; almost 8% of the entire region is vulnerable to cyclones; and 68% of the area is vulnerable to drought. (Rao et al. 2017) highlighted between 1990 and 2000, there were approximately 4344 fatalities and 30 million annual disaster-related injuries. The Himalayas are a young, folded mountain range where stress release is a frequent occurrence. Situated in the Himalayas, Uttarakhand state is divided into four parallel zones: the Siwalik range, the greater Himalayas, the middle Himalayas (sometimes called the Inner or lesser Himalayas), and the Tarai Bhabhar zone (Sati 2023). The combination of the hostile terrain, different glaciers, rivers, other climatic and geological conditions makes Uttarakhand a vulnerable state (Sharma and Masiwal 2024).

Uttarkashi is a district of Uttarakhand State, India, located in the Himalaya. It is categorized under Seismic Zone IV and V for Earthquake zonation (Fig. 1a) and is also prone to frequent landslides (Fig. 1b). In recent years, this area has seen both minor and large earthquakes. The security of the region's natural, manmade, cultural, and financial resources is a top priority for the local government, citizens, and decision-makers. Over the past few decades, the population of Uttarkashi town has grown extremely quickly due to economic causes such the decline in rural areas' economic opportunities and the resulting migration to metropolitan areas (Kumar et al. 2022). Slums will most likely proliferate as a result of the town's increasing urbanization, which will also put a significant burden on its resources. Most newly constructed buildings are poorly planned and built thus, even if the older structures were built in accordance with the applicable standards at the time, they might not meet the more exacting requirements of the most recent standards.

24.1.3 Need for Disaster Management

The Super Cyclone in Orissa (October 1999) and the Bhuj earthquake in Gujarat (January 2001) highlighted the critical need for a comprehensive, multi-dimensional approach to disaster management. This approach must integrate diverse scientific, engineering, financial, and social processes, while emphasizing a multi-disciplinary, multi-sectoral strategy. Incorporating risk reduction into development plans and policies is essential for building long-term resilience (Rao et al. 2017). In recent years, the Government of India has ushered in a paradigm shift in disaster management. This new approach is rooted in the belief that sustainable development is only achievable when disaster mitigation is an integral part of the development process (Sinha 2013). A key pillar of this approach is that disaster mitigation must be multi-disciplinary,

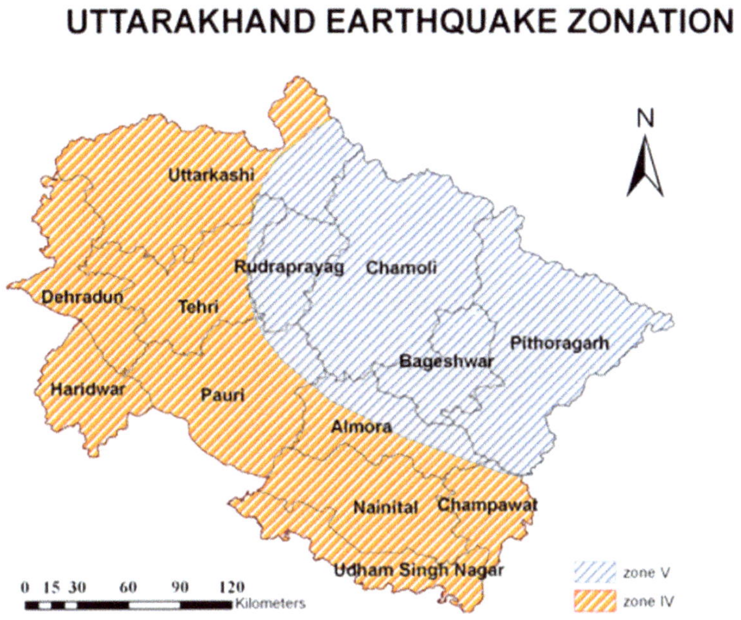

(a)

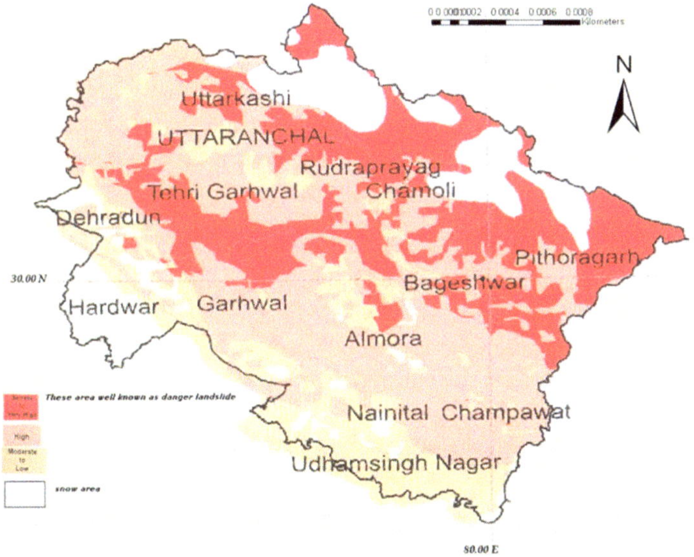

(b)

Fig. 24.1 **a** Uttarkashi earthquake zonation. *Source* http://dmmc.uk.gov.in/pages/display/95-earthquake-zone, **b** Uttarkashi landslide zonation. *Source* http://dmmc.uk.gov.in/pages/display/96-landslide-zone

encompassing all sectors of development. The policy is also driven by the belief that investing in mitigation is far more cost-effective than incurring the high costs of relief and rehabilitation efforts.

The groundwork laid by (Bolt, 1975) on the frequent occurrence of landslides in geodynamically sensitive areas initiated a deeper exploration into the relationship between seismic activities and geological disruptions. This early study underscored the necessity for ongoing geological assessments in earthquake-prone zones. Furthering this initial work, (G. Yu et al. 1994) delivered a comprehensive analysis of the Uttarkashi earthquake, detailing seismic behaviors and ground motions to predict the impacts of future earthquakes. Their research marked a significant advancement in the use of seismic data for disaster preparedness. (Cotton et al. 1996) expanded upon the 1991 Uttarkashi earthquake's analysis, offering insights into the faulting processes that are characteristic of major earthquakes in the Himalayas. This study contributed to a better understanding of the geological features that influence seismic risks. By the late 1990s, as noted by (Jain, 1998), there was an emerging concern over the complacency in earthquake preparedness in India, despite historical data showing frequent significant seismic events. This observation highlighted the need for enhanced structural standards and disaster readiness. In a comparative study, (Jain et al. 1999) examined differences in the impacts of the Uttarkashi and Chamoli earthquakes, attributing variations in damage and casualties to distinct site characteristics. This underscored the importance of localized geological and environmental studies in shaping effective disaster management strategies.

The early 2000s saw a shift toward understanding the environmental impacts of disasters, with (Narendran, 2001) and (Jigyasu, 2002) exploring how forest fires and local community knowledge play roles in disaster management. These studies emphasized the integration of environmental management with local community involvement in disaster preparedness plans. (Parashar Amit and Biswas Sas 2003) further investigated the ecological dynamics of forest fires in the Himalayas, pointing out the complex interplay between natural and anthropogenic factors in the degradation of forest ecosystems. The focus on the intersection of human activity and natural disasters continued with (Gupta and Bist, 2004) discussing the causes and consequences of landslides in Uttarkashi. Their findings stressed the combined impact of natural events and human interventions on disaster frequency and severity. By mid-2000s, (Negi, 2004) highlighted the critical role of forest cover in stabilizing the fragile ecosystems of the Himalayas, directly linking deforestation and land management practices with increased landslide risks.

As the decade progressed, the scope of research expanded to include more detailed assessments of disaster impacts on soil and ecosystems. (Sinha and Goyal, 2004) provided an overview of the seismic history of India, advocating for a comprehensive national earthquake disaster risk management policy based on their findings. Continuing from the assessments of seismic risks and environmental impacts, further research by (Kimothi et al. 2005) focused into the specific effects of landslides on Uttarkashi, examining the direct impact on the local community and infrastructure. This study brought to light the severe consequences of landslides exacerbated by poor urban planning and the lack of effective disaster management strategies. In subsequent years, research by (Pandey and Verma, 2007) and (G.S.I., 2008) focused on the agricultural impact and geological assessments related to landslides and floods. These studies were instrumental in understanding the broader environmental and economic implications of natural disasters in the Himalayas. A significant advancement in disaster research came with the comprehensive report on the Uttarakhand disaster by (Disaster Nature Fury, 2009). This report cataloged a variety of natural calamities, emphasizing the cumulative effect of multiple disaster types on the region and the urgent need for integrated disaster management plans.

By 2012, (Kumar et al. 2013) conducted a focused study on the effects of forest fires on soil nutrient status in the Chir Pine forests of the Garhwal Himalayas. Their findings highlighted the critical environmental impacts of forest fires on soil quality and vegetation, informing strategies for forest management and fire control. That same year, (Nautiyal and Bhandari, 2012a, b) presented a district-level disaster management action plan, showcasing how localized planning can address the unique challenges posed by the geological and climatic conditions of the Himalayas. Their work underscored the importance of geographic, climatic, and administrative factors in disaster response and preparedness. Further expanding the research landscape, (Verma and Jayakumar, 2012) explored the effects of forest fires on forest soil, emphasizing the dual nature of fire as both beneficial and destructive. Their work contributed to a nuanced understanding of fire management in ecosystem stability and health. By 2013, (Kumar A 2013) and (Kumar Govind et al. 2013) addressed the socio-economic and infrastructural challenges posed by disasters in Uttarkashi. These studies pointed out the necessity of integrating disaster risk reduction into the developmental plans to enhance resilience and ensure sustainable development. Lastly, by 2014, (Eapen, 2014) critically analyzed the exacerbated impact of unchecked tourism and rapid infrastructure development on the natural calamities in Uttarakhand. This study highlighted the complex interplay between development and disaster risk, stressing the need for sustainable planning and strict regulatory frameworks to mitigate disaster impacts. The research journey has seen further refinement with recent contributions like (Jhariya et al. 2014) and (Rai et al. 2014), which have expanded our understanding of the ecological impacts of natural disasters. Jhariya and Raj emphasized the significant long-term effects of fires on both flora and fauna, exploring how repeated fires alter habitat structures and community compositions. This work is vital for comprehending the broader ecological impacts of disaster-induced changes in habitat and for formulating strategies to mitigate these effects.

Meanwhile, (Rai et al. 2014) used remote sensing technology to study landslides, illustrating how satellite imagery and GIS technologies have become indispensable tools in the modern disaster management toolkit. Their research leverages these technologies to create detailed landslide inventories that are critical for hazard analysis and mitigation planning. (Khakhi et al. 2015) focused on the specific impact of forest fires on nutrient dynamics in forest soils, particularly in the Blue Pine forests. Their findings on nutrient depletion and alteration highlight the importance of managing fire regimes to maintain soil health and forest ecology, stressing the nuanced balance needed in forest management practices. Furthering the integration of disaster management with forest management, (Kaur and Gupta, 2016) conducted a vulnerability analysis of forest fires in Uttarakhand, emphasizing the role of spatial and climatic variables in fire dynamics. Their research underscores the importance of using scientific data to inform fire management strategies and adapt to the changing patterns influenced by climate change. (Murthy and Devgiri, 2015) also contributed to this body of knowledge by studying the chemical properties of soil post-forest fires. Their findings, which indicate changes in pH and nutrient availability due to fire, provide critical information that can guide post-fire recovery and reforestation efforts.

Disaster management holds a vital position in India's policy framework, as the poor and underprivileged are often the most severely impacted by calamities. The government's initiatives stem from this approach, which has been formalized into a comprehensive National Disaster Framework—a strategic roadmap. This framework addresses institutional mechanisms, disaster prevention strategies, early warning systems, mitigation efforts, preparedness and response measures, and human resource development, all aimed at reducing disaster risks across various states in the country.

The increasing frequency and intensity of disasters have become a matter of concern for Govt. and NGO's all over the area. Huge amount of money that has been spent on relief and rehabilitation which could otherwise be spent for development, the loss of property, lives, livelihood, and environment is also a major concern. There is hardly a year when some parts of the region do not face the specter of earthquake, landslide, drought, flash flood, forest fire, cloud burst, avalanche, crop disease, etc. Along with these disasters, manmade disasters such as road accidents, building collapse, fire hazards, bush fire, and epidemic now become a part of life. Having all these copious disasters in the region, it is important to have a strong disaster management plan which works on every phase of disaster.

(a) **Pre-disaster Phase-**Pre-disaster phase is the before disaster stage. In factual sense, this, the real time for disaster management plans, mainly includes the preparedness, mitigation, prevention, and other pre-disaster planning strategies.
(b) **During Disaster Phase-**During disaster phase is the current stage of any disaster. It mainly includes the response, community participation, rescue, relief, and other during disaster planning strategies.
(c) **Post-disaster Phase-**Post-disaster phase is the after disaster stage. It mainly includes the damage assessment, community health, restructuring, rehabilitation, and other post-disaster planning strategies.

The study of these disaster threats is very important from the ecological and economical point of view and for the same, the analysis of changes that occurred in the area due to earthquakes, landslides, flash floods, and forest fires are depicted so that the danger threat zone areas are demarcated for planning disaster risk reduction techniques/policies. CBDM planning is an approach intended primarily by and for the community in disaster-prone areas. Incorporating community-based approaches in disaster management planning is found to be more effective as the local communities are knowledgeable about the disasters happening in their environment, they are the first populace who are facing the hazard and also first group of people who respond. Uttarkashi is a disaster-prone district, and because of the existence of different disasters, there is an urgent need of community-based disaster management planning.

Major research gaps identified are divided into the following issues:

Communities-related issues, disasters and their impact-related issues, past disasters issues, current hazard issues, future issues, and other issues (e.g., political, social, religious, economic, disaster funding, role of government, NGOs, and other aspects that influence the hazard and disasters).

To solve these issues, it is felt that CBDMP should be included in four administrative levels:

- State level unit, district level unit, block unit, and village level unit.

The role of these units is important because community alone cannot implement community-based disaster management. It will take concerned efforts at different levels and across different sectors to improve outstanding linkages between community and disasters. There are some common failures in pre-, during, and past disasters phases:

(a) **Pre-disaster Failure**

A good administrative setup, a well-formulated disaster management plan, and appropriate position of local community participation in different phases of disaster management can mitigate the influence of disaster very effectively, but in reality the composition of above these are rare to see.

(b) **During Disaster Failures**

A major shortcoming in disaster response plans is the delayed activation of government machinery, which often begins operating only 48 h after a disaster strikes. Time is a critical factor in mobilizing resources and administrative systems to address the immediate needs of affected populations. This delay results in significant losses for those impacted. There is an urgent need to focus on improving disaster management programs by ensuring timely action and fostering active participation from the affected communities themselves.

(c) **Post-disaster Failures**

Maskrey (1989) pointed out 'Many vulnerable communities revert to coping with risk, often in the same or worse conditions than before the disaster actual struck'. In actual life, post-disaster management efforts lack community participation. There is a wide gap between what is needed by the affected community and what is being provided by the authorities.

The main objectives of this study are to assess earthquakes, landslides, floods, and forest fires in the study area and to evaluate their physical, socio-cultural, economic, and ecological impacts. The study also focuses on planning disaster risk reduction through risk identification, prevention, mitigation, and preparedness. It aims to improve systemic responses to pre-and post-disaster scenarios by considering the developmental needs of Himalayan states, preserving natural resources, promoting sustainable pilgrimage growth, implementing timely warning systems, and enhancing disaster management communication among stakeholders for sustainable development in Uttarakhand.

24.2 Materials and Methodology

24.2.1 Study Site

Uttarkashi district lies between latitude 30°28' and 31°28' North and longitude 77°49' and 79°25' East. Its maximum length from east to west is about 154 kms and maximum breadth from north to south is about 109 kms. Uttarkashi is a border district of the state of Uttarakhand. The total area of the district is 8016 sq.km and area-wise it is the largest district of the state Uttarakhand. Uttarkashi town is present at an elevation of 1150 mts above sea level on the bank of Bhagirathi River, and most of the terrain is hilly. As of 2011 India census, Uttarkashi District had a population of 330,086. **Males** constitute **51.1%** of the population and **Females 48.9%**. Out of total population 92.6% people live in rural area. Covering an area of about 12.02 sq.km., the Township of Uttarkashi has a total population of about 18,220 approx (2011 census). On the other hand, 7.4% people stay in urban area. Uttarkashi has an average literacy rate of 75.81% out of which male literacy is 88.79%, and female literacy is 62.35%. In district, 14.03% of the population is under 6 years of age.

The socio-economics of an area plays a vital role in determining its risk exposure to disasters and the community's ability to cope with hazards. The unique geography of hill regions like Uttarkashi, combined with their floral and faunal diversity, further defines these risks and resilience factors. Land use and natural vegetation of the area include various categories mentioned in Table 24.1. Floral species such as Aconitum heterophyllum, Asparagus racemosus, and Taxus baccata contribute to the ecological balance, while faunal species like Panthera pardus, Herpestes edwardsi, and Ursus thibetanus are crucial to the biodiversity of the region [Source-Forest Department Kotbangla, Uttarkashi District]. Settlements in these areas are typically built using local materials like stones and clay bricks, with livestock housed nearby. Livelihoods primarily depend on agriculture, livestock rearing, and the collection of medicinal herbs, although the steep slopes, limited fertile land, and poor infrastructure pose challenges. Despite the potential for horticulture, growth is hindered by marketing and logistical issues. Animal husbandry supplements income but is not a full-time occupation. The district lacks significant industrial development, relying on cottage industries like wool and woodcraft. However, improving infrastructure, especially in horticulture, tourism, and agriculture, remains a priority for sustainable growth.

Uttarkashi town, situated in the most seismically sensitive Zone V of the Garhwal Lesser Himalaya, lies just south of the Main Central Thrust (MCT) near Sainj. This

Table 24.1 Land use/land cover distribution

Category	Sub-category	Area in sq.km
Agriculture	Cropland	523.13
	Plantation	94.44
	Fallow	0.04
	Current shifting cultivation	
Barren/uncultivable/wasteland	Salt affected land	
	Gullied/Ravinous land	2.01
	Scrub land	17.05
	Sandy area	
	Barren rocky	1991.88
	Rann	
Built-up	Urban	3.17
	Rural	6.29
	Mining	
Forest	Evergreen/semi-evergreen	2951.84
	Deciduous	87.10
	Forest plantation	0.61
	Scrub forest	162.49
	Swamp/mangroves	
Grass/grazing	Grass/grazing	871.63
Snow and glaciers	Snow and glaciers	1190.01
Wetland/water bodies	Inland wetland	0.05
	Coastal wetland	
	River/stream/canals	109.03
	Water bodies	**5.29**

Source http://bhuvan.nrsc.gov.in/gis/thematic/tools/document/LULC502/MAP/UK.pdf

region falls within a highly seismotectonically active area (seismic zone V) and has experienced two major earthquakes since 1991. The epicenter of the 1991 Uttarkashi earthquake was in Agoda Village, near Uttarkashi town. Tectonic movements have led to the development of multiple faults and weak planes throughout the region. The Uttarkashi formation consists of Netala Quartzite, Lower Uttarkashi Limestone, Pokhri Slate, Upper Uttarkashi Limestone, and Bareti Quartzite (Jain, 1971). The rocks of Varunavat Parvat are primarily composed of quartzite and phyllites from the Damta Group, which are highly shattered, fragmented, fractured, and thinly jointed. These formations are covered by 20–25 m of slided material, with beds dipping at 30–35° toward N350° into the hill slope. The area is marked by three major joint sets, and the soils are relatively new, generally thin, porous, and unstable.

The drainage near Uttarkashi includes the Bhagirathi River, originating from Gaumukh, and its major tributary Asi Ganga, which begins at Dodital Lake. Significant lakes in the region are Dodital, linked to Lord Ganesha, and Maneri, formed by the Maneri-Bhali hydroelectric project. Other notable lakes include Nachiketa Tal and Kedar Tal. Main springs in Uttarkashi are Gangnani and Surya Kund.

24.2.2 Methodology

The methodology used in data collection, selection process, questionnaire process, field work, data analysis, and other work has been contained in the following steps:

For the present study of earthquakes, landslides, flash floods, and forest fires, the data set and type of activity used are:

24.2.2.1 Collection and Study of Basic Information

First step of any research study is to collect the basic information and data of the region and after that enhance the knowledge before starting the field or further research work. For the present study, basic information and data are mainly collected from institutes, organizations, libraries, and other reliable sources. The main findings after considering the information are selection and identification of sensitive zones and Himalayan ranges. Existence of many hostile elements makes the region more vulnerable to disasters like earthquake, landslides, flash flood, and forest fire. So, it is important to give eco-sensitivity zone (IV, V) an appropriate consideration in the base study.

Outcome-This study helped to analyze the most disaster-affected areas in study site.

24.2.2.2 Selection and Questionnaire Process

A focus group has been established to incorporate audience perspectives into the study. This focus group serves as a qualitative research tool, where participants are asked to share their perceptions, opinions, beliefs, and attitudes regarding a disaster, concept, idea, service, or awareness. The questions were posed both individually and in an interactive group setting, encouraging open dialogue and exchange of ideas among participants.

After getting the base identification list, second step is the selection of sample blocks and villages for further study. Many villages of Uttarkashi region subsists in zones IV and V, but for the present study selected 12 villages (Bhatwari, Jamak, Rawada, Gangori, Pata, Sangrali, Mandon, Tiloth, Joshiyara, Pokhri, Dharasu, and Nalupani), on the basis of its hazard status, disaster history and existence of MCT zone (Fig. 24.2).

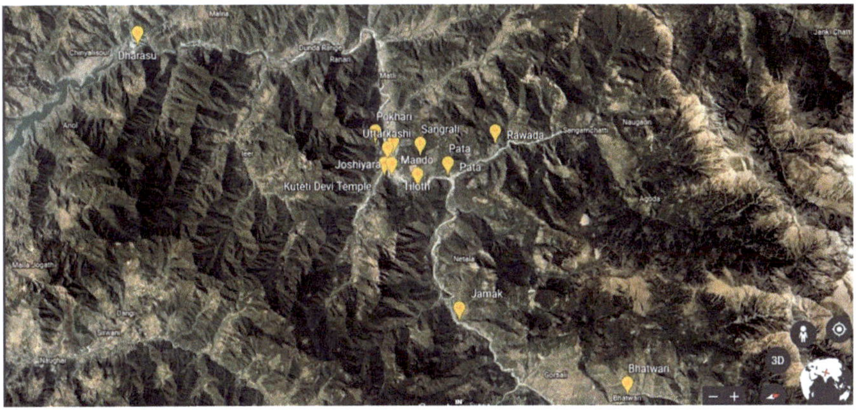

Fig. 24.2 Geographical map of the sample villages. *Source* Google Earth

For the selection process:

- Different villages of two blocks were selected which was through proportionate random and purpose sampling.
- Different stakeholders were involved for the collection of information.

In CBDM planning community is in the core so it's very essential to communicate with the people. PRA tools are the best to connect with the rural community. For PRA, it's very necessary to make a considerable questionnaire so that it can be understood by the local community easily and on the other hand also fulfills the prerequisite of the research work. According to the need of planning risk reduction, village-based, household-based, and other stakeholders-based questionnaires have been prepared.

A. Village-Based Questionnaire

The village-based questionnaire aims to gather comprehensive information about the village, past disasters, active hazards, and vulnerabilities. It is completed by community leaders and includes six main multiple-objective questions covering demographics, natural hazards, disaster preparedness, pre-disaster planning, during disaster health response, and post-disaster policies and training.

2 Household-Based Questionnaire

The household-based questionnaire collects family-specific information, focusing on their views and knowledge about disaster phases and community management. It includes two multiple-objective questions addressing general family information and disaster-related knowledge, such as pre-disaster preparedness, during disaster emergency handling, and post-disaster material understanding and training.

3 Government and Non-government Questionnaire

This descriptive questionnaire assesses disaster management plans and strategies of district authorities and NGOs. It involves input from key stakeholders like the District

Disaster Management Authority, hospitals, police, forest departments, government colleges, and NGO contributions.

There are three main reasons for making different questionnaires for different stakeholders, first is to collect the individual information of village, household, and ecology of the study site, second is to compare different stakeholders' point of view and contradiction about hazard and disaster, and third is to find out the community participation echelon with the help of different queries.

Outcome-Questionnaire survey helps to analyze the effects of disasters on the socio-cultural and economical aspects of the area and awareness created among villagers will further help them to take the precautionary steps before and during the upcoming disasters.

24.2.2.3 Field Work

The third and most important step in the study is fieldwork, conducted in the selected blocks of Dunda and Bhatwari in Uttarkashi. A color scheme is used for charts representing the visited areas: BLUE for District, YELLOW for Blocks, and RED for Villages. During the fieldwork, sketch maps of 12 sample villages were drawn, providing a visual layout of house locations, schools, roads, and key infrastructure, which can be easily understood by both officials and the public. Basic information diagrams were created using data from local institutions such as district headquarters, hospitals, schools, and police stations, including their distances from villages. Lab work followed, focusing on cartographic mapping and data analysis.

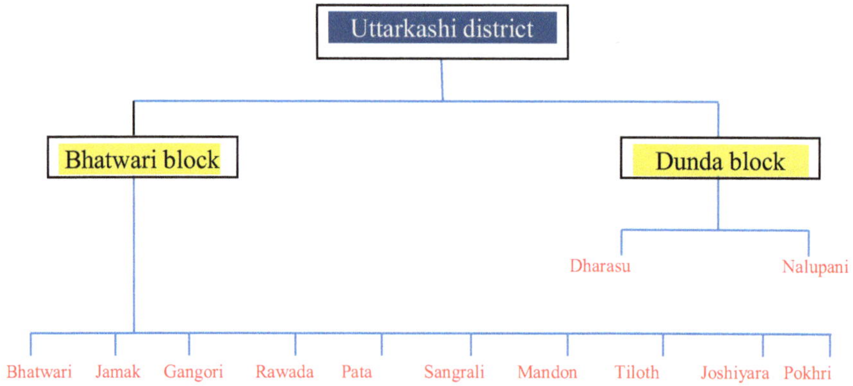

24.3 Results and Discussion

A. **Village-Based Questionnaire-Multiple-Objective-Based Questionnaire**

1. **Demographics**

(a) **Village Name and Population**

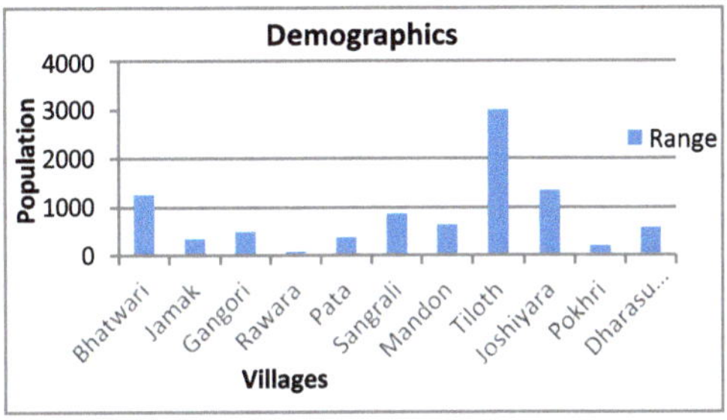

Out of 12 villages

- Max. population- Tiloth
- Min. population- Rawara

(b) **Male–Female Ratio**

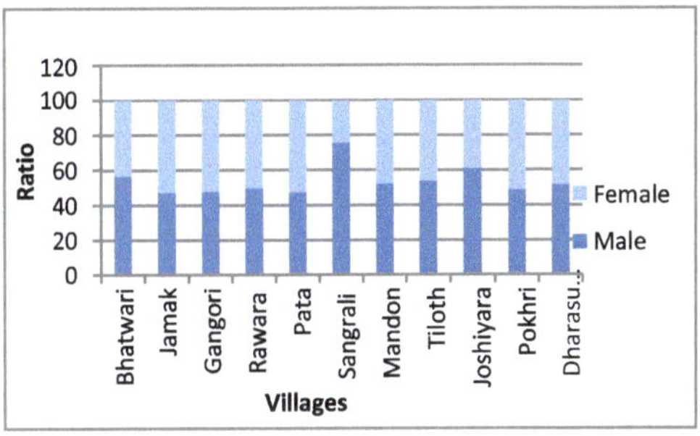

- Max. male ratio- Sangrali
- Max. female ratio- Pata
- Equivalent male female ratio- Rawara and Dharasu/ Nalupani

(c) **Access to Internet**

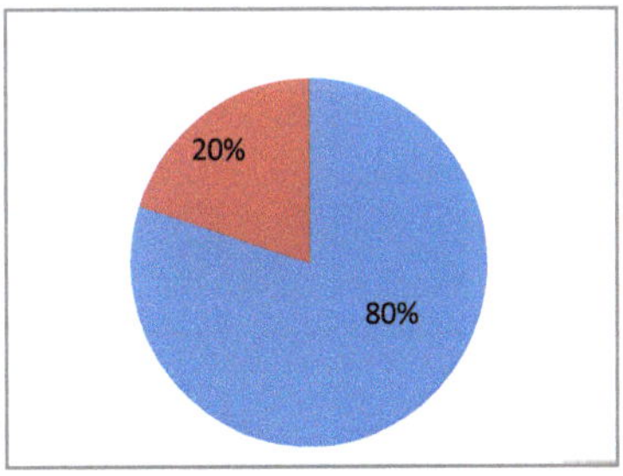

Blue colour with 80% represents- YES to internet access

Red colour with 20% represents- NO to internet access

Access to internet helps the community to be updated with the weather conditions

2. Natural Hazards

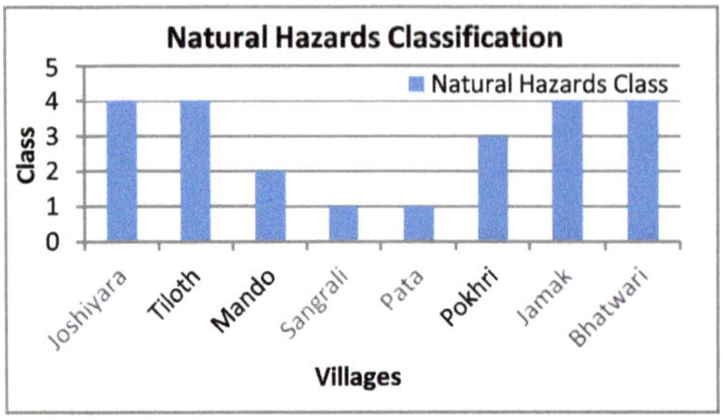

In the graph classes are represented on the basis of studied hazards-

1-Landslide

2-Flash flood and Landslide

3-Earthquake, Landslide and Forest fires

4-Earthquake, Landslides, Flash flood, Forest fire

3. Disaster Preparedness and Response (DPR)

(a) Responsibility

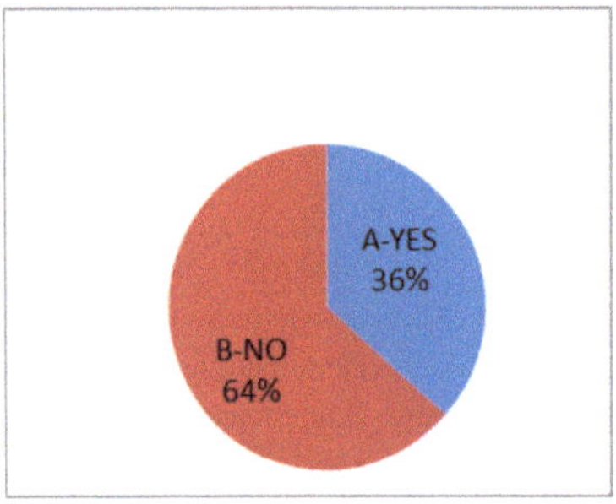

Stakeholder's response-

- 36%-'YES', there are responsible persons for DPR in the community
- 64%-'NO', there are no responsible persons for DPR.

(b) Interaction with the Officials

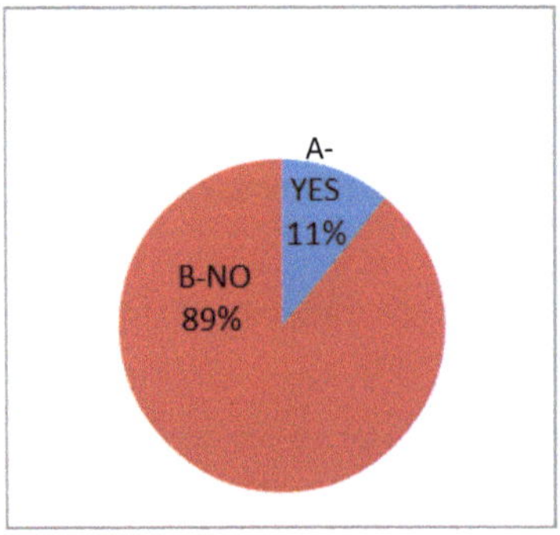

Stakeholder's response-

- 11%-'YES', Responsible person interacts with the Officials for suggestions regarding DM
- 89%- 'NO', Responsible person doesn't interact with the officials.

4. **Pre-disaster**

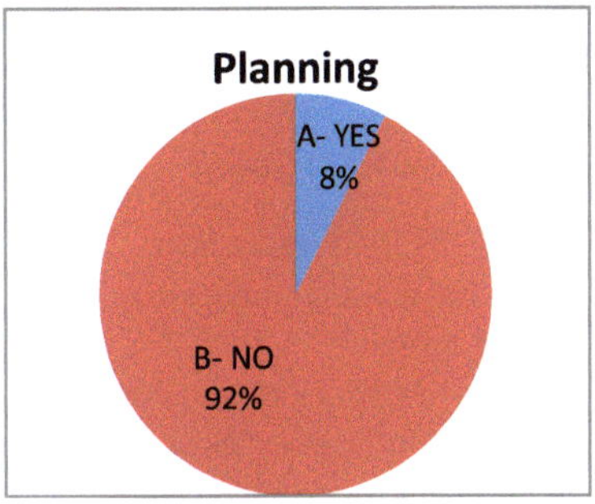

Stakeholder's response-

- 8%-'YES', Community have their own DM Plan
- 92%-'NO', there is no DM Plan

5. **During Disaster**

(a) **Specific Component on Health**

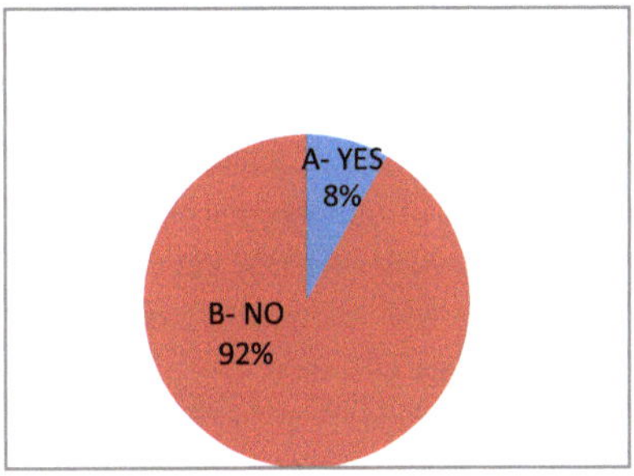

Stakeholder's response-

- 8%- 'YES', Community have specific component on health
- 92%- 'NO', Community don't have specific component on health

(b) **Helping Bodies**

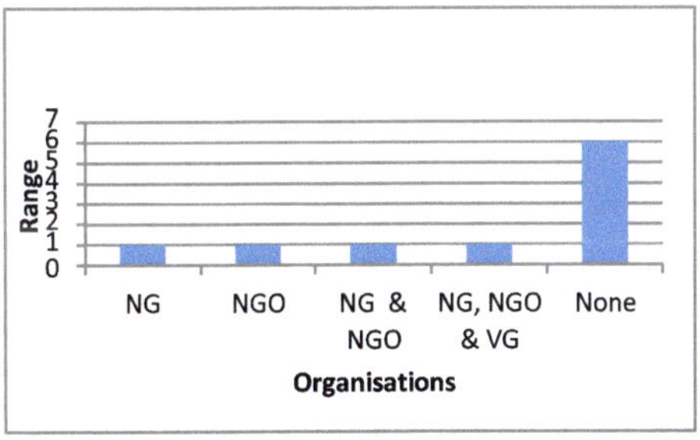

- Most of the community believes there are no helping bodies during disaster

6. **Post-disaster**

(a) **Laws/Policies/Regulations in the District**

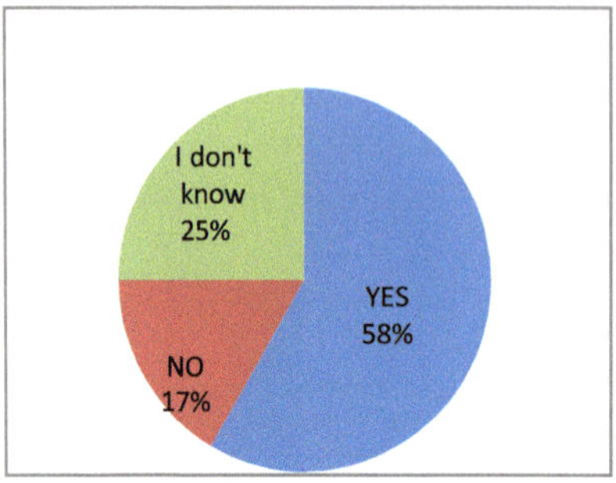

Stakeholder's response-

- 58%- 'YES', there are laws/ policies/ regulations regarding DM in the district
- 17%- 'NO', there are no laws for it
- 25%- 'Don't know', about DM laws in the district

(b) **Own Policies for Preparedness and Response**

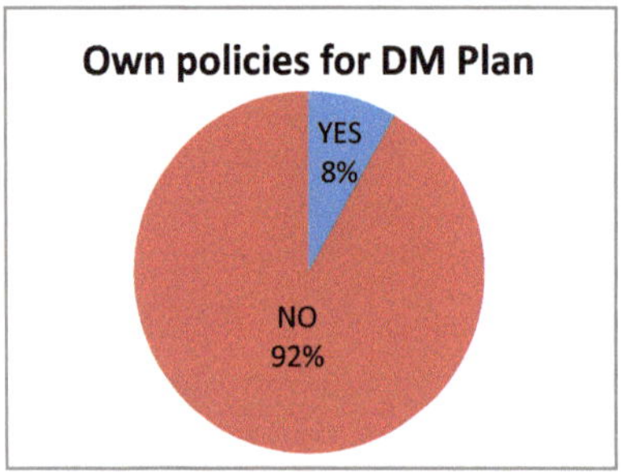

Stakeholder's response-

- 8%- 'YES', Community has their own policies for DM Plan
- 92%- 'NO', Community don't have any DM Plan

(c) **Reliability for Conducting Training programs**

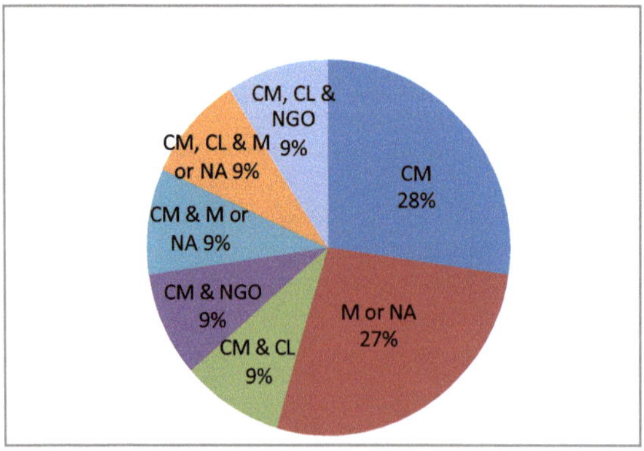

Stakeholders believe that Reliable Community Members (28%) must be trained by the Municipal or National Authority (27%) so that they will be able to execute the training programme themselves and respond effectively during disasters to reduce risk reduction

B. **Household-Based Questionnaire**

1. **Multiple-Objective-Based Questionnaire**

Bhatwari (Fig. 24.3a)

1. General information of families

(a) Income Range

Community of this village has an annual avg. income of Rs (3000–5000 and 10,000–12,000).

2. Disaster-related Knowledge

(a) Pre-disaster

i. Information on Health, Disaster Preparedness, and Response

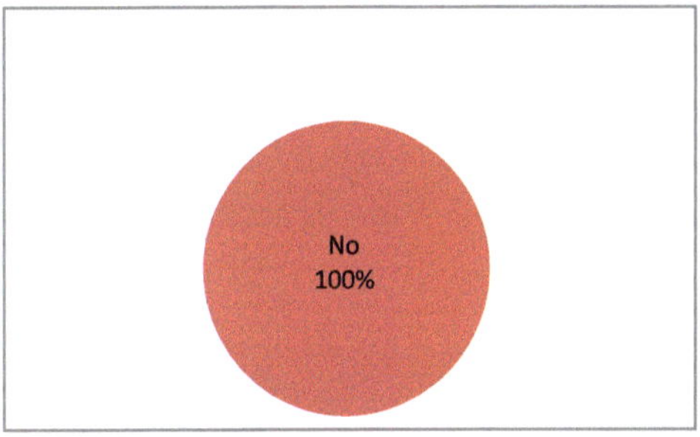

100% of the community reports that the information on health, disaster preparedness and response is not being shared

(b) During Disaster

i. Dissemination of Emergency Situations

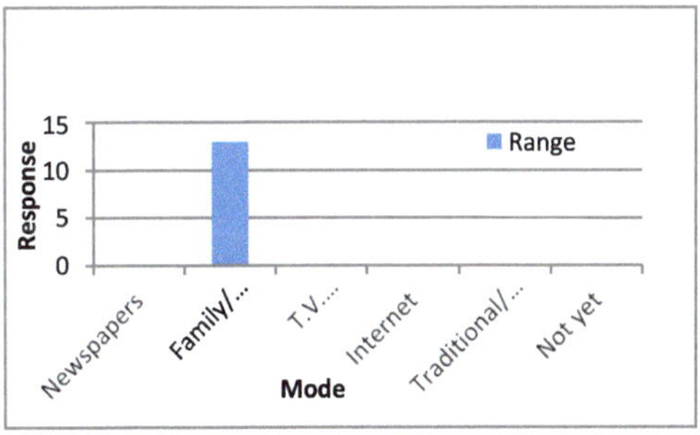

Family and friends according to the community are the mode of dissemination during emergency situation rather than Newspapers, T.V., Internet, and Traditional/announcement method

Fig. 24.3 **a** Bhatwari after Landslide of 2013, **b** Jamak after Earthquake of 1991, **c** Gangori after Asiganga (2012) and, **d** Rawada after Asiganga Flash Bhagirathi (2013) Flash flood, **e** Pata after Varunavat, **f** Sangrali after Varunavat Landslide (2013), **g** Mandon after Forest fire, **h** Reconstruction by community Asiganga and Bhagirathi Flash flood members of Tiloth after flash flood (2013), **i** Embankment formed in Joshiyara, **j** Landslide zone between after Flash flood (2013) Dharasu and Nalupani

Fig. 24.3 (continued)

ii. Health-related problems

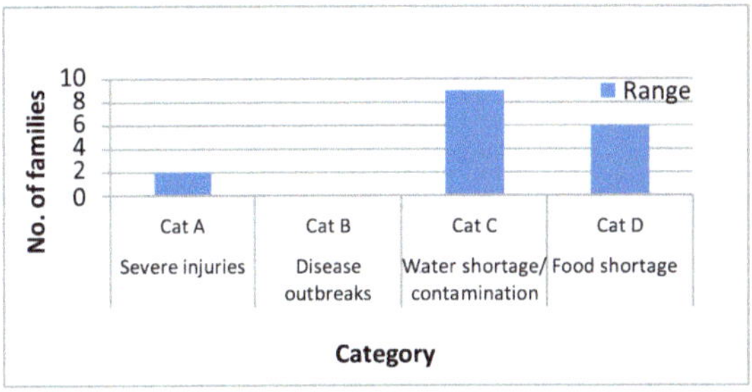

Major health related problem that can be observed is of Water shortage/ contamination

(c) Post-disaster

i. Understanding of Material provided

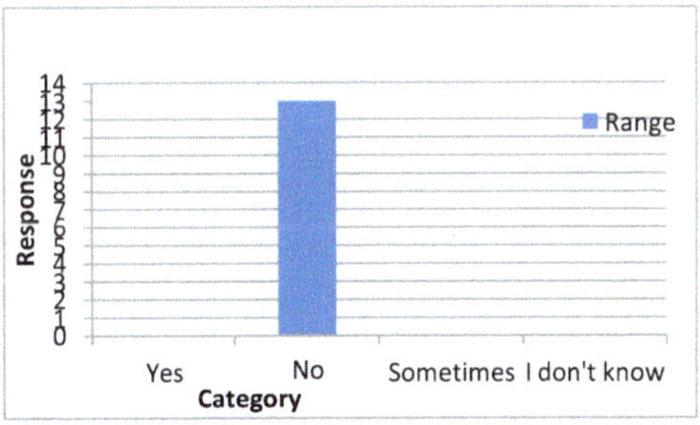

According to the community the materials provided in post disaster condition are not understood by most of the community which may be due to the language in which the instructions are given or due to lack knowledge of using it

ii. Reliability for conducting training programs

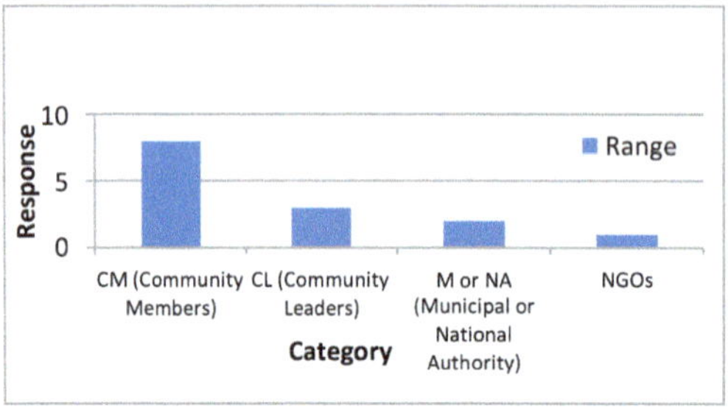

Graph shows community want the reliable community members are to be trained for conducting training programme

Jamak (Fig. 24.3b)

1. General information of families

   (a) Income Range

   Community of this village has an annual avg. income of Rs (10,000–20,000; 30,000–40,000 and 50,000).

2. Disaster-related Knowledge

   (a) Pre-disaster

   i. Information on Health, Disaster Preparedness, and Response

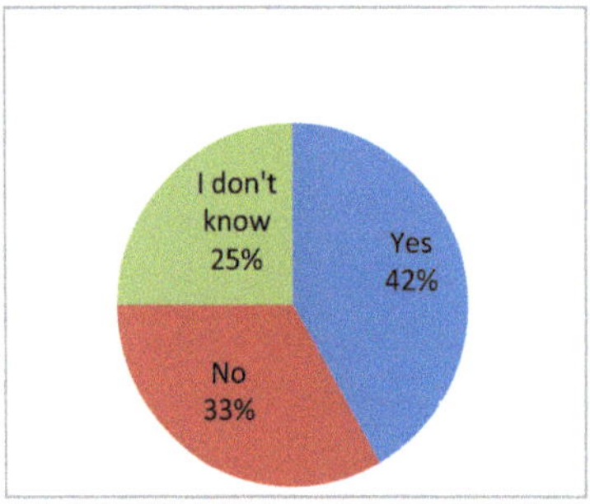

42% of the community agrees that the information on health, disaster preparedness and response is being shared

(b) During Disaster

i. Dissemination of Emergency Situations

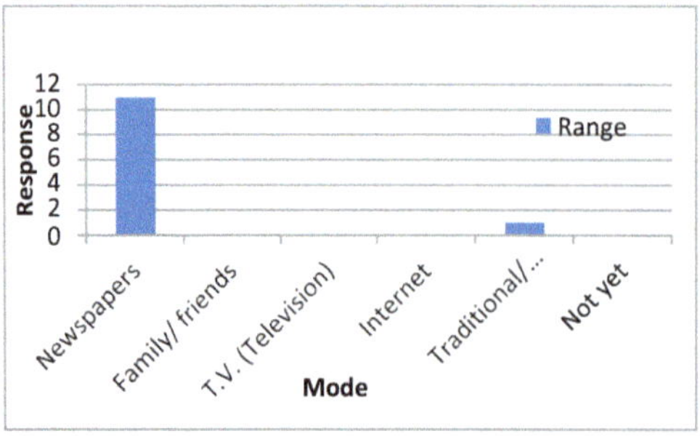

Newspapers according to the community are the mode of dissemination during emergency situation

ii. Health-related problems

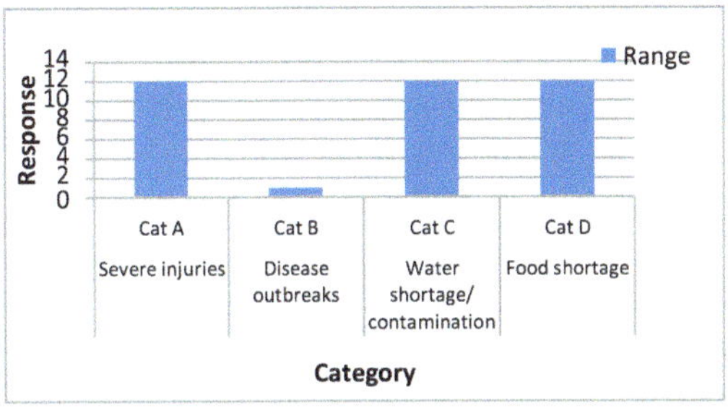

Major health related problem that can be observed is of Water shortage/ contamination, Severe injuries and Food shortage

(c) Post-disaster

i. Understanding of Material Provided

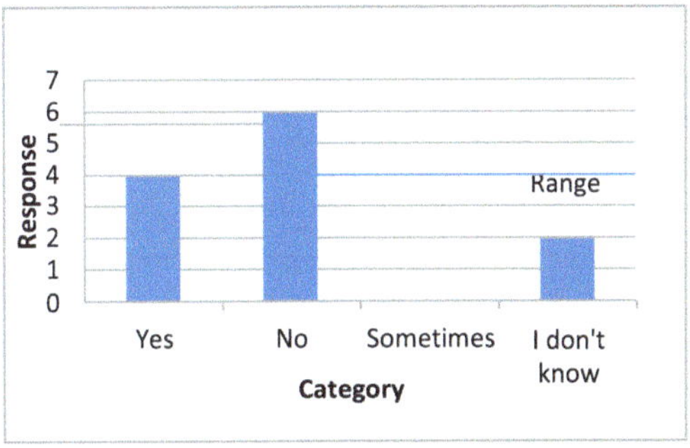

According to the community the materials provided in post disaster condition are not understood by most of the community

ii. Reliability for conducting training programs

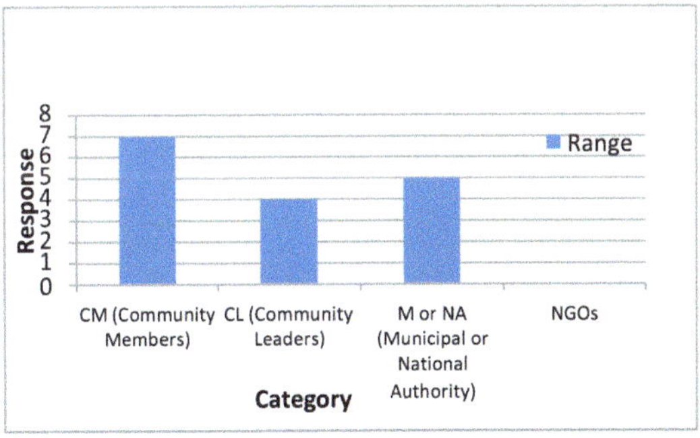

Graph shows community wants reliable community members are to be trained for response during disaster

Gangori (Fig. 24.3c)

1. General information of families

   (a) Income Range

   Community of this village has an annual avg. income of Rs (10,000–30,000; 40,000–60,000; 1 lakh-2 lakh and 5 lakh-6 lakh).

2. Disaster-related Knowledge

   (a) Pre-disaster

   i. Information on Health, Disaster Preparedness, and Response

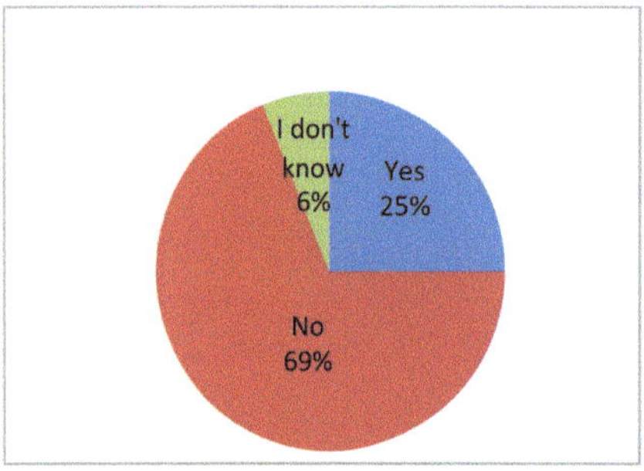

69% of the community reports that the information on health, disaster preparedness and response is not being shared

(b) During Disaster

i. Dissemination of Emergency situations

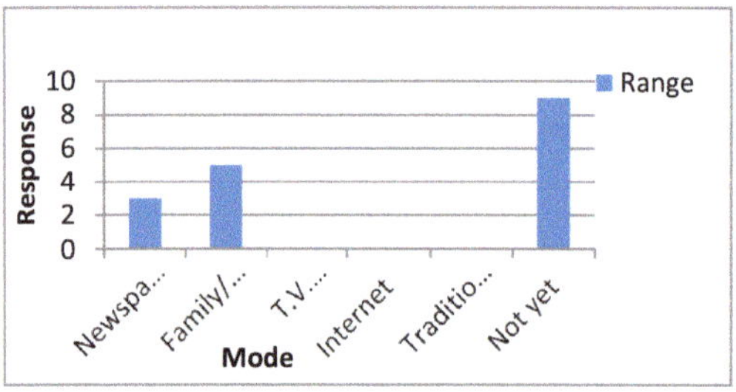

According to the community, no one except friends and family and sometimes newspapers are seen helpful for dissemination of emergency situation

ii. Health-related problems

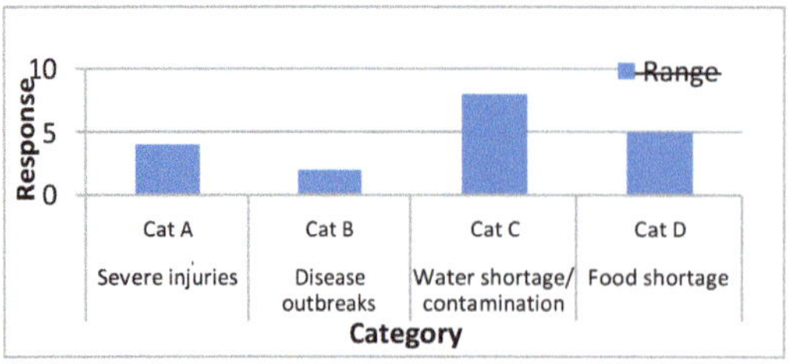

Major health related problem that can be observed is of Food and Water-shortage/ contamination

(c) Post-disaster

i. Understanding of Material provided

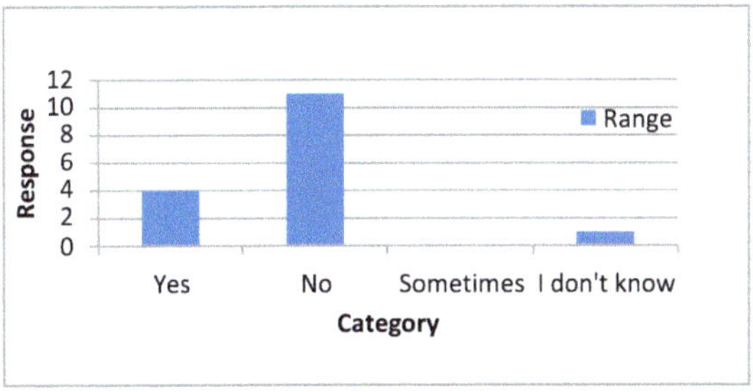

According to the community the materials provided in post disaster condition are not understood by most of the community

ii. Reliability for conducting training programs

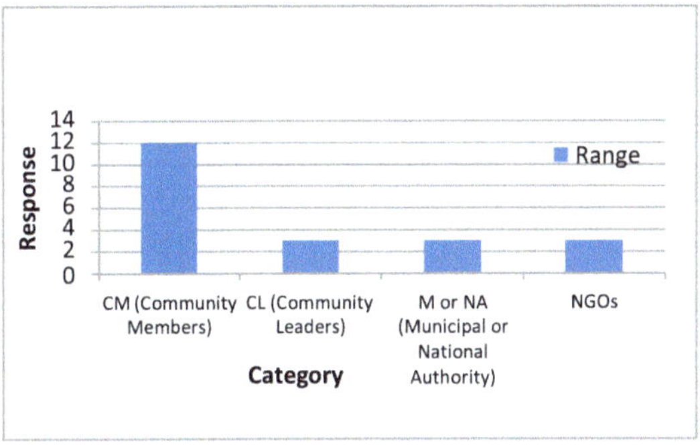

Graph shows community wants reliable community members are to be trained for response during disaster

Rawada (Fig. 24.3d)

1. General information of families

(a)Income Range

Community of this village has an annual avg. income of Rs (20,000–40,000).

2. Disaster-related Knowledge

(a) Pre-disaster

i. Information on Health, Disaster Preparedness, and Response

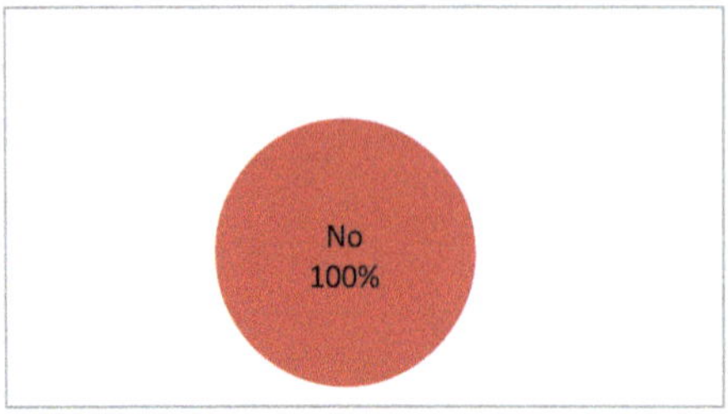

100% of the community reports that the information on health, disaster preparedness and response is not being shared

(b) During Disaster

i. Dissemination of Emergency situations

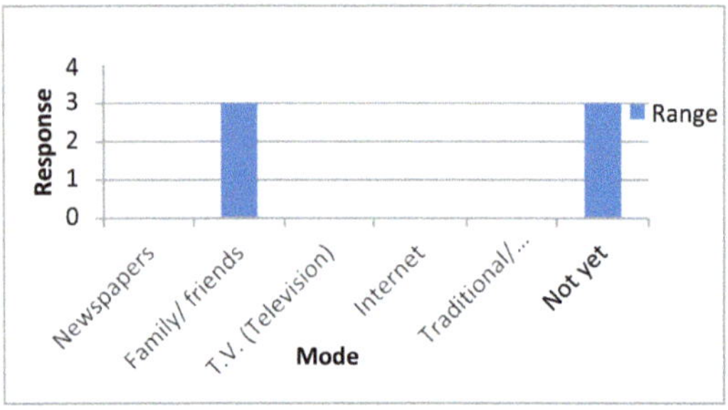

Either Family and friends or no one yet according to the community are the mode of dissemination during emergency situation

ii. Health-related problems

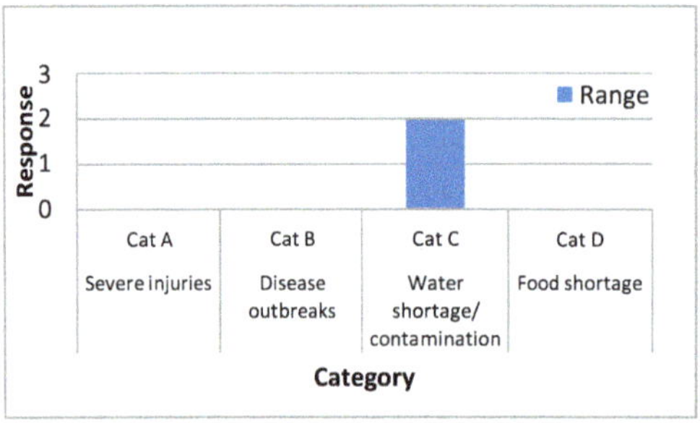

Major health related problem that can be observed is of Water shortage/ contamination

(c) Post-disaster

i. Understanding of Material provided

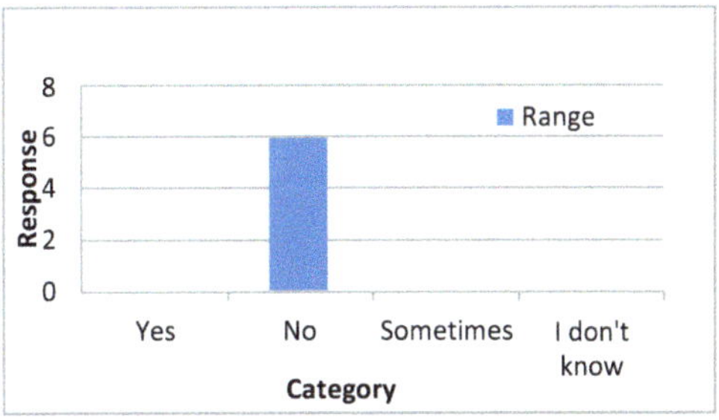

According to the community the materials provided in post disaster condition are not understood by most of the community

ii. Reliability for conducting training programs

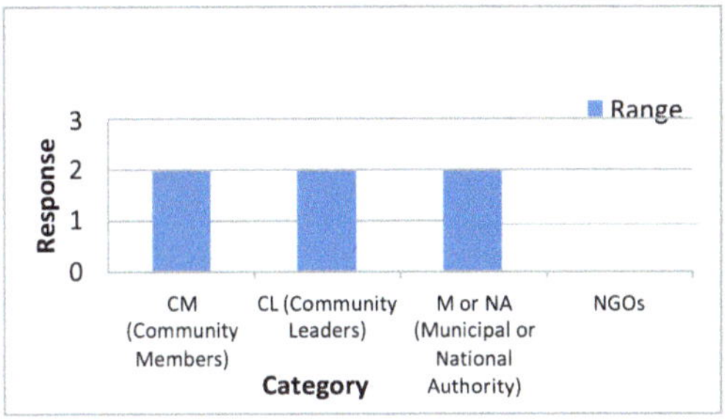

Graph shows community wants the Reliable Community members, Community leaders and Municipal or National Authority are to be trained for response during

Pata (Fig. 24.3e)

1. General information of families

    (a) Income Range

    Community of this village has an annual avg. income of Rs (10,000; 30,000–40,000 and 2 lakh).

2. Disaster-related Knowledge

    (a) Pre-disaster

    i. Information on Health, Disaster Preparedness, and Response

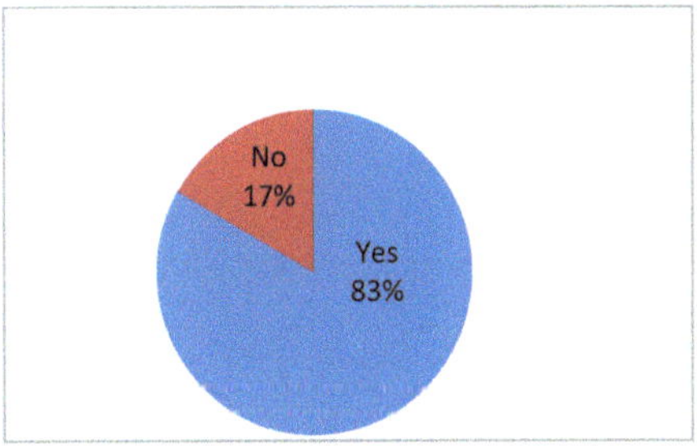

83% of the community agrees that the information on health, disaster preparedness and response is being shared

(b) During Disaster

i. Dissemination of Emergency situations

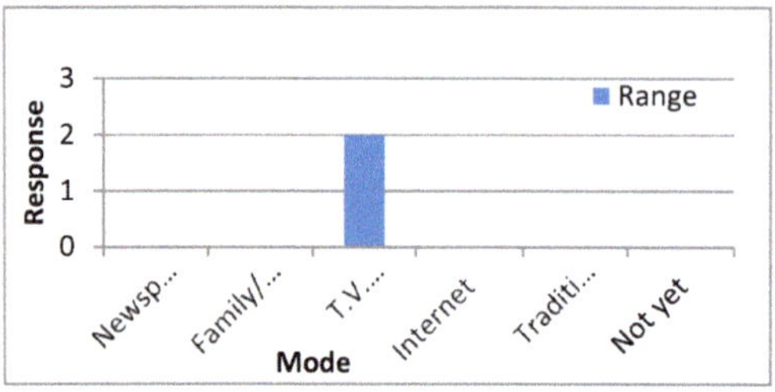

Television (T.V.) according to the community is the mode of dissemination during emergency situation

ii. Health-Related Problems

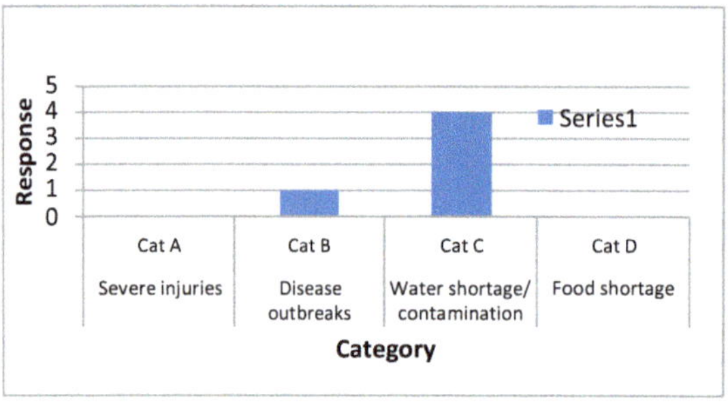

Major health related problem that can be observed is of Water shortage/ contamination

(c) Post-disaster

i. Understanding of Material provided

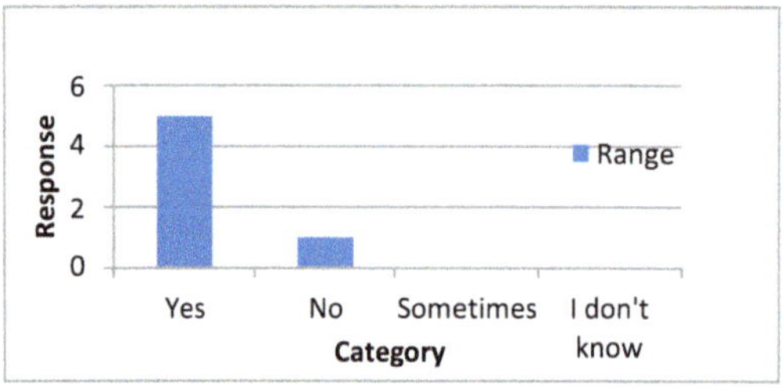

According to the community the materials provided in post disaster condition are understood by most of the community

ii. Reliability for conducting training programs

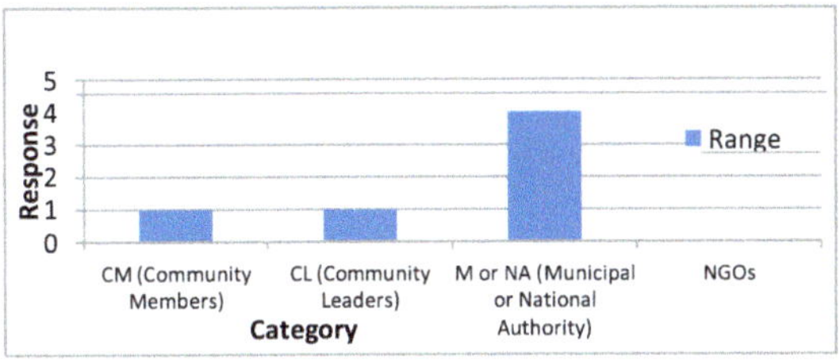

Graph shows community want the reliable community members and leaders are to be trained by Municipal or National Authority for response during disaster

Sangrali (Fig. 24.3f)

1. General information of families

(a)Income Range

Community of this village has an annual avg. income of Rs (5000; 10,000 and 2 lakh).

2. Disaster-related Knowledge

(a) Pre-disaster

i. Information on Health, Disaster Preparedness, and Response

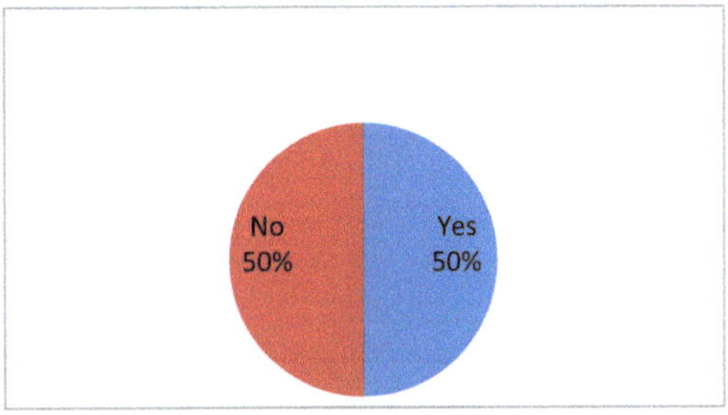

50% of the community agrees that the information on health, disaster preparedness and response is being shared

(b) During Disaster

i. Dissemination of Emergency situations

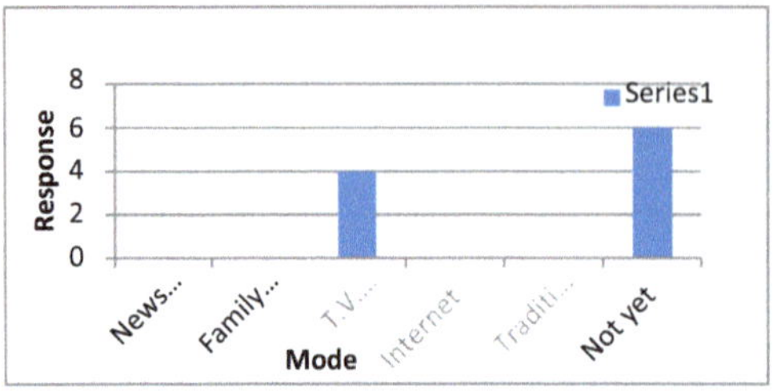

According to the community no one yet except television sometimes is the mode of dissemination during emergency situation

ii. Health-related problems

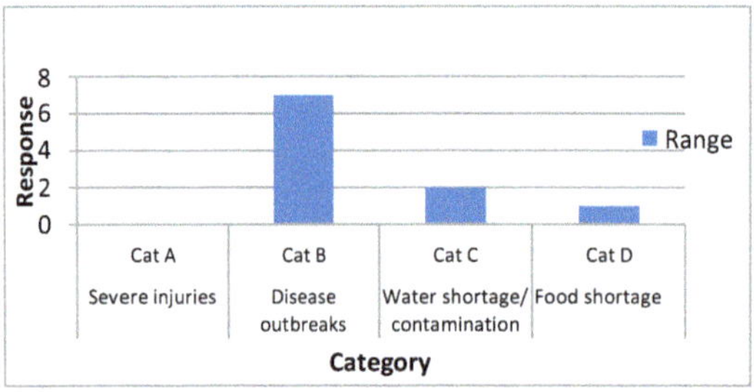

Major health related problem that can be observed is of Disease outbreaks like itching, allergy, cough and asthma due to dust from developmental works carried out for landslide

(c) Post-disaster

i. Understanding of Material provided

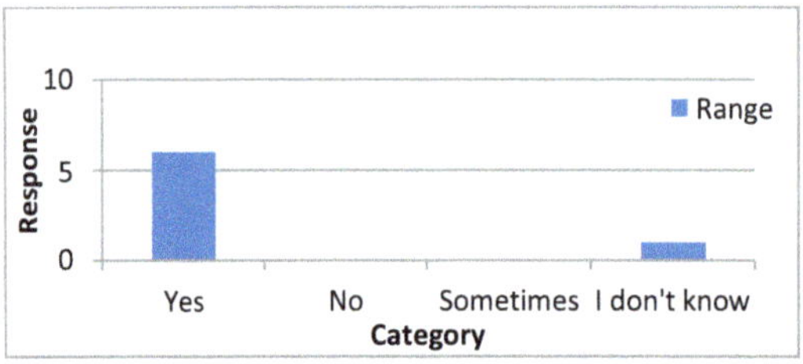

According to the community the materials provided in post disaster condition are understood by most of the community

ii. Reliability for conducting training programs

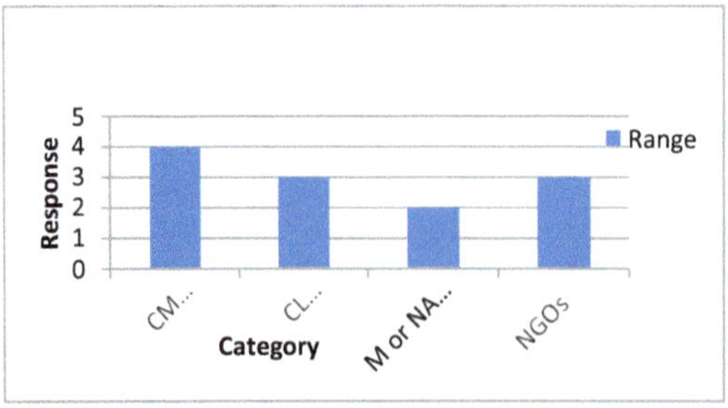

Graph shows community want the reliable community members and leaders are to be trained by Municipal or National Authority for response during disaster

Mandon (Fig. 24.3g)

1. General information of families

   (a) Income Range

   Community of this village has an annual avg. income of Rs (10,000–40,000; 50,000 and 3 lakh).

2. Disaster-related Knowledge

   (a) Pre-disaster

   i. Information on Health, Disaster Preparedness, and Response

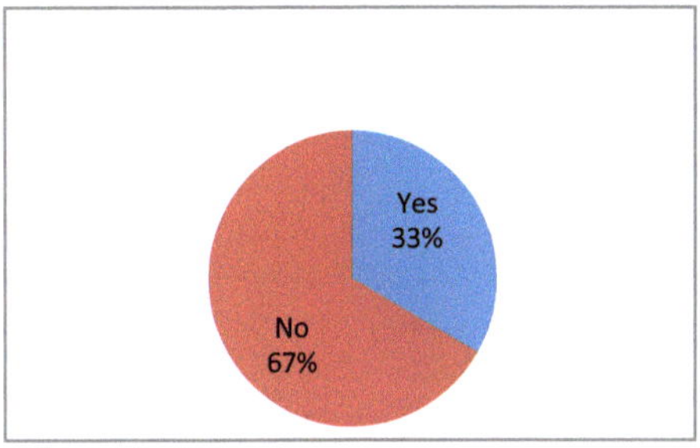

67% of the community reports that the information on health, disaster preparedness and response is not being shared

(b) During Disaster

i. Dissemination of Emergency situations

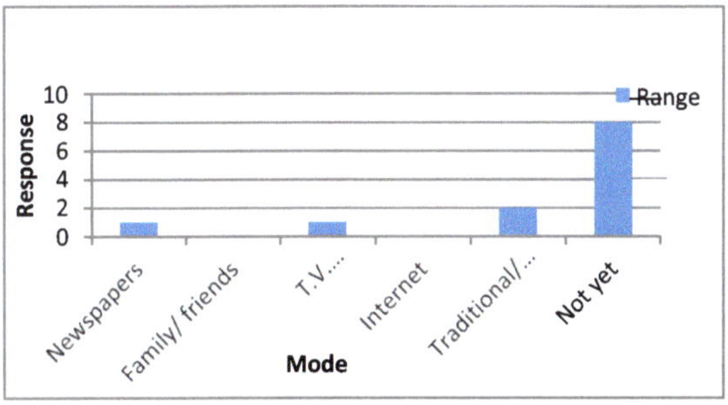

According to the community majorly no one is noticed yet for the dissemination during emergency situation

ii. Health-related problems

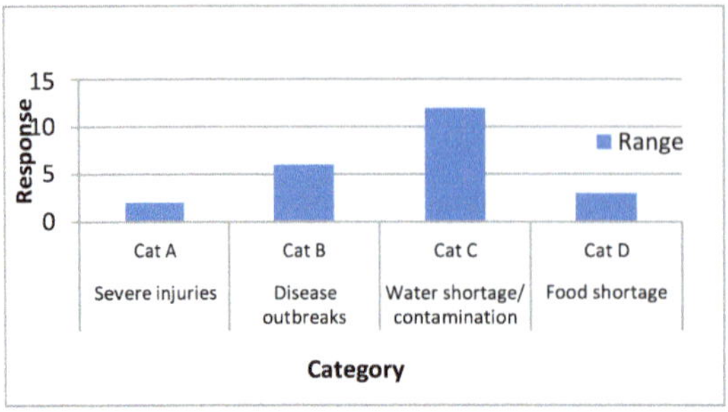

Major health related problem that can be observed is of Water shortage/ contamination

(c) Post-disaster

i. Understanding of Material provided

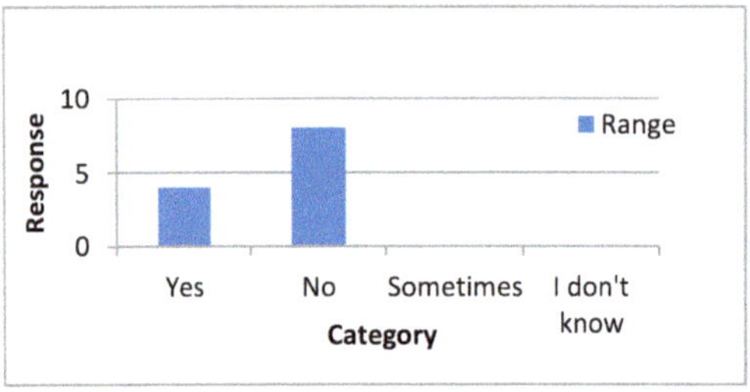

According to the community the materials provided in post disaster condition are not understood by most of the community

ii. Reliability for conducting training programs

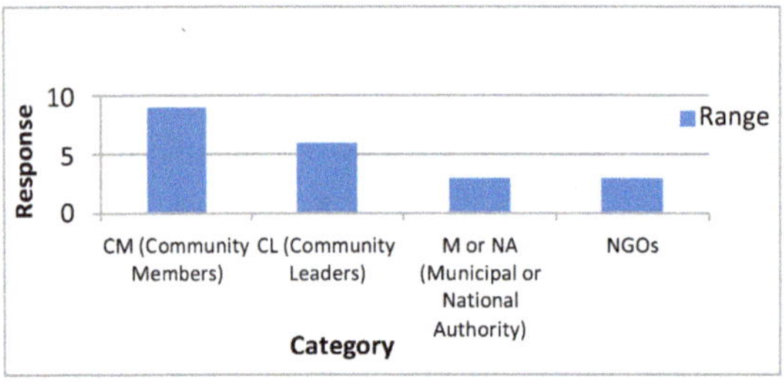

Graph shows community want the reliable community members and leaders are to be trained for response during disaster

Tiloth (Fig. 24.3h)

1. General information of families

(a) Income Range

Community of this village has an annual avg. income of Rs (10,000; 60,000 and 1 lakh).

2. Disaster-related Knowledge

(a) Pre-disaster

i. Information on Health, Disaster Preparedness, and Response

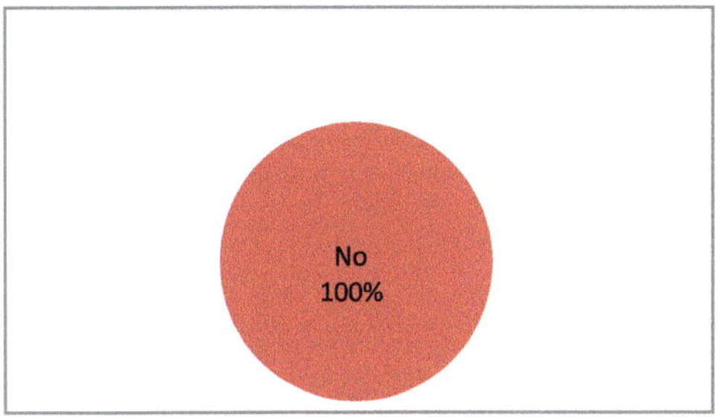

100% of the community reports that the information on health, disaster preparedness and response is not being shared

(b) During Disaster

i. Dissemination of Emergency situations

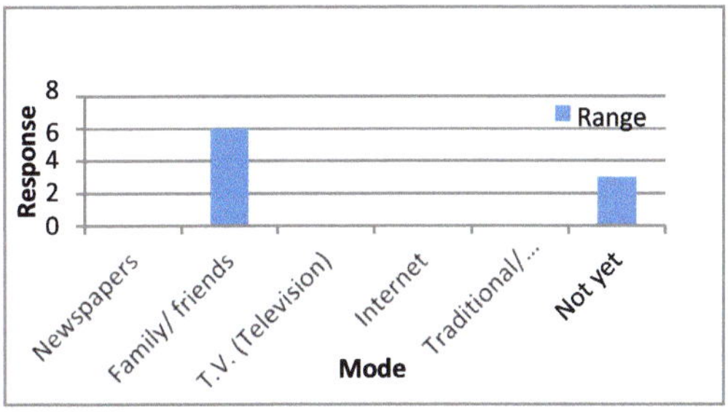

Family and friends according to the community are the mode of dissemination during emergency situation

ii. Health-related problems

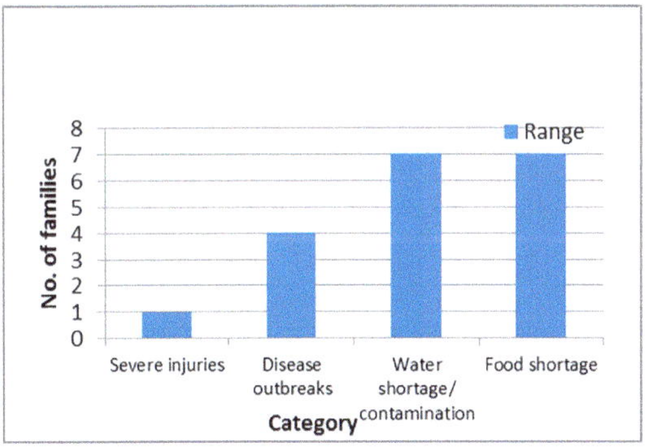

Major health related problem that can be observed is of Food and Water shortage/ contamination

(c) Post-disaster

i. Understanding of Material provided

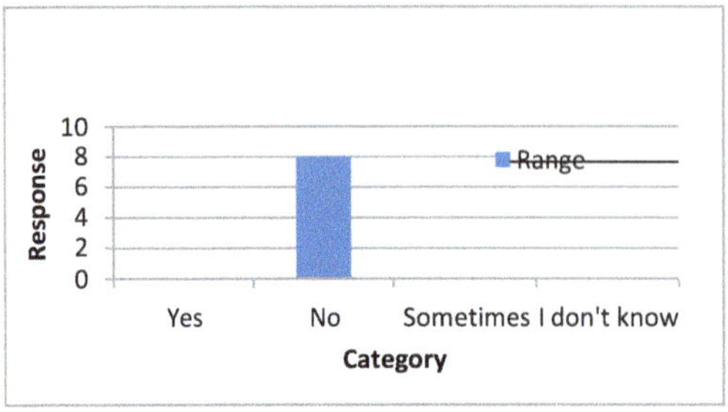

According to the community the materials provided in post disaster condition are not understood by most of the community

ii. Reliability for conducting training programs

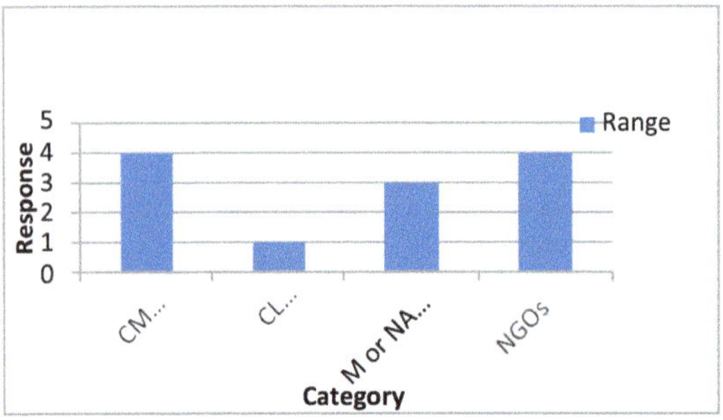

Graph shows community wants the reliable Community Members and NGOs are to be trained for response during disaster

Joshiyara (Fig. 24.3i)

1. General information of families

   (a) Income Range

   Community of this village has an annual avg. income of Rs (10,000–30,000; 31,000–60,000 and 1–2 lakh).

2. Disaster-related Knowledge

   (a) Pre-disaster

   i. Information on Health, Disaster Preparedness, and Response

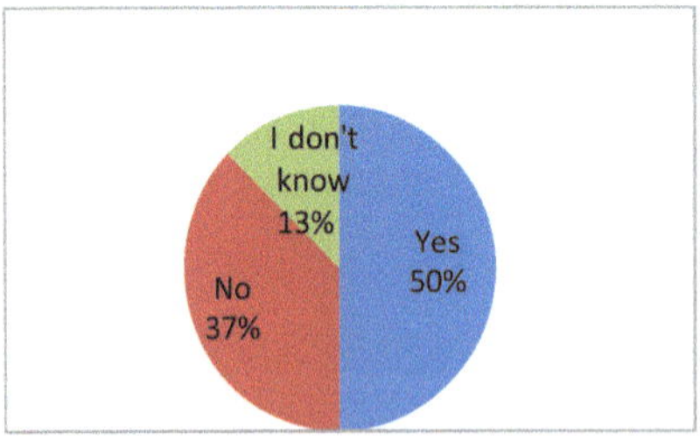

50% of the community agrees that the information on health, disaster preparedness and response is being shared

(b) During Disaster

i. Dissemination of Emergency situations

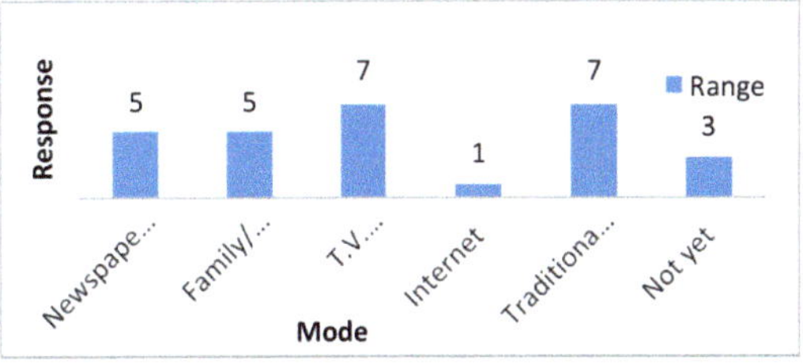

T.V. and Traditional methods according to the community are the mode of dissemination during emergency situation

ii. Health-related problems

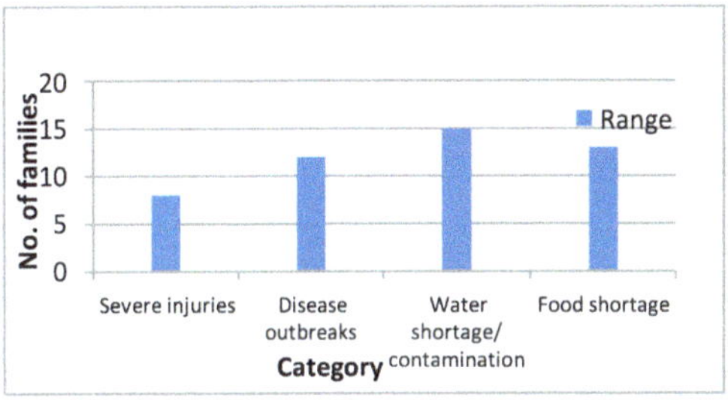

Major health related problem that can be observed is of Water shortage/ contamination

(c) Post-disaster

i. Understanding of Material provided

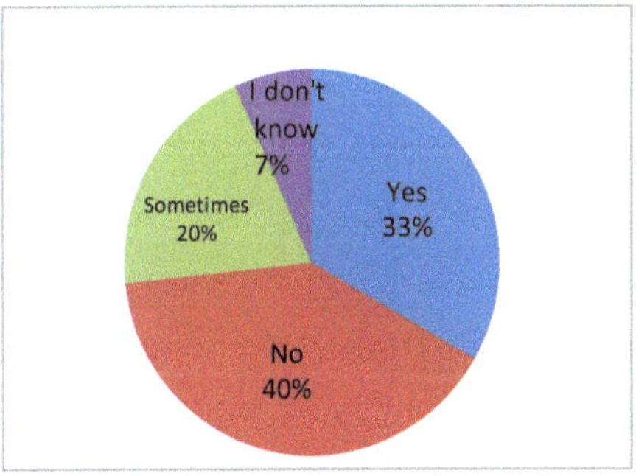

According to the community the materials provided in post disaster condition are not understood by most of the community

ii. Reliability for conducting training programs

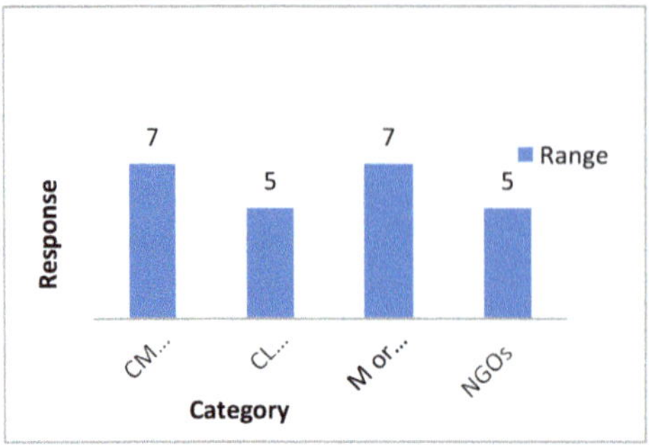

Graph shows community wants the reliable Community Members are to be trained by M or NA for response during disaster

Pokhri

1. General information of family

   (a) Income Range

   Community of this village has an annual avg. income of Rs (20,000–30,000; 40,000–50,000 and 1 lakh).

2. Disaster-related Knowledge

   (a) Pre-disaster

   i. Information on Health, Disaster Preparedness, and Response

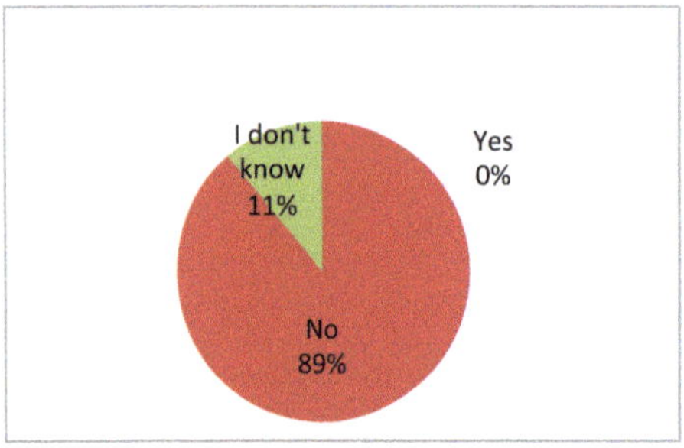

89% of the community reports that the information on health, disaster preparedness and response is not being shared

(b) During Disaster

i. Dissemination of Emergency Situations

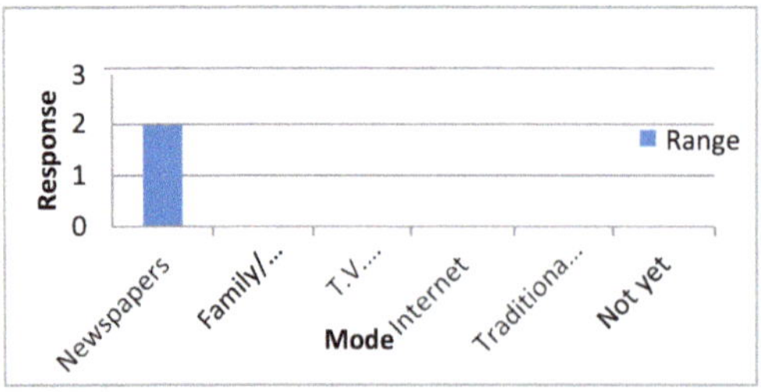

Newspapers according to the community are the mode dissemination during emergency situation

ii. Health-related problems

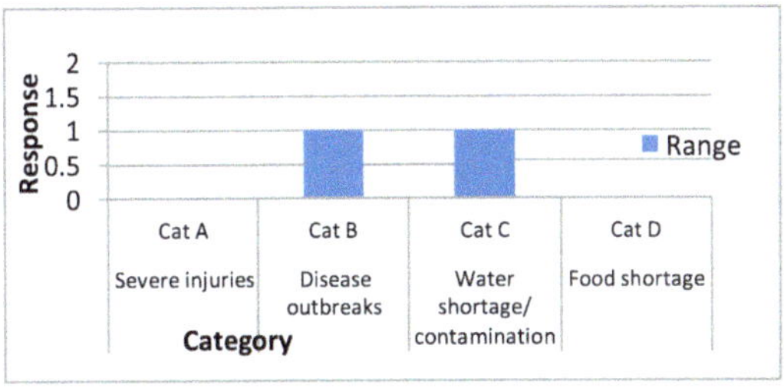

Major health related problem that can be observed is of Disease outbreaks and Water shortage/ contamination

(c) Post-disaster

i. Understanding of Material provided

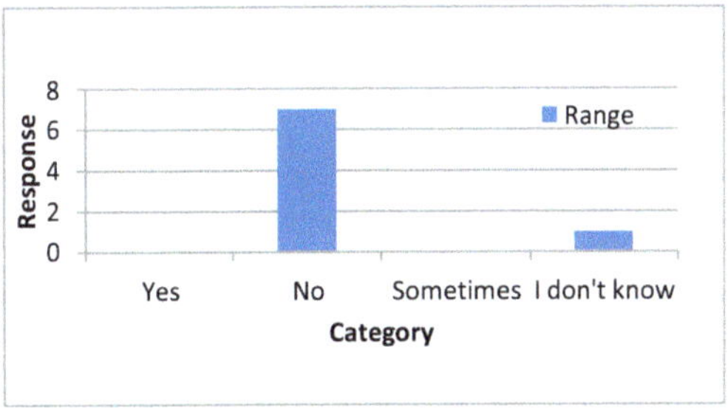

According to the community the materials provided in post disaster condition are not understood by most of the community

ii. Reliability for conducting training programs

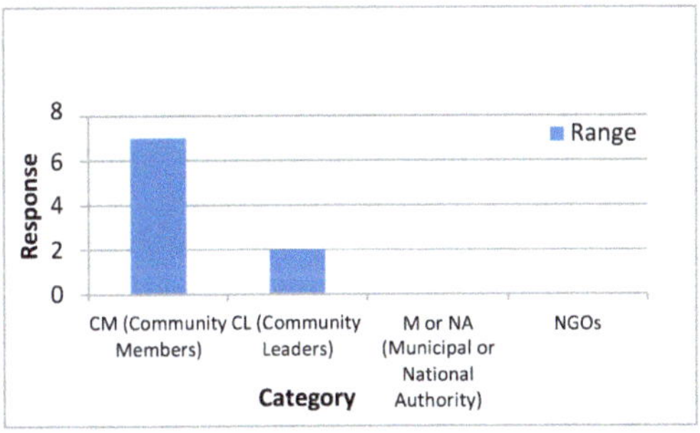

Graph shows community want the reliable community members are to be trained for response during disaster

Dharasu and Nalupani (Fig. 24.3j)

1. General information of families

(a) Income Range

Community of this village has an annual avg. income of Rs (20,000–30,000 and 2 lakh-3 lakh).

2. Disaster-related Knowledge

(a) Pre-disaster

i. Information on Health, Disaster Preparedness, and Response

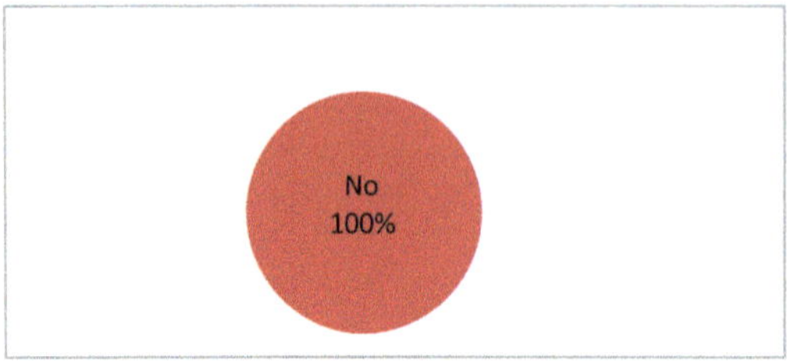

100% of the community reports that the information on health, disaster preparedness and response is not being shared

(b) During Disaster

i. Dissemination of Emergency situations

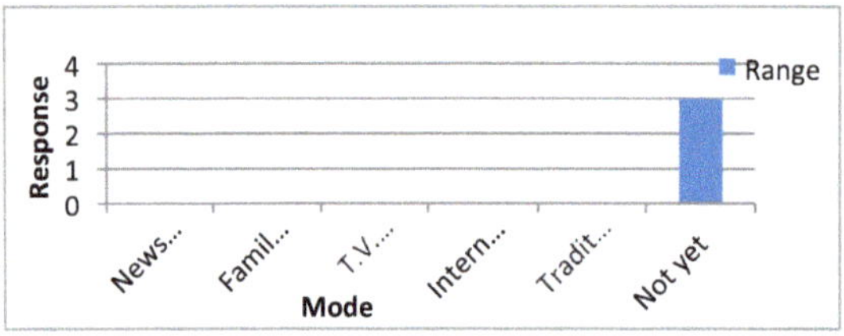

No one yet according to the community is the mode of dissemination during emergency situation

ii. Health-related problems

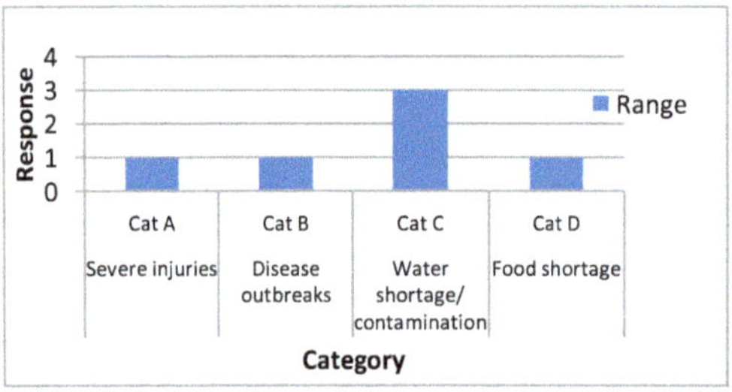

Major health related problem that can be observed is of Water shortage/ contamination

(c) Post-disaster

i. Understanding of Material provided

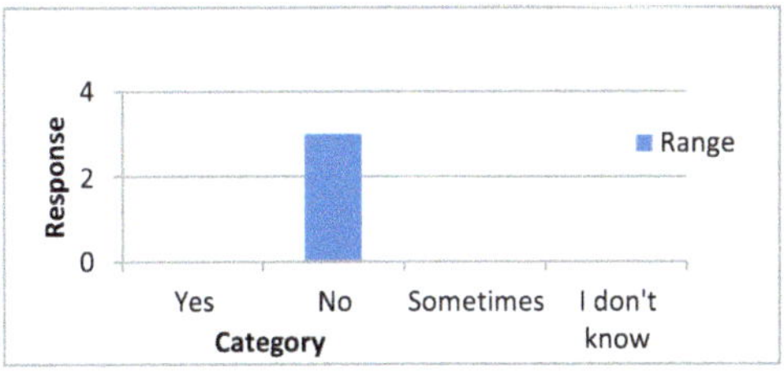

According to the community the materials provided in post disaster condition are not understood by most of the community

ii. Reliability for conducting training programs

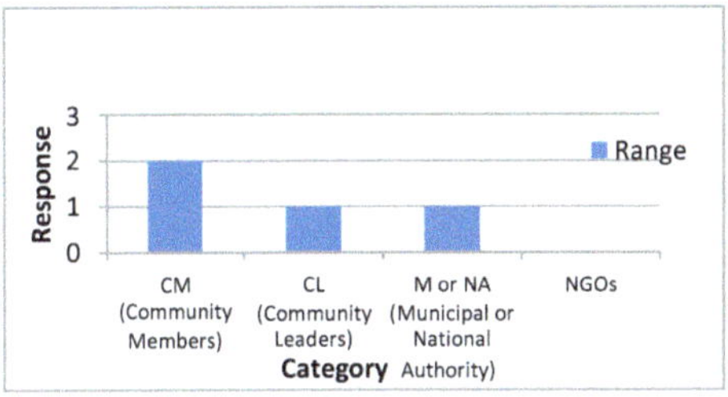

Graph shows community want the reliable community members are to be trained for response during disaster

2. **Descriptive Disaster-Related Questionnaire**

   (a) Human Loss/Livestock Loss/Infrastructural Loss/Field Loss

   According to the random sampling survey, (Fig. 24.4) many people since year 1978 lost their family members, livestocks, houses, cowsheds, fields, trees, plots, buildings, shops, stores, and offices due to different disasters from time to time which affected their livelihood and economy to the great extent.

   (b) Developmental Activities

   Developmental activities like construction of dams, bridges, good education facility in town, and construction of embankments affected the region both negatively and positively. Impacts of the developmental activities are:

   Positive impacts-Embankments, dams (Fig. 24.5a, b), roads, and bridges were beneficial to some extent.

   Negative impacts-After the flash flood disaster in Asi Ganga (2012), the embankments were not properly constructed which lead to the major cause of destruction in Bhagirathi flash flood disaster (2013). Community believe that the increasing developmental activities has changed the climate of the region which causes loss of natural resources, uneven rainfall pattern, uneven and low snowfall density and increased hailstorms resulting in lower production of native crops like Manduwa, Cheena, Koni, Jhangora, Makki, Dhan, Gehu, Tor, Gaith, Soyabean, Dal, Rai, Sarson, Aloo, and Pyaz.

   Community suggests that 2% benefit should be given to those farmers whose fields were used in developmental activities as later they don't have

Fig. 24.4 Random sampling survey in different villages of study site

any source of economy and education should be good at village level itself in order to prevent unemployment and migration respectively.

(c) Tourism Activities

Uttarkashi region is biodiversity-rich and a Hindu Pilgrimage site which attracts tourists not only from India but also from other countries of the world. Though tourism increases the economy of the district and creates awareness among local people via different means (environmental awareness and new technological interventions) but major effect of tourism viewed among the community is of socio-cultural impact.

According to the community during tourist season, the region is majorly affected by pollution, infections, sanitation problems, increased marketing

Fig. 24.5 **a** Embankment near Joshiyara, **b** Dam near Gyansu

value of goods and services, and heavy transportation. Elder community members have their perception that tourism activities have brought a great change in their traditional food and customs and their whole traditional living style.

(d) Challenges Faced by Community and How They Are Prepared

Community along with disasters faces some major challenges of fear, psychological problems, severe damages, loss of basic amenities, electricity facilities, medical facilities and transportation facilities, food, water and shelter shortages and migration. According to the survey, the most affected groups are labors, children, and women.

Effect on labors-majority of labors in town are from different states or places which leave them behind from being benefited by the relief provided during or post-disaster.

Effect on children-due to disasters' impact children lost their parents, and education facilities like schools and books.

Effect on women-majority of women affected are either pregnant or women with infants. Community suggests these three groups must be specially focused from the protection planning point of view.

(e) Disaster-Related Issues that are not Covered by National Level/Community Leaders

National Authority and Community leaders try their level best to achieve the goals of Disaster Management but there are many other disaster-related issues reported by the communities that are not been focused by them.

According to the community, relief was unequally distributed; i.e., unaffected people were benefited instead of affected ones.

Community expects-Proper survey of relief distribution should be done, there should be backup plan for education, food, shelter, and alternative transportation facilities in case any disaster occurs in the region, insurance awareness must be there, survey of eco-sensitive zones must be done timely, early warning systems should be improved and debris collected on the river bed after the flash flood disaster should be equally labeled in order to prevent another disaster to occur.

(f) Recommended Next Steps to Reduce Health Risks in Emergencies

To reduce the health risks, community suggests-

Good medical facilities with sufficient staff should be present in the region during the suspected duration of occurrence of disaster, water storage facilities should be done prior to the disaster situation, proper drainage system, installment of sewerage plant, and influx of huge amount of tourists in one go must be limited as it may cause pollution and also may affect the carrying capacity of the area; fiber structures should be made instead of concrete to prevent from injuries caused by earthquakes and community should be involved at each plan and rescue trainings.

A. **Government and Non-government Authority/Organization/Institute Descriptive-Based Questionnaire:**

1. **District Disaster Management Authority**

Basic Profile and Information from DDMA Uttarkashi:

Name-Mr. Devendra Singh Patwal; **Position-**DDM Officer Uttarkashi.

Duties-During and in post-disastrous condition, their duty is to provide relief, to community, documentation to DDMA, to work as Liaison Officer (Incharge), to be updated with the site idrn.gov.in, Sendai Framework and modification of DDMA Plan yearly.

Responsibility of Disaster Preparedness/Response-Responsible bodies for preparedness is not only DDMA but this work is done with collaboration of different stakeholders like NGOs, CBOs, PRIs, and Govt. bodies (SDRF, Forest, Irrigation, UJVNL, etc.) while DDMA during response provides the relief like tents, food, shelter (schools, govt. buildings), blankets, and lights as it has storage of various equipments (search and rescue equipments, ropes, shoes, stretchers, generators, mic, shovel, blower fire line).

DDMA is interacting with the community at three levels (Block level, Tehsil level, and Gram Panchayat level) with the help of updated contact numbers of the concerned leaders on regular basis. Dissemination of information regarding disasters and emergency situations is done with the help of Gram Samiti, traditional methods, awareness programs, search and rescue training programs.

Responsibilities and Resilience Measures Undertaken During Disaster and in Post-disastrous Conditions

Earthquakes-Responsibilities-Providing first aid and medical facilities, conducting relief program, colonies setup, food and blanket distribution, rehabilitation programs (for infrastructural losses), and ODCH program. **Measures-**Search and rescue operations, restoration and infrastructural maintenance.

Landslides-Responsibilities-Opening of roads, arrangement of labors and excavators by BRO and PWD. **Measures-**Removal of debris, maintenance and restoration of pipelines and road maintenance.

Flash floods-Responsibilities-Channelization, create protection wall, irrigation management, and providing shelter. **Measures**-Prevention by embankments and planning for urbanization.

Forest fires-Responsibilities-Providing equipments to the fire fighters. **Measures-**To form of forest fire lines and trenches.

Steps DDMA have introduced regarding

Earthquakes (Fig. 24.6a)

(a) **Disaster Risk Assessment-**Assessment of buildings in structural form, carrying capacity of the area and settlement.
(b) **Disaster Risk Prevention-**Awareness, training, and earthquake proof building construction techniques.
(c) **Disaster Risk Mitigation-**New construction activities and earthquake shaped building techniques are suggested by Geologists and experts in urban areas, but for rural areas construction depends upon awareness.
(d) **PDNA-**Collection of information (damage and property loss) > IMD informs the time, location, and magnitude > Field information (damage and property loss) > Immediate Relief programs (Medical, food, and rescue) > Damage assessment (infrastructure) by concerned depts. > further loss assessment is done by SDMA and forwarded to NDMA.

Fig. 24.6 **a** Effects of earthquakes, **b** Loss of biodiversity due to landslide, **c** Damages caused due to flash flood, **d** Ecological disaster due to forest fires

Landslides (Fig. 24.6b)

(a) **Disaster Risk Assessment-**History of disasters is an important aspect to assess the disaster, and concerned departments like BRO, PWD, and DDMA are the monitoring department for this.
(b) **Disaster Risk Prevention-**Creating awareness among community, treatment of the area, and Rehabilitation plan
(c) **Disaster Risk Mitigation-**Detailed geographical study, treatment measures, by avoiding construction/settlement in that particular area and construction activities are further approved by specialized suggestions.
(d) **PDNA-**Field survey by Revenue dept., NGOs, local public > Collection of information by local govt. representatives, public representatives, and others > all line depts. makes report in prescribed format and > final submission to DDMA, SDMA (regarding funds and necessary needs) > forwarded to NDMA.

Flash floods (Fig. 24.6c)

(a) **Disaster Risk Assessment-**History of disaster, river morphology study, flood discharge data, and flood zonation mapping.

(b) **Disaster Risk Prevention-**Awareness and training, dam regulation monitoring, and early warning system
(c) **Disaster Risk Mitigation-**Flood protection works in affected areas, flood zonation mapping for construction, and flood fold data.
(d) **PDNA-**Different Dams and CWC provides information to DDMA (hydrology project) > field survey > immediate relief program > damage assessment > collection of data to SDMA > after coordination with other bodies (Govt., NGOs) > rehabilitation programs

Forest fires (Fig. 24.6d)

(a) **Disaster Risk Assessment-**Livelihood, dependency in forest, and habitation in forest area.
(b) **Disaster Risk Prevention-**Community awareness and training, increasing dependency of forest to local people
(c) **Disaster Risk Mitigation-**Community participation, administrative participation, mixed- forest development, development of control rooms 24×7, development of fire crews station.
(d) **PDNA-**Collection of information from forest supervisor or forest of community > launching of relief program (search and rescue) > damage assessment by forest dept. and other concerned depts. and village community > rehabilitation programs (Fig. 24.6).

2. **District Hospital**

Name-Mr. Ram Sanjeevan Nautiya; **Position-**Accountant IDSP-District Hospital Uttarkashi CMO of District Hospital in collaboration with DDMA is responsible to conduct Disaster Preparedness/ Response meetings. Knowledge/ Training provided is through RED Cross Society, FMR and Mock Drill in pre-disaster situations (2–3 months before) by RBSK and ANM cell.

District hospital has updated contact details of concerned persons to be contacted during and post-disaster at DDMA level, Fire brigade, Police ITBP, and Extra Medical Staff (FMR = RED Cross + Health Dept.), which help the community for Ortho facilities, First Aid, and Medicines like (Diarrhea, Chlorine tablets, Electrol, and Blood Pressure machines).

3. **District Police Line**

Name-Mr. Mukesh Kumar and Mr. C.P. Joshi; **Position-**Constable (Trainer at DM cell).

The line constables are guided by DDMA to carry out Awareness and training programs (Fig. 24.7) in 7 disaster management stations in the district in which 80–90% programs are carried out within the duration of 2–3 months.

Disaster Risk Prevention/Mitigation Steps

Sign boards in landslide zone and damage due to earthquakes after confidential reports are addressed and announcements are made to prevent from disaster and to

Fig. 24.7 Disaster management training carried out by constables of Police line-Uttarkashi

mitigate, reporting hospital, making general diary of equipments required, provided, and damaged videography is done further steps to be taken.

Relief Provided During and After Disaster

Rescue projects, sign out symptoms, pre-hospital treatment are carried out In case of blood loss (direct pressure, pressure point, elevation, tonigation).

In case of dead (identified but not declared dead, EMS, 4 ribbons are used (WHITE-light injuries, BLUE-light critical, RED-critical, BLACK-dead or no check).

In case of huge disaster, 7 Desk meeting is carried out by various depts. (DM, PWD, Food, Law/ order, Communication, Medical, Transportation).

Logistic-provides equipments, search and rescue team, first aid team, first responder (first FR) and information.

Medical FR-Emergency medical service (EMS), network, 108 ambulance, police and fire service alters the team members

Trained FR-Examines the condition of disease, Critical Condition of live Victim (CCAV) and Critical Air wave Breathing (CAB).

4. **District and Range Forest Department**

Name and Positions

Dr. Sandeep Kumar IFS (DFO-Uttarkashi Dist.); Mr. N. S. Rawat (RO-Dharasu) and Mr. Khemraj S. Bigana (Office Incharge-Taknor-Bhatwari Range).

Forest fires Preparedness and Response

This is the responsibility of whole forest dept. and all duties are divided at division level. DFO supervises entire jurisdiction level of a district forest, RO who supervises the work done at range level. Range is undertaken by RO who controls different sections of range (Fig. 24.8a, b and c).

The dept. interacts with the community leaders at Gram Panchayat level and Van Panchayat level and has updated contact details of community leaders at Gram Panchayat level, Pradhan, Sachiv, and Van Sarpanch. Some of the equipments storage by Forest dept. are Water bottle, Pathal, Torch, Spade (Phawda), Belcha, Ration, Water, Vehicles Labors wireless walkie talkie sets.

According to the survey, Dunda block is more vulnerable to forest fires than Bhatwari block. The reason being less vulnerable is the region not covered with Pinus sp. but id mixed forest with species like Pangar, Kharsu, Moru, Deodar, Rahi, Murainda, and Khail.

Disaster Risk Assessment/Mitigation

Landslides-Seeds of grasses are broadcasted like Ringal, Jingal, Uttes, Leamon grasses, etc. Risk is less found in Gully plugging and generally brush wood check dam or stone check dam is created at the top of the hill, but if slope is more, then stone check dams are preferred.

Forest fires-When litter fall is more in any area during Jan–Feb then the area is marked as prone to forest fire. Control burning is done in Dec-Jan-Feb and leaves are collected and burned. For the prevention of forest fire, sapling, ground flora, birds and animals which are highly affected due to forest fire, forest dept. is prepared prior to the disaster for which mock drill, fire drill, community awareness, arrangement of vehicles, equipments, and basic amenities is done and collected NWFPs are distributed to the community for daily basis.

For mitigating measures, plantation of Quercus, Deodar, Mango, Medicinal plants (Amla, Reetha, Harad, Bahera) is started after month of July. It is an established fact that Pine needles are highly inflammable and are of the main cause of forest fire but on the other side this species is useful for the production of Resin (Terpentine, Leesa) which helps the revenue dept.

Fig. 24.8 **a** Forest Choki-Kuteti depicting equipments for response during Forest fire, **b** Range Office-Dharasu and **c** Nursery-Dharasu

Resilience Measures for Quicker Post-disaster Recovery

Landslides-Awareness and rescue programs are carried out; food and shelter are provided to the needed ones, temporary bridge formation (woods), fund, estimate, and budgets are managed.

Forest fires-All possible measures are applied to combat the situation. Rescue operation are carried out, food water, shelter, and first aid kits are distributed by the staff members, and after receiving the funds from the govt. preparedness measures are carried out.

PDNA

Post-disaster-Contingency rehabilitation plan is prepared. In addition to soil and water conservation measures, trenches are created and grass spp. are planted in the affected area.

Disaster Preparedness- is monitored by satellite Bhuwan ISRO, NASA, and the status is uploaded/updated and can be seen on website of UK. SMS alert system is sent to all officials of UK. Social Media group is created by govt. of UK on which forest fire location of respected divisions are updated daily.

5. **Government Degree College**

Name—Dr. Arun Agarwal.
Position-Head Botany, Coordinator with IGNOU-Disaster Management Course.

No one in the college came for Disaster Preparedness/Response Program so self-initiative was taken for knowledge/training to 6 blocks for 2 months May–June through project of IGNOU 2012–13 by Dr. D.C. Goswami.

College is not updated with numbers of concerned persons to be contacted during disaster or post-disaster condition instead NSS and NCC students are helping for the relief camps by DDMA.

6. **Contribution of NGOs**

Contribution of an NGO-Shri Bhuwaneshwari Mahila Ashram (SBMA) in Uttarkashi Disaster 2012–13

Name of the Organization-Shri Bhuwaneshwari Mahila Ashram, Anjanisain, District-Tehri Garhwal (Uttarakhand).

Date of Registration-23/09/1978; Registration No.–2426–1978-79;

Founder-Late Swami Manmathan. This foundation aims to work with communities, its organizations/institutions, government, technical, and academic institutions, and other civil society where every child is happy, its wellbeing assured.

The main funding body is PLAN INDIA.

1. Tmabakhani and Ganga Nagar:
   Dry ration packing kit 46, Anapai kit 55, non-food item 55.
2. Balmiki Basti:
   Dry ration pack, NFI kit, Sanitary kit and Shelter kit 89 families.
3. Sangamchatti 2012:
4. European commission sponsored:
   NFI kit 1250,
5. Unconditional cash transfer concert:
   Widow, disabled, single women who were above 60 and can't work for MNREGA were paid by UCT-1000.
6. 2012 Cash for work:
   Damage shop, house kitchen, cowshed, agricultural field restoration 209 families were supported-600 families.
7. Joshiyara:

Shelter, NFI kit, Sanitary kit, Kitchen kit, Food kit-500 families and 300 shopkeepers.

(Jain et al. 1999) mentioned in a report that the extensive and well-developed river terraces on which the villages of Uttarkashi are situated may have significantly influenced the ground shaking during the 1991 earthquake. However, according to my survey, the local community believes that the primary cause of ground shaking was not the river terraces, but rather anthropogenic factors, such as developmental projects along the Bhagirathi River, which weakened the stability of the surrounding hills. This highlights the differing views between scientific assessments and community perceptions, emphasizing the need for a more holistic understanding of disaster causality. Further supporting the region's vulnerability a study (Kimothi et.al. 2005) observed that Uttarkashi, located along the Bhagirathi River in the upper Lesser Himalayan zone of the Garhwal Himalayas, has a long history of disasters. One of the most recent was the Varunavat Parvat (mountain) landslide in September 2003, which affected over 5000 people. The landslide caused widespread damage to buildings and roads, clogged sewage and drainage systems, and disrupted transportation, communication networks, drinking water, and electricity supplies. However, the study highlights that proper mitigation measures implemented at the site have been effective in reducing risks indicating the importance of proactive disaster management. In addition to earthquakes and landslides, natural events like forest fires also pose significant risks to both the environment and the local economy. (Certiny 2005) explored the impact of forest fires on physical, chemical, mineralogical, and biological soil properties. The effects are primarily determined by the severity of the fire, including peak temperatures and duration. Forest fires also have economic consequences, as they reduce crop yields, non-timber forest products (NTFPs), and other forest resources, ultimately affecting district revenue. This further illustrates how different forms of natural disasters, from seismic to environmental, have far-reaching effects on both communities and ecosystems.

An extensive literature review reveals that the majority of mitigation measures were implemented after the 2013 flash flood in the Bhagirathi River catchment. However, according to the community survey, the most significant impact was felt during the 2012 flash flood in the Asi Ganga River catchment, which received little attention. Proper planning and preparedness post-2012 could have mitigated the effects of the 2013 disaster.

The pattern of disaster impacts continues with flash floods. A review of existing literature reveals that most mitigation measures were implemented after the 2013 flash flood in the Bhagirathi River catchment. However, community feedback from the survey suggests that the 2012 flash flood in the Asi Ganga River catchment had a greater effect, which went largely unnoticed. The lack of sufficient focus on this earlier event highlights the critical need for comprehensive disaster preparedness and planning, as timely interventions after the 2012 flood might have lessened the damage from the 2013 disaster. This underscores the importance of learning from past events to improve future resilience.

24.4 Conclusion

The present study clearly demonstrates that while development is essential for every region, special caution must be exercised when modifying Himalayan slopes. These slopes, shaped by both exogenic and endogenic processes, are precariously balanced and highly sensitive to disruption. Although planners and policymakers are aware of the fragile nature of the Himalayan ecosystem, this awareness is often overshadowed by the pressure to exploit the region's resources for growth and to create rapid access through road networks. Rapid, unplanned urbanization is significantly destabilizing this delicate balance. To mitigate further harm, there is an urgent need to discourage the establishment of settlements on steep mountain slopes and near riverbanks. A comprehensive assessment and monitoring plan for urban growth in high-risk and vulnerable disaster zones should be a core part of any strategy to manage and control disasters in these ecologically sensitive areas. Effective disaster management requires proactive measures before, during, and after an event to predict, prevent, and mitigate the impact of disasters. These actions should be taken at the community or household level. Community-based disaster management programs (CBDMPs), which have not yet received adequate focus, should be prioritized at the appropriate levels—district, block, and village. These programs should include a survey of hazardous areas, vulnerability assessments, community participation evaluations, disaster risk zoning, and raising awareness among stakeholders through information and education campaigns.

Additionally, the formation of disaster management committees, task forces (for search and rescue, first aid, sanitation, and shelter management), community fund creation, and mock drills are essential to sustaining preparedness. The government's focus should also prioritize physical, social, and economic vulnerability, budget availability, and accessibility. Responders believe that beyond the initial rescue and relief efforts, the most significant and challenging issue for the local community is rehabilitation. Disaster victims often lose their villages, homes, and properties, and it can take months or even years to recover. In such situations, the local community is frequently left abandoned, which is why timely relief and rescue operations, along with a coordinated collective approach, are essential for recovery. 'It is, therefore, critical to strictly regulate development near streams and rivers. Legislative action is necessary to formulate policies that address this issue, and strong executive enforcement of these policies is required to ensure compliance, safeguarding the region from further harm'.

Strategy for Rehabilitation

The rehabilitation of affected communities is the most critical and challenging task ahead. A comprehensive strategy for rehabilitation should begin with a close survey of the impacted villages, assessing the nature and extent of loss. Prioritizing areas in need of urgent help and support, ensuring the fair and transparent distribution of relief, and providing knowledge and technological assistance are essential components. The community's rehabilitation must also include training and educating them on how to prepare for future disasters. The strategy can be divided into two phases: immediate plan and long-term plan.

Immediate Plan

The immediate plan focuses on addressing the basic needs for survival in the aftermath of a disaster. The primary goal is to provide essential supplies like food, utensils, stoves, milk for children, and clothing to ensure the community can sustain itself during the initial recovery phase. Providing these necessities promptly helps alleviate suffering and stabilizes the situation while preparing for further intervention.

Long-Term Plan

The long-term plan is vital for sustainable rehabilitation. This phase focuses on ecological and economic recovery, including rebuilding shelters, houses, agriculture, and livestock. Livelihood support through skill development and sustainable resource use is crucial to ensure the community can rebuild its economy. Additionally, providing training on disaster management practices will empower the community to be better prepared for future disasters and more resilient in the face of future challenges.

The potential benefits of implementing this strategy are substantial. NGOs focusing on environmental safety will become more active, and the local community will be more aware and proactive regarding environmental and disaster safety. Socio-cultural and economic life will be less affected by disasters if people take preventative measures. Moreover, the government will likely enhance its focus on addressing both natural and manmade disasters in hilly areas, and fair and honest surveys will ensure that relief and rehabilitation efforts are accurately directed. The increased awareness and proactive measures from NGOs, the community, and the government will directly support our mission to reduce vulnerability to all hazards. This collaborative approach strengthens our efforts to integrate prevention and mitigation into everyday life, making the region more resilient to future disasters. Our mission is to reduce vulnerability to all types of hazards, whether natural or manmade. This is no small task, considering the district's large population and its exposure to multiple natural hazards. However, it is imperative that we learn from past mistakes and strengthen grassroots mechanisms during the preparedness phase. The recent Uttarakhand disaster, of such severe magnitude, should serve as a wake-up call for all of us. This is just the beginning—the ultimate goal is to integrate prevention and mitigation into our everyday lives.

We are confident that with the right measures in place, we can significantly reduce the risk of disasters like landslides and floods from occurring again, especially not at the devastating cost the district has already endured in terms of human lives, livestock, property, and livelihoods. Our vision ahead is to create a safer and more secure community through sustained collective efforts, the synergy of national resources, and active participation from the public. What may seem like a distant dream today will be realized within the next two decades. This is our goal, and we are committed to pursuing it with unwavering dedication. While the path ahead may appear challenging, it will become much easier as we move forward together, united in our efforts to build a resilient future.

References

A preliminary report Uttarakhand disaster (2009). Disaster Nature Fury

A status report- disaster management in India (2004) GOI, Ministry of Home Affairs, NDMA Division

Amit P, Biswas S (2003) The impact of forest fire on forest biodiversity in the Indian himalayas (Uttaranchal)

Bolt BA (1975) Landslide hazard, geological hazard, Springer Verlog, New York 150

Catherin N, Kiran P, Joseph B (2017) Assessment of post-traumatic stress disorder among disaster affected children in a high school in Uttarkashi district. Uttarakhand, India

Certiny G (2005) Effects of fire on properties of forest soils: a review 143(1):1–10

Cotton F, Campillo M, Deschamps A, Rastogi BK (1996) Rupture history and seismotectonics of the 1991 Uttarkashi, Himalaya earthquake. Tectonophysics 258

Disaster: natural fury: a preliminary report on Uttarakhand Disaster, a report prepared by Himalayan environmental studies and conservation organization (HESCO)

Eapen A (2014) Uttarakhand disaster—a wake up call: a case study on Uttarakhand disaster response–2013. Indian J Public Admin 302 60(2):301–312

Ecology. J Indian Res 1(3):106–116

Yu G, Khattri KN, Anderson JG, Brune JN, Zeng Y (1994) Strong ground motion from the Uttarkashi, himalaya, India, earthquake: comparison of observations with synthetics using the composite source model. Seismological Society of America

Geological survey of India. Treatment and stabilization of Varunavat Parvat Landslide, Uttarkashi District, Uttarkhand

GSI (1992) Uttarkashi earthquake: October 20, 1991, Special Publication No. 30, Geological Survey of India, Calcutta

Gupta V, Bist KS (2004) The 23 September 2003 Varunavat Parvat landslide in Uttarkashi township, Uttaranchal. Current Sci 87(11)

Jain SK (1992) On better engineering preparedness: lessons from the 1988 Bihar earthquake. Earthquake Spectra 8(3):391–402

Jain SK (1998) Indian earthquakes: an overview. Indian Concrete J 72

Jain SK, Das S (1993) Analysis of strong motion records from Uttarkashi earthquake for assessment of code provisions for different seismic zones. Earthq Spectra 9(4):739–754

Jain SK, Murty CVR, Arlekar JN, Rajendran CP, Rajendran K, Sinha R (1999) Chamoli (Himalaya, India) earthquake of 29 March 1999. EERI Special Earthquake Report, EERI Newsletter, vol 33

Jhariya MK, Raj A (2014) Effects of wildfires on flora, fauna and physico-chemical properties of soil–an overview. J Appl Nat Sci 6(2):887–897

Jigyasu R (2002) Reducing disaster vulnerability through local knowledge and capacity. The Cace of Earthquake Prone Rural CommuSnities in India and Nepal

Kaur T, Gupta S (2016) Forest fire vulnerability analysis in India–a case study of Uttarakhand state. Int J Adv Sci Techn Res 2

Khaki BA, Singh VRR, Wani AA, Thakur RK (2015) Effect of forest fire on soil nutrients in blue pine (*Pinus wallichiana* A.B. Jackson) Ecosystems. Indian Forester 141

Kimothi MM, Joshi V, Garg JK, Ajai (2005) Ancient religious Uttarkashi town (Garhwal Himalayas, Uttaranchal, India) in the grip of fear of landslide: observations from IRS–P6 (Resourcesat-1) high resolution LISS-IV data. Map India. Geomatics

Kumar Govind VV, Jain K, Gairola A (2013) A study and simulation of cloudburst event over Uttarkashi region using river tool and geomatic techniques. Int J Soft Comput Eng (IJSCE) 3(1)

Kumar A (2013) Demystifying a himalayan tragedy: study of 2013 Uttarakhand disaster

Kumar KJA (2013) Flash floods in Uttrakhand–lessons for disaster management. India Disaster Report 2012. National Institute of Disaster Management 2013. Science Reporter, pp 28–29

Kumar M, Sheikh MA, Bhat JA, Bussmann RW (2013) Effect of fire on soil nutrients and under storey vegetation in Chir pine forest in Garhwal Himalaya, India. Acta Ecol Sin 33:59–63

Kumar S, Kumar P, Singh A, Ashwani, Kumar M (2022) Socio economic livelihood vulnerability to mountain hazards: a case of Uttarakhand himalaya, India. In: Remote Sensing and Geographic Information Systems for Policy Decision Support, pp 169–190. Singapore: Springer Nature Singapore
Maskrey A (1989) Disaster mitigation: a community based approach. Oxfam GB
Mausam (2015) Cloudburst and landslides in Uttarakhand: a nature's fury? 66(1):139–144
Murthy MM, Devgiri GM (2015) Assessment of soil chemical properties as affected by forest fire in Rajiv Gnadhi National Park, Karnataka. Adv App Res 7
Murty CVR, Jain SK (1997) Seismic performance of bridges in India during past earthquakes. The Bridge and Structural Engineer, ING-IABSE 27(4):45–79
Naithani S (2013) Disaster mitigation using participatory learning action for a village, Uttarkashi, Uttarakhand. Scholarly Res J Interdisciplinary Stud 1
Narendran K (2001) Forest fires origins and ecological paradoxes
Nautiyal H, Bhandari SP (2012) Ddmap, district disaster management action plan for hill district Uttarkashi, (Uttarakhand). Int J Sci Technol Res 1(7)
Negi CVS (2004) State of environment report for Uttaranchal. NIDM Uttarakhand (2009). National Disaster Risk Reduction Portal
Pandey AC, Verma A (2007) Geoinformatics based landslide hazard zonation in Tehri dam environs with special emphasis on spatio-temporal dynamics of landslide (Unpublished M.Tech thesis, BIT Mesra, Ranchi)
Pant V (2012) Thesis on community based disaster management planning (CBDMP) in Uttarakhand Kumaun University, Nainital
Pearce L (2003) Disaster management and community planning, and public participation: how to achieve sustainable hazard mitigation. Nat Hazards 28:211–228
Prakash S (2012) NIDM. Memorandum submitted by the Government of Uttarakhand
Prakash S (2013) NIDM. A study on flash floods and landslides disaster on 3rd August 2012 along Bhagirathi Valley in Uttarkashi District, Uttarakhand
Rai PK, Mohan K, Kumra VK (2014) Landslide hazard and its mapping using remote sensing and GIS. J Sci Res 58:1–13
Rao LM, Mohan JR, AppaRao APV, Kumar ARS, Rao PRRCK (2017) Disaster management: prevention, mitigation and preparedness. Int J Multidisciplinary Edu Res 6(4):206–231
Sati VP (2023) Natural and cultural diversity in the Himalaya. Springer
Sharma A, Masiwal R (2024) Climate crisis in the Indian Himalayas: an introduction. In: Addressing the climate crisis in the Indian Himalayas: can traditional ecological knowledge help? Cham: Springer Nature Switzerland, pp 1–30
Sinha D (2013) Disasters and disaster aid 9 disasters conceptual paradigm and implications for social work intervention, with special reference to the Indian scenario. Governance, Development, and Social Work 162
Sinha R, Goyal A (2004) A national policy for seismic vulnerability assessment of buildings and procedure for rapid visual screening of buildings for potential seismic vulnerability
Sitlhou A, Singh TB (2014) Post-fire nutrient availability in the sub-tropical forest ecosystem of the Koubru Hills, Manipur.f1000research. 3–30
Verma S, Jayakumar S (2012) Impact of forest fire on physical, chemical and biological properties of soil: a review. Int Acad Ecol Environ Sci 2(3):168–176

Websites

http://uttarkashi.nic.in/pages/display/241-district-disaster-management-authority%28ddma%29 http://www.census2011.co.in/census/district/574-uttarkashi.html http://www.peasantautonomy.org/uttarkashi-forest-fire.html http://www.villageprofile.in/uttarakhand/uttarkashi https://en.wikipedia.org/wiki/Uttarkashi_district

https://sandrp.wordpress.com/2013/09/25/uttarakhand-rainfall-since-1901-and-in-the-light-of-2013-disaster/
https://www.citypopulation.de/php/india-uttarakhand.php
Location map of Uttarkashi district. www.gov.ua.nic.in
http://uttarakhandtourism.gov.in/utdb/?q=uttarkashi
http://www.hindustantimes.com/photos/india/uttarkashi-under-nature-s-fury/photo-eKFbcUGx5doEvBRZeNaPqK.html
http://indianexpress.com/article/india/india-news-india/uttarakhand-forest-fire-claims-5-lives-ndrf-forces-deployed-to-control-flames-2776785/

Chapter 25
Global Disaster Risk Reduction Strategies

Kopal Verma Saini and Anil K. Gupta

25.1 Introduction

Globally, the disaster risks are evolving at all levels with the changing climate and growing urbanization, increasing the vulnerability of the marginalized population. The pattern of such risks is becoming uncertain and haphazard, thereby making it much more difficult to mitigate these risks from turning into disasters (Thomas and López 2015). Therefore, the international community is recognizing the critical need for an efficient and effective disaster risk reduction (DRR) strategy which can address the threats posed by natural and human-induced disasters globally. The nations are also now compelled to reassess and strengthen their disaster resilience approaches because the pace of development can be largely hindered or even reversed by one or many of these disasters. Climate change has perplexed the disaster risk management by introducing new dimensions to the risks and escalating the incidences of unprecedented and unpredictable disasters, posing serious threat to the sustainable development (Zimmerman and Keiler 2015). The close interconnection between disasters and climate change can impact unevenly across geographies and communities and so demands a careful consideration. The advent of the Paris Agreement underscored the urgency of addressing climate change and mitigating disaster risks, and bringing it to the forefront of decision-makers' agendas to catalyse efforts to reverse its effects (Field 2012).

This chapter delves into recent international developments in DRR strategies, examining collaborative efforts, frameworks, and multi-lateral platforms on the global stage. This will help gain insights on the evolving panorama of international DRR efforts to create a resilient and secure world for communities. Over recent decades, several global frameworks have been formulated for DRR, including the

K. V. Saini · A. K. Gupta (✉)
Coalition for Disaster Resilient Infrastructure, Delhi, India
e-mail: anilg.gov.in@gmail.com

N. A. Siddiqui et al. (eds.), *Advances in Disaster Management, Volume 2*, Springer Transactions in Civil and Environmental Engineering,
https://doi.org/10.1007/978-981-96-4047-8_26

Yokohama Strategy for a Safer World, the Hyogo Framework for Action 2005–2015, and its successor, the Sendai Framework for Disaster Risk Reduction. While these frameworks provide valuable guidance for mitigating disaster risks, their adaptation to local contexts is imperative due to changing risk patterns induced by climate change (Zimmerman and Keiler 2015). Figure 25.1 gives a sequence of major international events that took place in last three decades which bolted the efforts in the direction of DRR globally.

25.2 International DRR Policy Frameworks

25.2.1 Yokohama Strategy for a Safer World: Guidelines for Natural Disaster Prevention, Preparedness and Mitigation

In the decades of 1970s and 1980s, the world was in the process of a major developmental sprint on one hand and on the other, it witnessed increasing incidences of disasters such as floods and earthquakes which caused significant economic and human losses. And so, the need for the efforts was realized to reduce the impacts of disasters on lives, livelihoods and infrastructure. The decade from 1990 to 1999 was marked as the **International Decade for Natural Disaster Reduction** by the UN General Assembly. So far, the activities were response-driven and addressed the post-disaster scenarios, but with the rising concern, this focus started to shift towards disaster risk reduction and prevention. This transition from a reactive to proactive approach was addressed during the first World Conference on Natural Disaster Reduction where the '**Yokohama Strategy for a Safer World**' was adopted in 1994.

Yokohama strategy outlined ten principles as mentioned in Fig. 25.2, which demonstrated that an integrated and advanced approach is essential in all aspects of an effective disaster management. This strategy directed the flow of efforts towards a culture of prevention. The world is interdependent, and hence, a collaborative effort is required to create a safer world which should transcend borders. Developing countries along with Landlocked Developing Countries (LLDCs), Least Developed Countries (LDCs), and Small Island Developing States (SIDS) are the most vulnerable to disasters, and the ones who suffer the most are the poor and socially disadvantaged section of the society. They should be enabled by providing scientific and technical assistance as well as access to data and technology so that they can revive, apply, and share their traditional wisdom on reducing the impacts of natural disasters (United Nations International Decade for Natural Disaster Reduction 1994).

At the mid-term of the International Decade for Natural Disaster Reduction, Yokohama Strategy urged the United Nations and the international community to hasten the implementation of policies and goals during the remaining decade and prepare the ground for the future generations. This strategy identified that it is imperative to bring

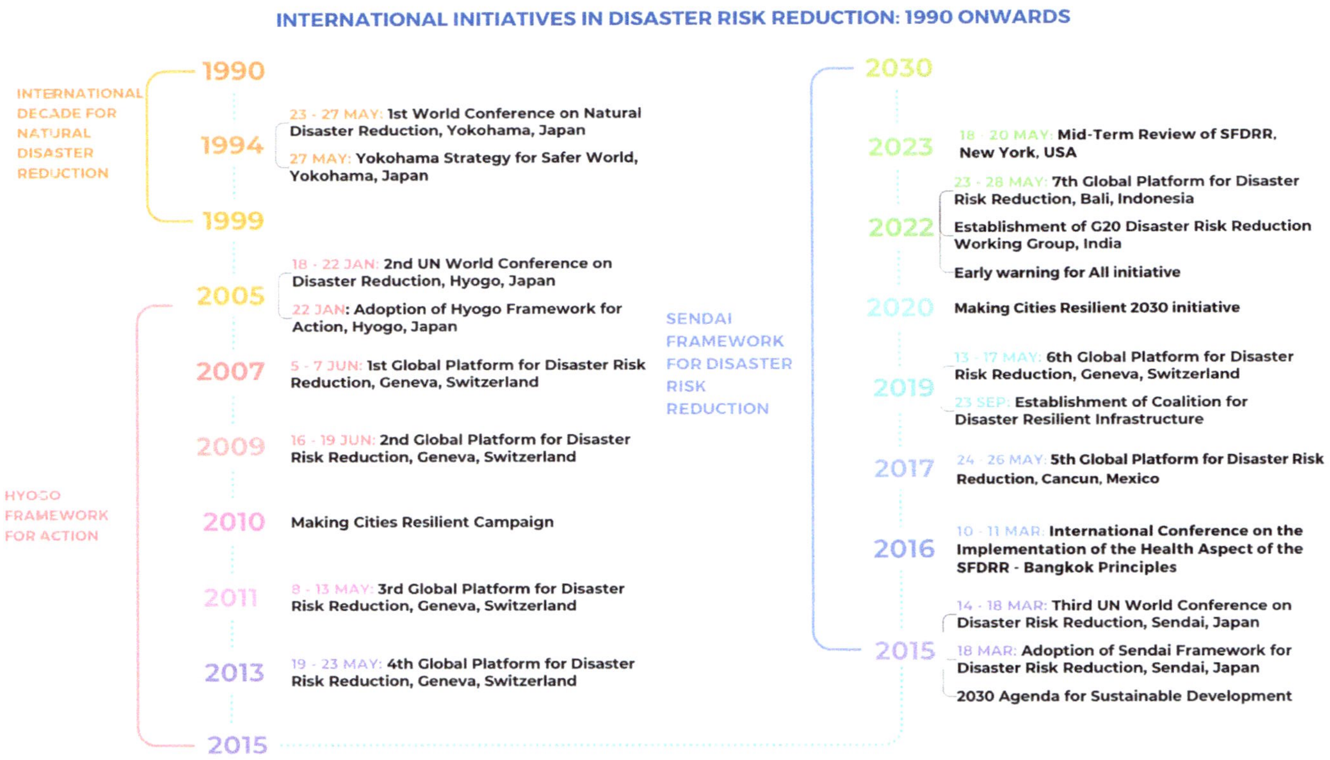

Fig. 25.1 International initiatives in disaster risk reduction: 1990 onwards

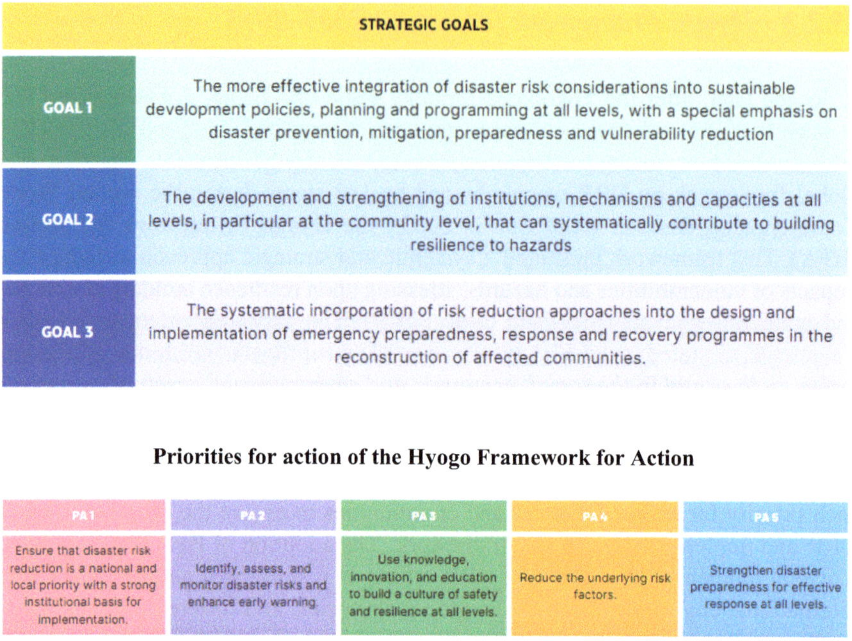

Fig. 25.4 Strategic goals and priorities for action of the Hyogo framework for action 2005–2015. *Source* HFA (2005)

framework fosters shared responsibility while ensuring adequate means of implementation through efficient resource allocation. Analysing the gaps observed in HFA underscored the need of an action-oriented framework which can be collaboratively implemented by the international bodies, governments, and various stakeholders while identifying and managing disaster risks and guiding investments to enhance resilience (SFDRR 2015).

Maintaining continuity in the previous DRR endeavours is very important so that the efforts don't go waste, and at the same time, it is also imperative to inculcate latest innovations and emerging technologies. This has been astutely balanced in the Sendai Framework. It aspires to shift focus towards disaster "risk" management, with the establishment of seven global targets and four priorities outlined in Fig. 25.5. The framework outlines a goal centred on preventing new risks, reducing existing risks, and fortifying resilience. It also defines guiding principles which emphasized states' primary responsibility to prevent and reduce disaster risk with all-of-society and all-of-state institutions involvement. This framework expands the scope of DRR by encompassing both natural and man-made hazards, alongside related environmental, technological, and biological risks. To bring out the most effective results, it is important that this framework operates in coherence with other global policy frameworks such as the Paris Agreement and 2030 Agenda on Sustainable Development (Box 1) (SFDRR 2015).

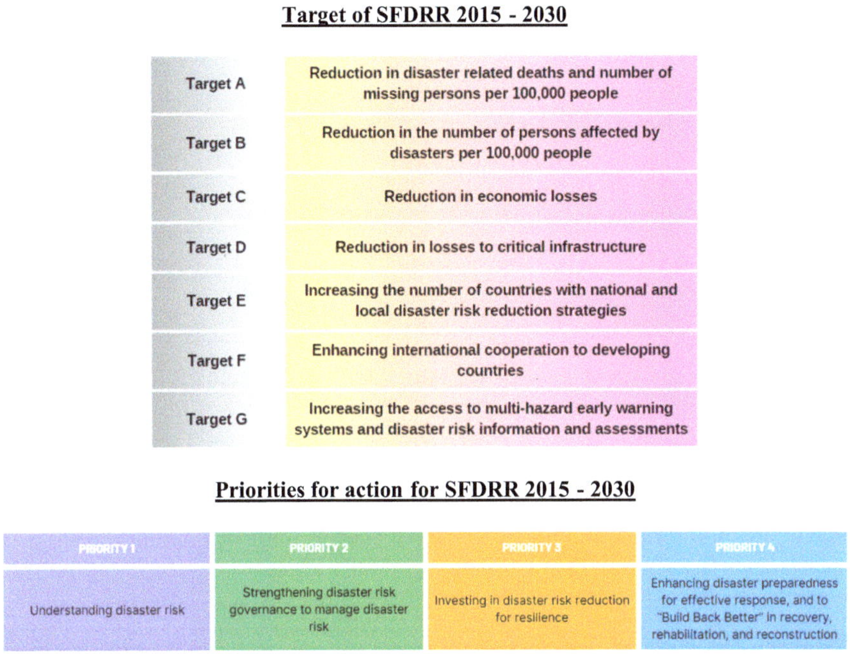

Fig. 25.5 Global targets and priorities for action in the Sendai Framework for Disaster Risk Reduction. *Source* SFDRR 2015

Box 1. 2030 Agenda for Sustainable Development

In September 2015, United Nations member states adopted a global initiative—2030 Agenda for Sustainable Development. It sets out a comprehensive and ambitious framework to address a wide range of global challenges and promote sustainable development over the next 15 years, until 2030. At the core of the agenda are the 17 Sustainable Development Goals (SDGs), which are an integrated set of targets and indicators covering social, economic, and environmental dimensions of development. The goals aim to eradicate poverty, ensure inclusive and equitable education, promote gender equality, achieve sustainable economic growth, combat climate change, and foster peaceful and inclusive societies, among other objectives. The 2030 Agenda emphasizes a holistic approach, recognizing the interconnectedness of various global issues and the need for collective action to achieve a more sustainable and equitable future (SDG 2015).

The mid-term review of the SFDRR which took place from 18 to 20 May 2023 aimed to assess the progress towards its 2030 goal and targets and identify challenges and gap areas to expedite efforts in the right direction. The review recommended a comprehensive and people-centric approach that operates in alignment with the 2030 Agenda for Sustainable Development. Various challenges were identified by the mid-term review like the uneven pace of implementation, limited data and financing access, and coordination gaps and silos between various sectors. In the first half of the SFDRR, various positive developments were observed including enhanced risk understanding, regional cooperation, and recognition of traditional knowledge. The review suggests that the support for local authorities to carry out the DRR initiatives should be increased, institutional silos should be minimized, and sustainable investments in DRR should be promoted. Along with these, it also urges to emphasize inclusive early warning systems, innovative financing instruments, and nature-based solutions in order to protect the vulnerable populations (MTR 2023). The review stressed upon a participatory and all-inclusive approach to ensure that no one is left behind in undertaking DRR initiatives. The initiatives like "Making Cities Resilient 2030," "Early Warnings for All," and the "Bangkok Principles" (Boxes 2–4) were also recognized by the review, and it also acknowledged the establishment of the G20's Disaster Risk Reduction Working Group during the Indian presidency, committed to advancing the SFDRR agenda.

Box 2. Making Cities Resilient 2030

The Making Cities Resilient (MCR) Campaign was launched in 2010 to increase awareness towards disaster risk reduction and gather the commitment of local governments and political leaders in this respect. Various consultations conducted in 2018–2019 resulted into shaping this campaign into its successor which is now known as Making Cities Resilient 2030 (MCR2030) initiative. This MCR2030 initiative highlights the urgent need to build resilience at local levels beyond 2020 and accelerate the progress in this direction which is essential to guide cities onto the resilience trajectory, aligning with the objectives of the Sendai Framework for Disaster Risk Reduction, the New Urban Agenda, the Paris Agreement, and the Sustainable Development Goals by 2030.

Box 3. Early Warnings for All

In 2022, the United Nations Secretary-General issued a call to safeguard all individuals on the planet by providing a universal coverage of early warning systems by the end of 2027. This is a very ambitious initiative that has been taken up recently and requires an urgent expansion of multi-hazard early warning systems (MHEWSs) globally. This revolutionary effort aims to protect every individual on Earth from the impacts of disasters and climate-related extreme events.

Box 4. Bangkok Principles

The International Conference on the Implementation of the Health Aspect of the SFDRR 2015–2030, held on 10–11 March 2016, in Bangkok, Thailand, recommended the following measures:

1. Integrate health in disaster plans.
2. Strengthen health cooperation for resilience.
3. Invest in health disaster reduction.
4. Include disaster reduction in health education.
5. Integrate health data in warnings.
6. Support cross-sector collaboration.
7. Develop coherent policies and strategies.

25.2.4 United Nations Framework Convention on Climate Change (UNFCCC)

The Conference of the Parties (COP) represents gatherings conducted within the framework of the United Nations Framework Convention on Climate Change (UNFCCC), a multinational treaty instituted in 1992. Throughout these sessions, participating countries (Parties) assess global efforts in alignment with the primary objective of the Paris Agreement, which seeks to limit global warming to around 1.5 °C above pre-industrial levels. This accomplishment is crucial, given the increasing influence of climate change on recent disasters, characterized by unpredictability and disruption in their occurrence patterns. Notably, COP 27 established

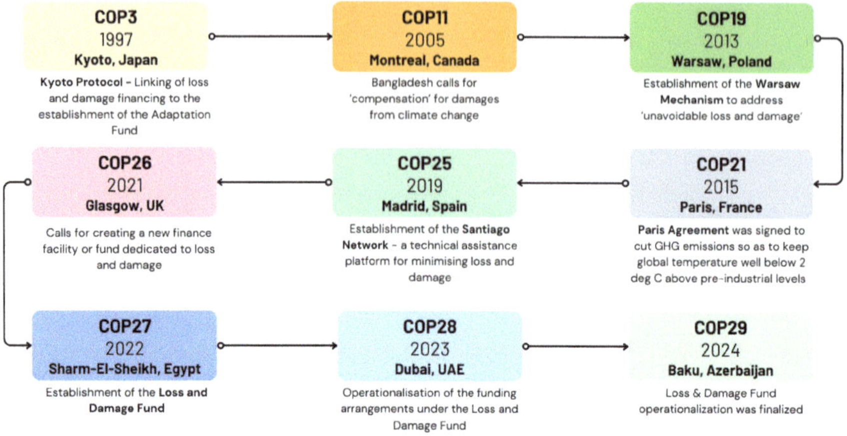

Fig. 25.6 Timeline of the important DRR-related developments that took place in various COPs

a dedicated Loss and Damage Fund. The COPs function as the principal decision-making body of the UNFCCC, adopting decisions and resolutions across various dimensions of climate action, including mitigation, adaptation, finance, technology, and transparency. Figure 25.6 represents a timeline of important developments pertaining to disaster risk reduction that took place in various COPs (Bodansky 1993).

25.2.4.1 The Santiago Network

The Santiago Network was established during the COP 25, Madrid, Spain, in December 2019, with an objective of "*catalysing technical assistance of relevant organisations, bodies, networks and experts, for the implementation of relevant approaches for averting, minimise and addressing loss and damage at the local, national and regional level, in developing countries that are particularly vulnerable to the adverse effects of climate change.*" The institutional arrangements for Santiago Network were established at COP26 and COP27, paving the way towards its operationalization. The functions of the Santiago network are aimed at addressing loss and damage associated with climate change, especially in vulnerable developing countries including the effective implementation of the Warsaw International Mechanism, catalysing demand-driven technical assistance, promoting collaboration and coordination among organizations and experts, and facilitating the development and dissemination of knowledge on comprehensive risk management approaches (Åberg and Jeffs 2022). As of 2023, pledges to the Santiago Network include commitments worth approximately USD 40.7 million and its secretariat will be hosted by UN Office for Project Services (UNOPS).

25.2.4.2 Loss and Damage Fund

The "Loss and Damage" (L&D) fund is a financial mechanism established during COP 27, held in November 2022, to address the irreversible consequences of climate change that cannot be addressed through adaptation efforts. The agreement to set up the L&D fund was endorsed during the COP27 held in November 2022. The L&D fund acknowledges and seeks to compensate for real losses incurred by communities, countries, and ecosystems due to climate change, extending beyond monetary value to impact human rights, well-being, and environmental sustainability (Alayza et al. 2022). In 2023, COP 28 adopted decisions on the fund's operationalization, approving its governing instrument, establishing a dedicated and independent secretariat, and placing it under the supervision of a governing board. Pledges, amounting to around USD 700 million to date, have been made by various countries.

25.2.5 G20 Disaster Risk Reduction Working Group

The Sendai Framework along with its predecessors has been considerably successful in bringing the global attention towards this very important issue of DRR, and this was a high time when this issue is also discussed at a multilateral forum like the Group of 20 (G20). The efforts on bringing DRR to the forefront started long back in the journey of G20 summits. Figure 25.7 presents a timeline when emphasis on disasters was made within the G20 deliberations. India's proactive initiative during its G20 Presidency to add a new working group on DRR has received a widespread support and appreciation worldwide. This working group presents an opportunity to bridge the capacity gap between various economies through regional, national, and international collaborations. The G20 DRR Working Group aims to integrate risk reduction measures into public and private sector investment decisions and policy-making to reduce existing risk, prevent the creation of new risk, and building resilient economies, societies, and natural systems. The five priorities outlined in the DRR working group are mentioned in Fig. 25.8. Current economic choices and development pathways are not in line with the commitments made in the Sendai Framework as well as the 2030 Agenda and the Paris Agreement, and the COVID-19 pandemic has even stalled or reversed the progress in some of its targets (Shaw and Kishore 2023). The G20 DRR Working Group is committed to changing this through the development of policy guidance, exchange of technical knowledge, promotion of capacity building, and identification of new approaches to financing that prioritize risk reduction and resilience. Since this was the first time this working group was formed in the G20, India in consultation with other G20 members, has created an elaborate three-year roadmap for the upcoming presidencies to carry forward the extensive work done and efforts made under this working group (G20 DRRWG 2023).

The G20 New Delhi Leaders' Declaration acknowledged that G20 can play a crucial role as a major development partner in DRR, highlighting the mitigation of

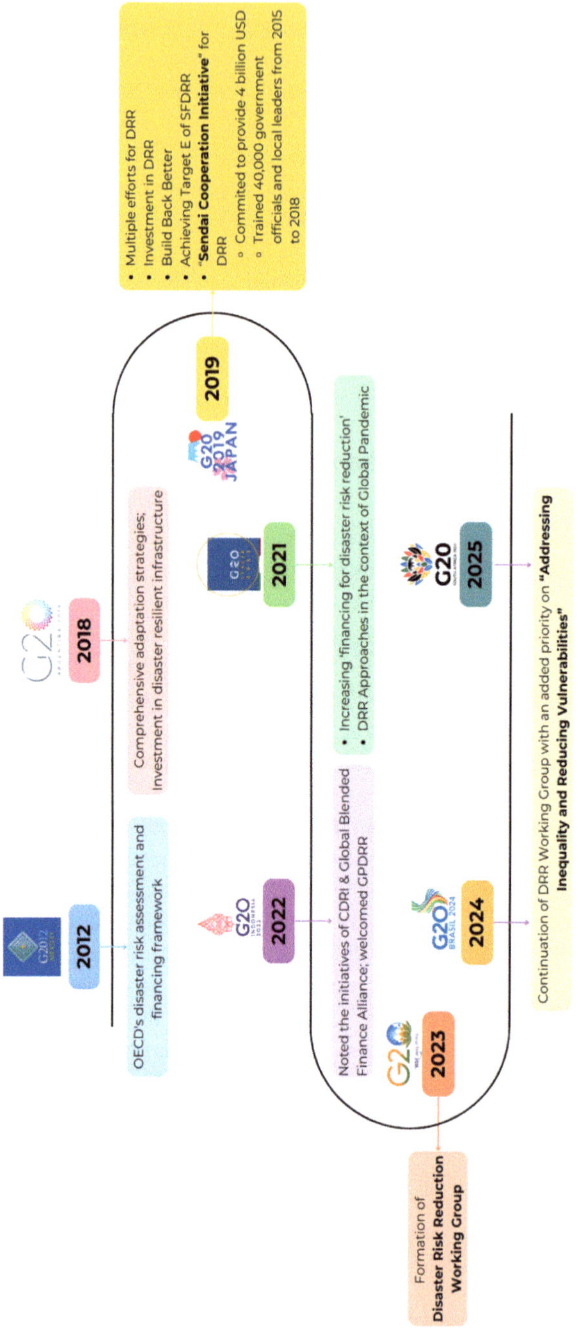

Fig. 25.7 DRR initiatives taken up by G20 Presidencies *Source* G20 DRRWG (2023)

Priority 1	Global coverage of Early Warning Systems for all hydro-meteorological disasters
Priority 2	Increased commitment towards making infrastructure systems disaster and climate resilient
Priority 3	Stronger national financial frameworks for disaster risk reduction
Priority 4	Strengthened national and global disaster response system to address the consequences of increasing frequency and intensity of disasters.
Priority 5	Increased application of ecosystems-based approaches to disaster risk reduction

Fig. 25.8 Priorities of the G20 disaster risk reduction working group in India's Presidency (G20 DRRWG 2023)

climate and disaster risks in LDCs and SIDS on priority. Through the working group's five priorities, the G20 countries committed to enhance visibility and advocacy for DRR and share global best practices. The establishment of the DRR working group coincided with the mid-term review of the SFDRR which further reinforced the commitments to achieve the targets set by the framework. In fact, the priorities of the DRR working group perfectly align with those of the SFDRR, which showcased and set a brilliant example for the forthcoming strategies to operate in synchronization with the existing frameworks. When it comes to effective sharing and collaboration, the DRRWG also noted the contribution of crucial global platforms and initiative like Global Platform for DRR (Box 5) and Coalition for Disaster Resilient Infrastructure (Box 6).

Box 5. Global Platform for Disaster Risk Reduction (GPDRR)

The Global Platform for Disaster Risk Reduction is a biennial event which functions as a collaborative platform for multiple stakeholders to assemble to exchange knowledge and discuss the recent advancements and trends to mitigate disaster risks. During these sessions, various stakeholders including national governments, the UN system, etc., come together to chalk out strategies and mechanism for assessing and accelerating the implementation of the Sendai Framework. So far seven sessions of GPDRR have occurred since its inception in 2007. The UN General Assembly recognizes the outcomes of these sessions which further contribute to the deliberations of the High-Level Political Forum on Sustainable Development (HLPF) which finally leads to a risk-informed implementation and monitoring of the 2030 Agenda for Sustainable Development. The outcomes of the 7th session of GPDRR, held in Bali, Indonesia, have been summarized as follows:

- *"Think Resilience" approach for investments and decision-making.*
- *Systemic changes required to address the real cost of disasters, to make proactive DRR investments.*
- *Mitigation becoming urgent with increasing greenhouse gas (GHG) emissions further intensifying catastrophic events.*
- *DRR planning requires a participatory, human rights-based approach, acknowledging diverse impacts on individuals.*
- *Inclusive Multi-Hazard Early Warning Systems (MHEWS) are essential, aligning with the UN Secretary-General's call for 'Early Warning for All' by* 2027.
- *Post-COVID-*19 *recovery demands a transformative, equitable, and community-driven approach, fostering resilience through gender-responsive and human rights-based strategies.*
- *Ecosystems offer environmental, socio-economic, and cultural benefits and so can become integral to infrastructure system.*
- *Aligning DRR and climate change adaptation enhances resilience, stressing the need for a comprehensive risk management approach, while acknowledging limited understanding of emerging hazards and reactive government policies.*

Box 6. Coalition for Disaster Resilient Infrastructure

In 2019, the Coalition for Disaster Resilient Infrastructure (CDRI) was established by Indian Prime Minister Shri Narendra Modi during the 2019 UN Climate Action Summit. It is an international collaboration that aims to enhance the resilience of infrastructure systems to disaster and climate risks, and it extensively engages with the UN bodies, national governments, academic institutions, development banks, and the private sector. It closely aligns with the targets/goals of SFDRR and SDGs and provides technical support to member countries, by emphasizing on standardized infrastructure design and management to reduce infrastructure losses from disasters. CDRI serves as a global platform to promote infrastructure resilience, stimulate knowledge exchange, bolster global planning, and expand universal access of disaster-resilient infrastructure by addressing the need to create hazard-resistant structures and fortifying the existing ones.

25.3 Conclusion

Disasters are a time which brings absolute unity in people because when a disaster strikes, nobody is concerned about the religion, gender, caste, nationality, etc., and no issue is bigger than saving lives. This kind of temperament and resolve is required to bring DRR to every nook and corner so that no one is left behind. We have come through a long journey to bring world's perspective from a response and relief centric approach to a mitigation and preparedness centric approach. But since disasters are taking an unpredictable and dubious shape with the changing climate and growing urbanization, the conventional approaches to disaster management may not be appropriate. Hence, effective implementation of DRR strategies requires standardized approaches, inclusive multi-stakeholder participation and opportunities for knowledge exchange and capacity development.

With every global strategy, plan or framework, significant lessons were learnt and efforts were made to address the gap areas in the subsequent one. This shows the commitment of the international community to fostering a more resilient and sustainable future. Also, looking at the interconnectedness and interdependence of different sectors, it is very important that all the global frameworks work in close coherence and synchronization with each other. By fostering collaboration, sharing best practices, and prioritizing resilience, nations can collectively build a more robust and all-inclusive framework to confront the evolving landscape of disaster risks.

References

Åberg A, Jeffs N (2022) Loss and damage finance in the climate negotiations

Alayza N, Bhandari P, Burns D, Cogswell N, de Zoysa K, Finch M, Díaz MJ, Fransen T, González ML, Krishnan N, Langer P, Larsen G, Srouji J, Warszawski N, Waskow D (2022) COP27: key takeaways and what's next

Bodansky D (1993) The United Nations framework convention on climate change: a commentary. Yale J Int L 18:451

Field CB (ed) (2012) Managing the risks of extreme events and disasters to advance climate change adaptation: special report of the intergovernmental panel on climate change. Cambridge University Press

G20 DRRWG 2023, G20 disaster risk reduction working group—outcome document and chair's summary 2023

HFA 2005, United Nations, Hyogo framework for action 2005–2015: building the resilience of nations and communities to disasters, 22 Jan 2005, A/CONF.206/6. Available at https://www.refworld.org/docid/42b98a704.html

McGlade J, Bankoff G, Abrahams J, Cooper-Knock SJ, Cotecchia F, Desanker P, Erian W, Gencer E, Gibson L, Girgin S, Hirsch F, Hirst L, Jordan S, Kopec E, Krausman E, Lopez-Ocon OG, Lorito S, Lovholt F, Moufouma-Okia W, Necci A, Peters K, Pulwarty R, Rush D, San Miguel J, Schneider J, Sengupta R, Twigg J, Vogt J, Walls R, Williams S, Wood M (2019) Global assessment report on disaster risk reduction 2019

MTR 2023, Political declaration of high-level meeting on the midterm review of the Sendai framework for disaster risk reduction 2015–2030

SDG 2015, UN General Assembly, Transforming our world: the 2030 agenda for sustainable development, 21 Oct 2015, A/RES/70/1. Available at https://www.refworld.org/docid/57b6e3e44.html
SFDRR 2015 (2015) Sendai framework for disaster risk reduction 2015–2030. In: Third UN world conference on disaster risk reduction, Sendai, Japan
Shaw R, Kishore K (2023) Disaster risk reduction and G20: a major step forward. Prog Disaster Sci 17:100274
Stanganelli M (2018) Hyogo framework for action: an analysis ten years later and a prospective for future decades. In: Handbook of disaster risk reduction & management, pp 499–521
Thomas V and López R (2015) Global increase in climate-related disasters. Asian Development Bank Economics Working Paper Series, No. 466 (https://www.adb.org/sites/default/files/publication/176899/ewp-466.pdf)
UNISDR (2005) Review of the Yokohama strategy and plan of action for a safer world
United Nations International Decade for Natural Disaster Reduction (1994) Yokohama strategy and plan of action for a safer world: guidelines for natural disaster prevention, preparedness and mitigation
Zimmermann M, Keiler M (2015) International frameworks for disaster risk reduction: useful guidance for sustainable mountain development? Mt Res Dev 35(2):195–202

Chapter 26
Disasters and NCD: Preparing for Health Emergencies

Atisha Sood and Anil Kumar Gupta

26.1 Introduction: The Looming Intersection of Disasters and NCDs in India

India, a vibrant nation steeped in culture and history, faces a burgeoning threat at the precipice of the twenty-first century: the escalating co-occurrence of natural disasters and non-communicable diseases (NCDs). This intersection poses a formidable challenge, jeopardizing the well-being of millions and necessitating proactive management strategies.

26.1.1 The Escalating Threat of Disasters:

India's vulnerability to natural disasters is alarmingly high. Ranked 6th among the most affected countries between 1993 and 2022 in the Climate Risk Index 2025 by Germanwatch, the nation grapples with a concerning rise in earthquakes, floods, cyclones, and droughts (Kumar et al. 2019). Climate change further exacerbates this trend, intensifying the severity and frequency of these events (World Bank 2020). The devastating consequences of these disasters ripple across communities, disrupting infrastructure, displacing populations, and disrupting essential services, including health care for NCD patients.

S. Atisha (✉) · A. K. Gupta
ECDRM Division, Centre for Disaster & Health, National Institute of Disaster Management, Ministry of Home Affairs, GoI, New Delhi, India
e-mail: soodatisha@gmail.com

N. A. Siddiqui et al. (eds.), *Advances in Disaster Management, Volume 2*, Springer Transactions in Civil and Environmental Engineering,
https://doi.org/10.1007/978-981-96-4047-8_27

26.1.2 The Burden of NCDs: A Ticking Time Bomb

Concurrently, India battles a burgeoning NCD epidemic. Data from the World Health Organization (WHO 2023) reveal that NCDs account for over 63% of all deaths in the country, with diabetes, cardiovascular diseases, respiratory illnesses, and mental health conditions leading the charge (Chatterjee et al. 2022). This burden is further amplified by social and economic inequalities, disproportionately impacting marginalized communities (Kumar and Banerjee 2018).

26.1.3 A Precarious Intersection

The convergence of these forces creates a perfect storm of vulnerability. Disasters disrupt NCD care, leading to medication shortages, inaccessible healthcare facilities, and compromised food security, all of which can trigger devastating health consequences for NCD patients (Ghazanchaei et al. 2021). The psychological trauma associated with displacement and loss amplifies existing conditions and contributes to a decline in mental well-being (Ghosh et al. 2021). This precarious intersection demands immediate attention and the development of robust frameworks for managing NCDs in the face of disaster.

26.1.4 Moving Forward

Ignoring this looming threat would be akin to turning a blind eye to an impending catastrophe. This chapter aims to provide a comprehensive framework for proactive NCD management in disaster situations, encompassing pre-disaster preparedness, effective response strategies, and resilient recovery plans. By acknowledging the vulnerability, equipping communities, and strengthening healthcare systems, India can build a future where NCD patients are not left at the mercy of these intertwined threats.

26.2 Disruptions Cascade: Unraveling the Impact of Disasters on NCD Management

When a disaster strikes, its impact ripples beyond immediate physical destruction, Table 26.1. For NCD patients, these ripples translate into a complex tapestry of vulnerabilities, disrupting every facet of their well-being.

Table 26.1 Overview of disaster impact on NCDs

Disaster type	Specific impact on NCD patients	Key challenges	Recommended interventions
Earthquakes	Disruption of healthcare infrastructure, medication shortages	Loss of access to care, increased risk of complications	Implement telemedicine, establish mobile medical units
Floods	Compromised food security, displacement	Uncontrolled blood sugar levels, exacerbation of respiratory symptoms	Create innovative medication distribution channels, prioritize NCD patients in evacuation plans
Cyclones	Destruction of homes, social fabric unraveling	Limited access to social support networks	Integrate mental health services, establish safe spaces for shared support
Droughts	Limited access to clean water	Increased risk of dehydration, exacerbation of health conditions	Prioritize NCD patients in resource allocation, ensure access to clean water in shelters

26.2.1 Lifeline Severed: The Crumbling Foundation of NCD Care

The very foundation of NCD management—access to medications and healthcare services—crumbles beneath the weight of a disaster. Medical facilities may be rendered inoperable, supply chains disrupted, and medication stocks depleted. This leads to uncontrolled blood sugar levels, uncontrolled hypertension, and exacerbation of respiratory symptoms, potentially triggering life-threatening complications for NCD patients (Ghazanchaei et al. 2021).

26.2.2 Beyond the Physical Toll: The Psychological Tsunami

The psychological impact of disasters can be equally devastating for NCD patients. The trauma of displacement, loss of loved ones, and uncertainty about the future can trigger anxiety, depression, and post-traumatic stress disorder (PTSD) (Mandal et al. 2022). These mental health consequences can further compromise their physical health and adherence to treatment regimens.

26.2.3 Social Fabric Unraveled: Exacerbating Existing Inequalities

Disasters often unravel the social fabric that supports NCD patients. Livelihoods may be lost, homes destroyed, and families separated. This can lead to decreased income, compromised food security, and limited access to social support networks—all vital factors for managing NCDs (Sharma et al. 2020). Existing social and economic inequalities further compound these vulnerabilities, disproportionately impacting women, children, older adults, and individuals from marginalized communities (Chakrabarti et al. 2018).

26.2.4 A Tapestry of Vulnerabilities: Recognizing the Nuanced Risks

Understanding the nuances of these vulnerabilities is crucial for designing effective interventions. Recognizing the unique challenges faced by different NCD groups, such as the increased risk of infections for patients with diabetes or the need for accessible oxygen for those with respiratory illnesses, is vital for targeted preparedness and response strategies.

26.2.5 Beyond the Immediate Storm: Embracing Proactive Preparedness

Ignorance breeds vulnerability. Acknowledging the multifaceted impact of disasters on NCD patients and proactively building resilience within communities and healthcare systems is the first step toward mitigating these risks. By building a robust framework encompassing pre-disaster preparedness, effective response strategies, and resilient recovery plans, India can empower NCD patients to navigate the storm's fury and emerge stronger from its aftermath.

Preparing for the unpredictable storms of natural disasters is crucial, particularly for mitigating the impact on vulnerable populations like NCD patients. The Kerala Floods and Cyclone Fani serve as stark reminders of this need. Yet, they also offer inspiration through two pioneering case studies. These initiatives demonstrate the power of proactive interventions like community health worker networks, telemedicine platforms, and pre-loaded patient data in disaster preparedness. By embracing such strategies across diverse regions and NCD types, we can ensure a more resilient healthcare ecosystem that stands strong even in the face of adversity.

Box Study 1: Community Resilience in Action: The Kerala Floods 2018

In August 2018, Kerala experienced devastating floods that displaced millions and disrupted healthcare access for countless NCD patients. Yet, amidst the chaos, the state's robust community-based NCD management system, built under the Kerala State NCD Cell, proved its worth.

Key initiatives:

- **Community health workers**, trained in NCD management by programs like "Snehajeevi," became the backbone of the response. They tirelessly identified and reached out to vulnerable patients in flood-affected areas, ensuring continuity of care.
- **Village health committees**, established under the "Aarogyasree" program, played a crucial role in refilling prescriptions, delivering medications, and setting up temporary care centers in shelters and public spaces.
- **Local peer support groups**, fostered by initiatives like "Samasthitha Kshetra Samithies," provided invaluable psychosocial support to NCD patients during this stressful time. These groups facilitated information sharing, emotional coping mechanisms, and a sense of community amidst the devastation.

Outcomes:

Despite the unprecedented challenges, the community-based network ensured NCD management continuity for over 80% of patients during the floods. This remarkable feat not only highlighted the effectiveness of existing programs but also strengthened the overall NCD management system in Kerala. The experience emphasized the vital role of community resilience, trained health-care workers, and strong social support networks in mitigating the impact of disasters on NCD patients.

Box Study 2: Technology Triumphs in Cyclone Fani 2019 (Odisha)

Cyclone Fani in 2019 ripped through Odisha, leaving a trail of devastation and severing communication networks. However, amidst the chaos, a pioneering technological intervention emerged, bridging the gap for isolated NCD patients and proving the power of innovation in disaster response:

Key Technologies:

- **Mobile Medical Units**: Equipped with telemedicine tools, diagnostic kits, and essential medications, these units, like the "Aarogya Rath" program, navigated impassable roads and reached remote villages, defying limited connectivity (Ghosh et al. 2019).

- **AI-Powered Triage**: Through a smartphone app like "Swasthya Mitra," healthcare workers conducted initial assessments and prioritized interventions, using AI algorithms to optimize resource allocation and ensure the most critical patients received timely care (Kumar et al. 2022).
- **Digital Health Records**: Pre-loaded patient data in platforms like "e-Aarogya" provided immediate access to medical history, allowing doctors to quickly understand medication regimens and adjust them as needed, even in offline settings (Bandyopadhyay et al. 2020).

Outcomes:

This tech-driven approach not only successfully reached over 85% of isolated NCD patients but also demonstrably improved resource allocation and treatment efficiency. The experience showcased the transformative potential of technology in disaster response, particularly for managing chronic conditions like NCDs.

26.3 Weaving a Safety Net: Building a Framework for Proactive NCD Management in Disaster Situations

In the face of the looming intersection of disasters and NCDs, India requires a robust framework for proactive management. This framework must encompass three crucial pillars: empowering communities, bolstering healthcare systems, and embracing technological advancements.

26.3.1 Empowering Communities: Grassroots Resilience

- ***Knowledge is Power***: Community-based awareness campaigns and education programs equip individuals with knowledge of NCD management during disasters. This empowers them to manage their conditions independently and create support networks within their communities (Rajagopalan et al. 2020).
- ***Mapping Lifelines***: Identifying and registering NCD patients, establishing accessible communication channels, and ensuring availability of emergency contact information are crucial for targeted outreach and support during disasters (WHO 2019). Integrating NCD management into local disaster preparedness plans and early warning systems further strengthens community resilience.
- ***Tech to the Rescue***: Leveraging telemedicine and digital health solutions provides remote consultations, medication refills, and health monitoring, enhancing access to care even in resource-constrained settings (Kumar et al. 2022).

26.3.2 Bolstering Healthcare Systems: A Foundation for Resilience

Fortifying the Foundation: Strengthening healthcare infrastructure, including primary care centers and specialized NCD clinics, to withstand disaster impacts is essential. Ensuring adequate supplies of essential medications and equipment, and establishing backup power and communication systems guarantees continuity of care even amidst disruptions (National Disaster Management Authority 2019).

Drilling for Resilience: Regular disaster drills and simulations test NCD management plans and identify areas for improvement. Training healthcare personnel in disaster response and NCD care, and fostering collaboration with community health workers and volunteers, creates a robust network for effective response (Sharma and Arora 2021).

Uninterrupted Care: Developing contingency plans for ensuring continuity of healthcare services during and after disasters is crucial. This includes establishing alternative care sites, deploying mobile medical units, and coordinating with local providers for referral networks (Sharma et al. 2020).

26.3.3 Embracing Technology: Transforming Preparedness

Bridging the Digital Divide: Investing in digital infrastructure and training healthcare personnel and communities in technology utilization ensure the effectiveness of digital health solutions. Addressing connectivity challenges in rural areas is crucial for inclusive preparedness (Bandyopadhyay et al. 2020).

Real-Time Data for Real-Time Response: Implementing data-driven systems for monitoring population health, resource availability, and disaster impact provides valuable insights for targeted interventions and resource allocation during response efforts (Ghosh et al. 2019).

Innovation for Resilience: Encouraging research and development of disaster-resilient healthcare technologies, such as portable medical equipment and disaster-resistant communication systems, can significantly enhance preparedness and response capabilities (WHO 2020).

By weaving these three pillars together, India can create a comprehensive framework for proactive NCD management in disaster situations. This framework empowers communities, strengthens healthcare systems, and embraces technological advancements, building a future where NCD patients are not victims of disasters, but active participants in building resilience and navigating the storms with greater preparedness and confidence.

26.4 Rising to the Challenge: Effective Response Strategies for NCD Management in Disasters

When disaster strikes, swift and effective response becomes paramount. For NCD patients, this translates into safeguarding their well-being amidst the chaos. This section outlines key strategies for navigating the storm's fury and ensuring continuity of care during emergencies.

26.4.1 Mobilizing the Specialized Forces

- Rapid Deployment: Immediately deploy specialized medical teams trained in NCD management and equipped with portable medication kits and diagnostic tools. These teams can provide on-site care in shelters and temporary healthcare facilities (WHO 2019).
- Harnessing Local Expertise: Partner with local healthcare providers and community health workers. Their knowledge of the affected area and established relationships with individuals are invaluable for reaching vulnerable NCD patients and tailoring interventions effectively.

26.4.2 Building Safe Havens for Care

- Shelters with a Difference: Establish temporary shelters and healthcare facilities equipped to cater to the specific needs of NCD patients. Ensure adequate space, temperature control, access to clean water and sanitation, and availability of essential medications and medical equipment.
- Prioritizing the Vulnerable: Prioritize NCD patients in evacuation plans and consider establishing dedicated shelters for high-risk individuals with specific conditions requiring oxygen support or intensive monitoring.

26.4.3 Maintaining the Lifeline of Care

- Creative Distribution Channels: Implement innovative distribution channels for medications and medical supplies. This may include mobile pharmacies, community distribution points, and partnerships with local shops and pharmacies.
- Embrace the Digital Bridge: Leverage telemedicine and digital health platforms to provide remote consultations, monitor health parameters, and offer psychosocial support to NCD patients, even in areas with limited accessibility.

26.4.4 *Mental and Emotional Needs*

- Integrate Mental Health: Integrate mental health services into disaster response efforts. Provide psychological first aid, trauma counseling, and access to mental health professionals to help NCD patients cope with the psychological stress and anxiety associated with the disaster.
- Safe Spaces for Shared Strength: Create safe spaces for peer support and community interaction where NCD patients can share experiences, offer mutual encouragement, and build resilience together.

26.4.5 *Continuous Adaptation and Improvement*

- Monitor and Refine: Continuously monitor the situation and adapt response strategies based on evolving needs and emerging challenges. Regularly communicate updates and disseminate vital information to NCD patients through accessible channels.
- Strengthening the Network: Ensure effective coordination between different stakeholders, including government agencies, healthcare providers, community organizations, and volunteers, to optimize resource allocation and maximize the impact of response efforts.

By diligently implementing these strategies, India can create a more effective and responsive system for ensuring NCD management during disasters.

26.5 Building Anew: Resilient Recovery for Sustainable NCD Management Post-disaster

The aftermath of a disaster presents a window not only for rebuilding, but also for strengthening resilience against future events. This section delves into strategies for ensuring long-term NCD management through resilient recovery plans, Table 26.2.

26.5.1 *Prioritizing NCD Needs in Recovery Plans*

- **Integrate and Allocate**: Integrate NCD management into overall recovery plans and allocate resources specifically for restoring and strengthening NCD healthcare services. This includes rebuilding damaged facilities, replenishing medication stocks, and providing ongoing support to healthcare personnel (National Disaster Management Authority 2019).

Table 26.2 Resilient recovery strategies for NCD management

Recovery strategy	Implementation steps	Key outcomes
Integration of NCDs in recovery plans	Allocate resources, rebuild damaged facilities, replenish medication stocks	Enhanced continuity of care, restored healthcare services
Disaster-proofing NCD care	Invest in resilient infrastructure, emergency power, and communication systems	Improved disaster resilience, reduced vulnerability in future events
Long-term community engagement	Implement NCD awareness campaigns, train community health workers	Sustainable community support, increased NCD awareness

- **Advocate for Change**: Advocate for policies and programs that address the social and economic determinants of NCDs, such as livelihood restoration, social protection schemes, and improved access to healthy food and sanitation. These elements are crucial for preventing NCD complications and promoting sustainable recovery (WHO 2024).

26.5.2 Building Back Smarter: Disaster-Proofing NCD Care

- **Resilient Infrastructure**: Reconstruct healthcare infrastructure with disaster resilience in mind. Invest in robust buildings, emergency power and communication systems, and disaster-resistant storage facilities for medications and medical supplies (WHO 2019).
- **Tech for the Future**: Leverage technological advancements to create a more resilient NCD management system. Utilize telemedicine infrastructure, digital health platforms, and real-time data monitoring to improve access to care and disaster preparedness in the long term (Ghosh et al. 2019).

26.5.3 Empowering Communities for Lasting Impact

- **Sustained Engagement:** Foster long-term community-based interventions. Implement NCD awareness campaigns, train community health workers in NCD management, and empower communities to advocate for their own health needs. This promotes ownership and ensures long-term sustainability (Rajagopalan et al. 2020).
- **Social Support Networks:** Build social support networks and peer support groups for NCD patients that can provide ongoing support, share coping strategies, and enhance resilience in the face of future challenges. These networks strengthen the social fabric and offer emotional well-being (Mandal et al. 2022).

26.5.4 *Sharing Lessons Learned: Paving the Way for Future Preparedness*

- **Post-disaster Assessments:** Conduct thorough post-disaster assessments to evaluate the effectiveness of response strategies and identify areas for improvement. Disseminate knowledge and best practices through research, training programs, and knowledge-sharing platforms (Ghazanchaei et al. 2021).
- **Collaborative Preparedness:** Collaborate with national and international stakeholders to develop evidence-based NCD management guidelines and disaster preparedness plans specifically tailored to the Indian context. This fosters knowledge exchange and enhances preparedness across regions.

Resilient recovery is not simply about rebuilding; it's about building back better. By prioritizing NCD needs, investing in resilient infrastructure, empowering communities, and sharing lessons learned, India can create a future where NCD patients are not only prepared for future disasters but also empowered to thrive in the face of them.

26.6 Contemplating the Horizon: NCDs and Disasters in the Indian Context

The convergence of natural disasters and non-communicable diseases (NCDs) in India paints a complex picture—one laced with vulnerabilities, disruptions, and resilience. This chapter has delved into this interplay, highlighting the cascading impact of disasters on NCD patients and emphasizing the urgent need for proactive measures. Here are a few lessons learned from post- disaster assessments, Table 26.3.

Standing at the precipice of this challenge, a few overarching truths emerge:

Table 26.3 Lessons learned from post-disaster assessments

Disaster event	Key lessons	Improved strategies
Earthquake	Importance of mobile medical units for on-site care	Incorporate mobile units into disaster response plans
Flood	Need for innovative medication distribution channels	Establish partnerships with local shops and pharmacies for distribution
Cyclone	Prioritizing NCD patients in evacuation plans	Develop dedicated shelters for high-risk individuals
Drought	Importance of addressing dehydration risks	Include clean water access in disaster response plans

- **Vulnerability is Multifaceted**: From disrupted medication access to psychological trauma and compromised social support networks, the impact of disasters on NCD patients extends beyond the physical. Understanding these diverse vulnerabilities is crucial for designing effective interventions.
- **Proactive Preparedness is Paramount**: Recognizing the looming threat is not enough. India must weave a tapestry of preparedness, strengthening communities, bolstering healthcare systems, and embracing technological advancements to mitigate the impact of disasters on NCD patients.
- **Community Resilience is a Bedrock**: Empowering communities through knowledge, communication channels, and technological solutions enables them to become active participants in their own well-being during disasters.
- **Adaptation and Learning Are Continuous**: No single framework can be foolproof. Continuous monitoring, refinement of response strategies, and sharing of lessons learned will be vital for navigating the evolving landscape of NCDs and disasters.

The road ahead demands concerted efforts from various stakeholders—government agencies, healthcare professionals, researchers, community leaders, and individuals themselves. By acknowledging the challenges, embracing best practices, and fostering a culture of proactive preparedness, India can mitigate the detrimental impacts of disasters on its NCD population, paving the way for a future where resilience reigns supreme.

References

Bandyopadhyay S, Dasgupta K, Ghosh S (2020) Bridging the digital divide to address healthcare and non-communicable disease challenges in rural India during natural disasters. Int J Disaster Risk Reduction 44:101580

Chakrabarti S, Dutta N, Chatterjee A (2018) Gender and non-communicable diseases: addressing inequities in access to healthcare services in India during disasters. J Gend Stud 27(3):346–360

Chatterjee A, Kumar A, Murthy M (2022) Social and economic inequalities in non-communicable disease prevalence in India: a review. Int J Public Health 37(1):1008620

Germanwatch (2024) Global climate risk index 2025. https://www.germanwatch.org/sites/default/files/Global%20Climate%20Risk%20Index%202025_1.pdf

Ghazanchaei E, Khorasani-Zavareh D, Aghazadeh-Attari J, Mohebbi I (2021) Identifying and describing impact of disasters on non-communicable diseases: a systematic review. Iran J Public Health 50(6):1143–1155. https://doi.org/10.18502/ijph.v50i6.6413

Ghosh S, Chakraborty S, Sharma A (2021) Mental health consequences of natural disasters in India: a review. Indian J Psychol Med 43(1):21

Ghosh S, Kumar P, Sharma A (2019) Utilizing a real-time data-driven approach for non-communicable disease management during natural disasters in India. Int J Med Informatics 126:146–152

Kumar PV, Menon A, Jayanthy A (2019) Non-communicable disease management during the Kerala floods: lessons learnt. J Postgrad Med 65(4):272

Kumar R, Banerjee A (2018) Non-communicable diseases and social inequalities in India. Lancet Glob Health 6(12):e1422–e1434

Kumar S, Gupta A, Yadav K (2022) Role of telemedicine in disaster management of non-communicable diseases in India: challenges and opportunities. Indian J Public Health 66(1):55–58

Mandal A, Das S, Sil A (2022) Mental health outcomes among non-communicable disease patients exposed to natural disasters in India: a systematic review. Int J Soc Psychiatry 68(3):226–236

National Disaster Management Authority (2019) National disaster management plan: chapter on health. National Disaster Management Authority, Government of India

Rajagopalan S, Sreedharan J, Panicker S (2020) Community-based approaches for non-communicable disease management during disasters: evidence from Kerala floods. J Fam Med Prim Care 9(4):1850–1855

Sharma U, Gupta T, Gupta S (2020) Impact of natural disasters on the management of non-communicable diseases and potential mitigation strategies: a case study of Uttarakhand floods in India. Disaster Med Public Health Prep 14(4):489–495

Sharma U, Arora S (2021) Training healthcare professionals for non-communicable disease management during disasters: a need of the hour. Indian J Med Educ 6(2):115–118

WHO (2019) Noncommunicable diseases and natural disasters: a toolkit for preparedness and response. World Health Organization

World Bank (2020) Climate change knowledge portal for India. https://climateknowledgeportal.worldbank.org/

World Health Organization (2020) WHO global strategy on health, environment and climate change: the transformation needed to improve lives and well being sustainably through healthy environments. World Health Organization

World Health Organization (2023) World health statistics: non-communicable diseases. Retrieved from 20 Dec 2023. From https://www.who.int/gho/ncd/en/: https://www.who.int/gho/ncd/en/

World Health Organization (2024) Tackling NCDs: best buys and other recommended interventions for the prevention and control of noncommunicable diseases. World Health Organization

MIX
Papier aus verantwortungsvollen Quellen
Paper from responsible sources
FSC® C105338

If you have any concerns about our products,
you can contact us on
ProductSafety@springernature.com

In case Publisher is established outside the EU,
the EU authorized representative is:
Springer Nature Customer Service Center GmbH
Europaplatz 3, 69115 Heidelberg, Germany

Printed by Libri Plureos GmbH
in Hamburg, Germany